HANDBOOK OF TRANSFORMER APPLICATIONS

OTHER McGRAW-HILL HANDBOOKS OF INTEREST

HANDBOOK OF TRANSFORMER APPLICATIONS

WILLIAM M. FLANAGAN

McGRAW-HILL BOOK COMPANY

New York   St. Louis   San Francisco   Auckland   Bogotá   Hamburg
Johannesburg   London   Madrid   Mexico   Montreal   New Delhi
Panama   Paris   São Paulo   Singapore   Sydney   Tokyo   Toronto

Library of Congress Cataloging in Publication Data

Flanagan, William M.
   Handbook of transformer applications.

   Bibliography: p.
   Includes index.
   1. Electric transformers—Design and construction—
Handbooks, manuals, etc. I. Title.
TK2791.F57 1986   621.31′4   85-5241
ISBN 0-07-021290-2

1234567890   DOC/DOC   898765

ISBN 0-07-021290-2

The editors for this book were Harold B. Crawford and Laura Givner, the designer was Judith Fletcher Getman, and the production
supervisor was Sally Fliess. It was set in Times Roman by University Graphics, Inc.

Printed and bound by R. R. Donnelley & Sons Company.

To my wife, Billie Morris, whose forbearance made this book possible.

Contents

Preface

Magnetic components are the product of a mature industry. Being away from the cutting edge of technology, they sometimes do not receive the attention which their economic significance requires. Misunderstanding and lack of information cause many magnetic applications to cost more than they should. This handbook is an attempt to help relieve this problem by providing updated information which heretofore has been widely scattered.

User and supplier frequently use different language and have opposing viewpoints. The resulting faulty communications are at the root of many of the difficulties encountered with magnetic devices. An objective of this book is to facilitate this communications process. Ultimately, the aim of all parties is to secure—in the shortest possible time and at the lowest possible cost—components which function correctly. Mutual understanding and a broad common technical base are important contributors to this aim. Because of the great importance of mechanical requirements to magnetic components, sections of this work are devoted to current mechanical and manufacturing practice as well as to electrical theory. Space is also devoted to specification preparation, testing, and quality control—factors which relate directly to both user and supplier.

The author is indebted to many people for assistance and encouragement in writing this book. His family, both immediate and extended, have provided indispensable support. His associates at Del Electronics

Corporation have been an inspiration. Walter Buchsbaum has provided wise counsel and encouragement. Hal Crawford and Laura Givner at McGraw-Hill have been most supportive and understanding. Finally, the author has been fortunate in having the help of Lois Porro in editing and preparing the manuscript.

William M. Flanagan

HANDBOOK OF
TRANSFORMER
APPLICATIONS

CHAPTER 1

Magnetic and Electrical Fundamentals

1.1 INTRODUCTION

Transformers are passive devices for transforming voltage and current. They are among the most efficient machines, 95 percent efficiency being common and 99 percent being achievable. There is practically no upper limit to their power-handling capability, and the lower limit is set only by the allowable no-load loss. Transformers and inductors perform fundamental circuit functions. They are a necessary component in electrical systems as diverse as distribution terminals for multimegawatt power-generating stations to hand-held radio transceivers operating on a fraction of a watt. This book is intended to ease some of the difficulties users and designers face from the limitations of these components, the resolution of which is still painfully evolving.

Transformers are the largest, heaviest, and often costliest of circuit components. The geometry of the magnetic circuit is three-dimensional. This property places a fundamental restraint on reducing transformer size. The properties of available materials limit weight reduction. The high cost of transformers is due to the impracticability of standardization, the materials needed, and the processes inherent in their manufacture. The problems associated with the use of magnetic devices can be minimized by the employment of astute application practices.

Transformers are indispensable for voltage transformation in power applications. Their ability to isolate circuits and to alter ground conventions can often be matched in no other convenient manner. They are needed in frequency selective circuits whose operation depends on the response of inductances. They are rugged, being capable of withstanding severe environmental conditions.

Transformers are essentially single-application devices. Designed for specific requirements, they do not offer optimum performance over a wide range of operation. They are not outstanding performers in applications requiring high-fidelity reproduction of audio or video signals. Wideband and high-impedance circuits often experience serious degradation when transformers are used. Transformers do not perform well in circuits which apply dc magnetization to the core. They are a problem in equipment in which size and weight must be kept to a minimum.

Transformers can sometimes be eliminated by circuit artifices. A bridge rectifier directly across the power line can replace a power transformer and rectifier if the voltage level and ground isolation can be accommodated. This is often done to obtain dc voltage for inverter circuits. Needed voltage transformation is then done at high frequency with a much smaller transformer. The direct coupling of semiconductor devices to loads eliminates audio transformers. Operation of driving circuits at voltages which will provide the desired output voltage by direct coupling may eliminate the need for voltage transformation at the output.

Closely related in both theory and construction to transformers are other magnetic devices which include inductors, saturable reactors, and magnetic amplifiers. Much of the discussion on transformers is applicable to them. The unique features of these devices are discussed separately.

1.2   FUNDAMENTALS OF MAGNETIC CIRCUITS

The literature on magnetic devices uses a vocabulary which may be unfamiliar to some readers. It is not difficult to learn these terms and to master the fundamentals of magnetic circuits. This section provides an introduction to the subject.

The study of magnetics begins with Coulomb's law for magnetic poles:

$$F = \frac{m_1 m_2}{\mu r^2} \tag{1.1}$$

where F = repulsive force between two magnetic poles of like polarity m_1 and m_2
  r = distance between the two poles
  μ = constant of proportionality called *permeability*

Units are chosen so that μ is equal to 1 in air. The poles m_1 and m_2 are mathematical fictions useful in defining relationships. If both sides of Eq. (1.1) are divided by m_2 and m_2 is assigned unit magnitude, then a new quantity, magnetizing force, symbol H, is defined:

$$H = \frac{F}{m_2} = \frac{m_1}{\mu r^2} \tag{1.2}$$

If m_1 and m_2 are both of unit magnitude, the magnetizing force will be of unit magnitude. The unit for H is called the *oersted* (abbreviated Oe).

Work is done in moving a unit test pole m_2 in the magnetic field associated with a magnetizing force. The integration of H with respect to r as r changes during the motion of m_2 gives the work done. The work done is called magnetomotive force (mmf):

$$\text{mmf} = \int_{r_2}^{r_1} H \, dr = \frac{m_1}{\mu r_1} - \frac{m_1}{\mu r_2} \tag{1.3}$$

The commonly used unit for mmf is the gilbert. An oersted is one gilbert per centimeter. Another convenient term for mmf is *ampere-turn,* and when it is used the magnetizing force is in ampere-turns per inch.

Permeability μ is a function of the medium. Since magnetic circuits frequently involve more than one medium, another term which is independent of the medium is used. This term, called *flux density,* symbol B, is obtained by multiplying both sides of Eq. (1.2) by μ:

$$B = \mu H = \frac{m_1}{r^2} \tag{1.4}$$

The unit for B is gauss. In air, where the permeability μ is 1, one oersted is equivalent to one gauss.

The flux density around a point pole m_1 will be uniform at a constant distance from the pole. The total flux emanating from the pole will be the flux density at a distance r from the pole multiplied by the area of a sphere of radius r with the pole m_1 at the center:

$$\phi = BA = \frac{4\pi r^2 m_1}{r^2} = 4\pi m_1 \tag{1.5}$$

The total flux is independent of the medium and will be continuous across spherical boundaries between media of different permeabilities surrounding the pole m_1. The unit of flux is the maxwell. One gauss is one maxwell per square centimeter. One kilogauss is one thousand gauss per square centimeter. One thousand maxwells per square inch, commonly referred to as *kiloline per square inch,* and kilogauss are both commonly used units for flux density.

Magnetic flux is analogous to current in an electric circuit, and magnetomotive force is analogous to voltage. This suggests Ohm's law for magnetic circuits:

$$\text{mmf} = \phi \mathcal{R} \qquad (1.6)$$

In magnetic circuits the proportionality constant in Ohm's law is called *reluctance,* symbol $\mathcal{R}$. Reluctance is directly proportional to the length of the magnetic path and inversely proportional to the area through which the flux flows, the proportionality constant being permeability:

$$\mathcal{R} = \frac{l}{\mu A} \qquad (1.7)$$

Reluctance does not have a unit assigned. The term gilberts per maxwell is sometimes used. In that system of units, the dimensions are in centimeters and μ is 1 for air.

Permeability is an important property of magnetic circuits. Ferromagnetic materials have permeabilities of 1000 or more. Air, insulating materials, and most electric conductors except ferromagnetic materials have permeabilities of approximately 1.

The permeability of ferromagnetic materials is not constant. It varies with frequency, waveshape of applied signals, temperature, and flux density. It is not even a single-valued function. The double-valued nature of permeability gives rise to the well-known hysteresis curve, which is only approximate, varying from batch to batch of the same material. The elusive nature of permeability is one reason that the study of magnetic devices appears somewhat esoteric. The analysis of magnetic circuits generally uses constant and single-valued approximations for permeability over a limited range. Selecting these approximations and their boundaries is a major task of magnetic design. See Secs. 1.6 and 6.4 for a further discussion of the permeability of magnetic materials.

1.3 CURRENT, FLUX, AND INDUCTANCE IN MAGNETIC CIRCUITS

A current flowing in a straight conductor generates a magnetic field around the conductor, the intensity of which is proportional to the magnitude of the current. It is convenient to think of the magnetic field as consisting of flux lines around the conductor. These lines are continuous and tend to repel each other. They form concentric circles with the conductor at the center of the circles. The plane of the circles is perpendicular to the conductor. The direction of the flux circles, either clockwise or counterclockwise, will reverse if the current reverses. The number of flux lines is proportional to the total flux around the conductor. The flux lines will be close together near the conductor and become farther apart as the distance from the conductor increases. Where the lines are close together, the flux density B is higher than where the lines are farther apart. The separation between adjacent flux lines increases indefinitely as the distance from the conductor increases, each circle of flux meanwhile remaining continuous around the conductor.

Imagine a plane passing through the conductor and its magnetic field. The flux lines will pass through the plane in one direction on one side of the conductor and in the opposite direction on the other side of the conductor. If this conductor is formed into a closed loop, all the flux lines which were on one side of the plane will now tend to be concentrated in the area enclosed by the conductor. If additional conductor turns are made around the area, forming a coil, additional flux lines are added within the enclosed area. If all the flux lines are contained in the area, the number of flux lines will be proportional to the product of the current and the number of turns.

The magnetic field associated with the turns of wire or coil is an energy reservoir. The current that will flow through the conductor when a voltage is suddenly applied across the coil must supply the energy contained in the magnetic field. The current will start at zero and rise at a rate proportional to the applied voltage. When the applied voltage is removed, the energy in the magnetic field is returned to the electric circuit in the form of a voltage appearing across the coil which is proportional to the rate of change of the current. Because the flux is proportional to the current, energy transfer between the circuit and the magnetic field will occur only when the current is changing. These considerations lead to the defining relationship for inductance:

$$e = L \frac{di}{dt} \tag{1.8}$$

TABLE 1.1 Maximum Flux Density Formulas for Commonly Occurring Functions

Function	Waveform	Formula
1. Sine wave voltage (steady-state)		$B_{max} = \dfrac{E_{rms} \times 10^8}{4.44NAf}$
2. Symmetrical square wave voltage		$B_{max} = \dfrac{E_{pk} \times 10^8}{4NAf}$
3. Interrupted symmetrical square wave voltage		$B_{max} = \dfrac{E_{pk} \times t \times 10^8}{2NA}$
4. Half sine wave voltage pulse		$B_{max} = \dfrac{E_{pk} \times 2 \times t \times 10^8}{\pi NA}$
5. Unidirectional rectangular voltage pulse		$B_{max} = \dfrac{E_{pk} \times t \times 10^8}{NA}$
6. Full-wave-rectified single-phase sine wave voltage (ac component only)		$B_{max} = \dfrac{E_{dc} \times 10^8}{19.0NAf}$
7. Half-wave-rectified three-phase sine wave voltage (ac component only)		$B_{max} = \dfrac{E_{dc} \times 10^8}{75.9NAf}$
8. Full-wave-rectified three-phase sine wave voltage (ac component only)		$B_{max} = \dfrac{E_{dc} \times 10^8}{664NAf}$
9. Current	Any	$B_{max} = \dfrac{LI_{max} \times 10^8}{NA}$

Note: In these formulas N is the number of turns in the winding across which the voltage is developed. A is the cross-sectional area of the core around which the winding is placed. If the area is expressed in square centimeters, the flux density will be in gauss. If the area is expressed in square inches, the flux density will be in maxwells per square inch. Time t is in seconds. Frequency f is in hertz. Voltage E is in volts. Current I is in amperes. Inductance L is in henrys.

where di/dt = rate of change of current, A/s

e = instantaneous voltage appearing across coil, V

L = constant of proportionality called *inductance* (unit of inductance is the henry, abbreviated H)

The proportionality between current and flux leads to Faraday's law:

$$e = N\frac{d\phi}{dt} \times 10^{-8} \qquad (1.9)$$

In Eq. (1.9) N is the number of turns of wire in the coil and $d\phi/dt$ is the rate of change of flux in maxwells per second. The multiplying factor derives from the system of units used. The results of integrating Faraday's law for various applied voltages are used extensively in magnetics. See Table 1.1.

1.4 AMPERE'S LAW

Ampere's law states that a current-carrying conductor placed in a magnetic field will have a force exerted upon it in accordance with the following relationship:

$$F = Bil \qquad (1.10)$$

where F = force on wire in dynes perpendicular to both direction of current flow and magnetic field

B = flux density in gauss perpendicular to the wire

i = current, A

l = length of wire, cm

The absence of a proportionality constant results from using Ampere's law to define the magnitude of the unit of current.

Consider a straight wire of infinite length as shown in Fig. 1.1 which

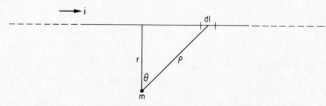

FIG. 1.1   Long current-carrying wire in a magnetic field.

is carrying a current i. A unit test pole m generates a magnetic field around the wire. The force exerted on an element of the wire will be

$$dF = Bi\,(\cos\theta)\,dl$$

The force exerted on the wire over its entire length will be

$$F = \int_{-\infty}^{+\infty} Bi\,(\cos\theta)\,dl$$

in which B is the flux density at the element dl. If the wire is in air, the flux density at any point on the wire will be

$$B = \frac{m}{\rho^2}$$

since the pole is of unit magnitude. The expression for the force exerted on the entire length of wire then becomes

$$F = \int_{-\infty}^{+\infty} \frac{im}{\rho^2}\,(\cos\theta)\,dl$$

$$= im \int_{-\pi}^{+\pi} \cos\theta\,d\theta$$

$$= \frac{2im}{r}$$

By Newton's law of action and reaction the force on the pole will be equal and opposite to the force on the conductor. Then by Eq. (1.2), since m is a unit pole,

$$\frac{F}{m} = H = \frac{2i}{r} \tag{1.11}$$

Equation (1.11) gives the magnetizing force generated by a current flowing in a long, straight conductor. By a similar procedure Ampere's law may be used to determine the magnetizing and magnetomotive forces in a coil of N turns. The derivation assumes that all the flux links all the turns. Figure 1.2 represents a cross section of a circular coil whose N turns occupy an infinitesimal area. The horizontal axis in the figure passes through the center of the coil. A unit magnetic pole is placed on the horizontal or x axis. A current i flowing in the coil of N turns will generate a magnetic field causing a force to act on the unit pole. That force, which is equal to the magnetizing force, is given by

$$F_x = Bil = \frac{mNil \sin \theta}{\rho^2}$$

$$\frac{F_x}{m} = H_x = \frac{2\pi Nir \sin \theta}{\rho^2}$$

From Eq. (1.3) the magnetomotive force at x resulting from moving the unit pole m from infinity to x will be

$$\mathrm{mmf} = -\int_{\infty}^{x} H_x \, dx$$

$$= -2\pi Nir \int_{\infty}^{x} \frac{\sin \theta}{\rho^2} \, dx$$

The total x component of the mmf set up by the coil will be obtained by integrating between the limits of $-\infty$ and $+\infty$. That integration can be more conveniently performed by changing the variable from x to θ. Then

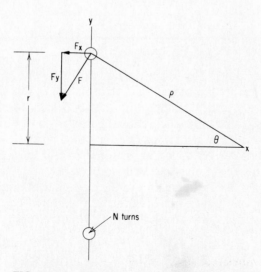

FIG. 1.2 Vector diagram and coil cross section used to determine magnetizing force.

the limits of integration will be 0 to π:

$$\text{mmf} = -2\pi Nir \int_0^\pi \frac{\sin^3 \theta \, (-\csc^2 \theta) \, d\theta}{r^2}$$

$$= -2\pi Ni \, [\cos \theta]_0^\pi$$

$$= 4\pi Ni$$

In this expression i is in abamperes and mmf is in gilberts. If i is expressed in amperes, the magnetomotive force in gilberts will be

$$\text{mmf} = 0.4\pi Ni \tag{1.12}$$

1.13

The y component of H has an external resultant force of zero. Internally the y component tends to crush the coil. This becomes an important consideration in coils carrying large currents or those subject to temporary overloads. Under such conditions the coil can develop shorts between adjacent turns and be destroyed by overheating. If the magnetic circuit is ferromagnetic, the magnetizing force will not extend to infinity in a straight line but will tend to follow the ferromagnetic core. Equation (1.12) is still a good approximation for the magnetomotive force and is regularly used for that purpose.

1.5 APPLICATION OF FARADAY'S LAW

The limitations imposed by ferromagnetic materials due to permeability, hysteresis, and saturation make the determination of flux density one of the most important calculations in magnetics design. See Secs. 1.6 and 6.1 for a discussion of these phenomena. Faraday's law, Eq. (1.9), is used to calculate flux density. The calculation is based on the assumption that the flux distribution over the cross-sectional area of the core is uniform. This assumption allows the flux term to be replaced by the product of flux density and cross-sectional area of the core. Since the applied voltage is usually a known quantity and the flux is unknown, Eq. (1.9) is transformed by integration into an explicit function of voltage:

$$B_{\max} = \frac{\phi_{\max}}{A}$$

$$= \frac{10^8}{NA} \int_0^t e \, dt \tag{1.13}$$

In Eq. (1.13) e is in volts, t is in seconds, and ϕ is in maxwells. If A is in square centimeters, then B will be in gauss. If A is in square inches, then B will be in maxwells per square inch.

In obtaining expressions for B_{max} from known voltages, the limits of integration and boundary conditions must be chosen with care. This is illustrated for the most commonly used formula for B_{max}, the case of a steady-state sine wave voltage. The value of the integral is positive for 1 half-cycle of voltage and negative for the other half-cycle of voltage. When the positive half-cycle of voltage begins, B will be at its maximum negative value. The total change in flux during the subsequent half-cycle will therefore be $2B_{max}$. The value of B_{max} is consequently only one-half of the value obtained from Eq. (1.13):

$$
\begin{aligned}
B_{max} &= \frac{E_{pk} \times 10^8}{2NA} \int_0^{\pi/\omega} \sin \omega t \; dt \\
&= \frac{E_{pk} \times 10^8}{2NA\omega} \left[\cos \omega t \right]_{\pi/\omega}^0 \\
&= \frac{E_{pk} \times 10^8}{NA\omega} \\
&= \frac{E_{rms} \times 10^8}{\sqrt{2} \; \pi f NA} \\
&= \frac{E_{rms} \times 10^8}{4.44 f NA}
\end{aligned}
\tag{1.14}
$$

Maximum flux densities for other waveforms for which analytical expressions exist may be determined in a similar manner through the use of Eq. (1.13). See Table 1.1 for maximum flux density formulas for some commonly occurring functions. If an analytical expression does not exist, the maximum flux density may be obtained graphically.

A magnetic device may experience a higher flux density under transient conditions than under steady-state conditions. If a magnetic core in a neutral state is suddenly subjected to a symmetrical magnetizing force, the first cycle of flux will start at zero, and so the maximum flux will be twice that for the steady-state condition. The device may saturate on the first cycle. A high inrush current may occur under these conditions.

The flux density in the core of an inductor may be determined from the current flowing through the inductor if the inductance is known. This is done by applying Eqs. (1.8) and (1.9):

$$
e = L \frac{di}{dt} = N \frac{d\phi}{dt} \times 10^{-8}
$$

By integrating both terms, the following is obtained:

$$Li = N\phi \times 10^{-8} = NBA \times 10^{-8}$$

from which the following expression for B is obtained:

$$B = \frac{Li}{NA} \times 10^8 \qquad (1.15)$$

The inductance L is a function of permeability. Equation (1.15) is, therefore, valid only over a range for which μ is constant and L is known.

Confusion sometimes arises over the origin of flux from a flow of current when the developed voltage is used to calculate the same flux. Flux is generated only by a flow of current. However, both current and flux are proportional to the time integral of voltage. Hence, flux density may be calculated directly from the time integral of voltage. The current that flows in response to a voltage applied to an inductor is that quantity needed to generate the flux calculated from the time integral of voltage. This current, in the case of a transformer, is the exciting current of the transformer, not to be confused with the load current which produces no net flux. The method selected to calculate flux density depends upon convenience and the known quantities.

In some applications the total flux in a magnetic device results from independently determined direct currents and ac voltages. If the device operates in the linear range, superposition holds. The flux from the direct current may be determined from Eq. (1.15), and the ac flux may be determined from Eq. (1.13) or Table 1.1. The total flux is then the sum of the ac and dc components. A typical example of this application is a filter inductor across which appears an ac voltage and through which direct current flows:

Illustrative problem

Determine the flux density in a filter inductor used in a single-phase full-wave rectifier with inductor input filter. The inductance is 1.0 H. The power frequency is 60 Hz. The output of the rectifier is 100 V dc. The direct current is 1.0 A. The number of turns in the coil is 1000. The effective area of the core is 10.9 cm^2.

Solution

It may be assumed that all the ac voltage from the rectifier appears across the input inductor because of the lower impedance of the filter capacitor. The resistance of the inductor is neglected. It is assumed that the inductor operates in the linear range so that superposition may be used.

The dc flux density is determined by the use of Eq. 9 in Table 1.1:

$$B_{dc} = \frac{LI_{dc}}{NA} \times 10^8$$

$$= \frac{1 \times 1 \times 10^8}{1000 \times 10.9}$$

$$= 9170 \text{ G}$$

Since the assumption is made that all the ac voltage appears across the inductor, Eq. 6 of Table 1.1 may be used to determine the ac component of flux density:

$$B_{ac,max} = \frac{E_{dc} \times 10^8}{19.0NAf}$$

$$= \frac{100 \times 10^8}{19 \times 1000 \times 10.9 \times 60}$$

$$= 805 \text{ G}$$

The total maximum flux density is the sum of the dc and ac flux densities:

$$B_{max} = 9170 + 805$$

$$= 9975 \text{ G}$$

1.6 MAGNETIC CIRCUITS

If the core of a magnetic device is continuous—as in a toroidal core, for example—the reluctance of the magnetic circuit will be constant along the magnetic path and will be represented by Eq. (1.7). The magnetizing force H will also be constant along the magnetic path so that from Eq. (1.12)

$$H = \frac{\text{mmf}}{l_i}$$

$$= \frac{0.4\pi Ni}{l_i} \tag{1.16}$$

In Eq. (1.16) i is in amperes, l_i is the length of iron path in centimeters, and H is in oersteds.

Most magnetic circuits are not continuous. They usually contain a small air gap in addition to the path in the core. When an air gap exists, the reluctance of the circuit is the sum of the reluctances of the iron core and the air gap:

$$\mathcal{R} = \frac{l_a}{\mu_a A_a} + \frac{l_i}{\mu_i A_i} \qquad (1.17)$$

where l_a = total air gap, cm
μ = permeability either of air or iron, as indicated by subscripts
l_i = total length of iron path, cm
A = cross-sectional area of iron path or air gap, as subscripts indicate

If fringing in the air gap is neglected, the two areas are equal.

From Eqs. (1.6) and (1.17):

$$\text{mmf} = \frac{l_a \phi}{\mu_a A_a} + \frac{l_i \phi}{\mu_i A_i}$$

Since $B = \mu H$, $\mu_a = 1$, and $\phi = BA$:

$$\text{mmf} = l_a B + l_i H_i \qquad (1.18)$$

In Eq. (1.18) mmf is in gilberts, l_a and l_i are in centimeters, B is in gauss, and H is in oersteds.

If both sides of Eq. (1.18) are divided by Bl_i, and if l_a is small compared with l_i, then the left-hand side of that equation will be approximately equal to the reciprocal of the effective permeability of the magnetic circuit. Equation (1.18) then becomes

$$\frac{1}{\mu_{\text{eff}}} = \frac{l_a}{l_i} + \frac{1}{\mu_i} \qquad (1.19)$$

In Eq. (1.19) l_a and l_i may be in any dimensions of length, provided they are the same, and permeability is a dimensionless number.

Equation (1.18) leads to an important inductance formula. First divide that equation by l_i:

$$\frac{\text{mmf}}{l_i} = \frac{l_a}{l_i} B + H_i$$

$$= B\left(\frac{l_a}{l_i} + \frac{1}{\mu_i}\right)$$

Then from Eqs. (1.12) and (1.19):

$$\frac{\text{mmf}}{l_i} = \frac{0.4\pi Ni}{l_i}$$

$$= B\frac{1}{\mu_{\text{eff}}}$$

Now solve for B.

$$B = \frac{0.4\pi Ni\mu_{\text{eff}}}{l_i}$$

Equate the above expression to Eq. (1.15).

$$\frac{0.4\pi Ni\mu_{\text{eff}}}{l_i} = \frac{Li}{NA} \times 10^8$$

Solving for L yields

$$L = \frac{4\pi N^2 \mu_{\text{eff}} A}{l_i} \times 10^{-9} \tag{1.20}$$

In Eq. (1.20) L is in henrys, A is in square centimeters, and l_i is in centimeters. This equation is widely used to calculate the inductance of ferromagnetic inductors. Equation (1.19) is frequently used to determine the permeability term in Eq. (1.20).

The energy contained in an inductance can be determined by multiplying both sides of Eq. (1.8) by $i\,dt$ and integrating:

$$\int ei\,dt = J = L \int \frac{di}{dt} i\,dt = \frac{1}{2}Li^2 \tag{1.21}$$

In Eq. (1.21) J is in joules, L is in henrys, and i is in amperes. Inductors with similar $\frac{1}{2}Li^2$ values are usually of similar size over a large range of inductances and currents. This energy term is sometimes called the *storage factor*. In this usage the ½ factor is sometimes dropped.

In Eq. (1.18) the mmf is determined from the current, often a known and constant quantity. When this condition exists, this equation contains two unknowns, B and H_i. The equation is linear. The unknown quantities are the same as those appearing on the magnetization curve. If the magnetization curve is approximated as a single-valued function passing through the origin, then Eq. (1.18) may be plotted on the magnetization curve and the unknowns determined graphically from the intersection of the straight line with the magnetization curve. This intersection is called the *operating point* because it frequently represents a position established by a dc magnetization around which a superimposed ac magnetization operates. The construction can be done by first letting H equal zero. Then the vertical intercept B_o is equal to mmf/l_a. Then the horizontal intercept can be determined by letting B equal zero. H_o is then equal to mmf/l_i. This process assumes that the air gap is known.

In a practical problem the air gap is not usually known, and hence the air gap must be selected for optimum performance. A number of factors converge on this problem. It is necessary to keep the operating point below saturation. If the current used to calculate the mmf is the peak current, then the operating point may be just below saturation. If the current used to calculate the mmf is a direct current and there are ac components of current superimposed, then the operating point must be selected to accommodate the ac component of flux which is usually calculated from Faraday's law (see Table 1.1). If maximum inductance is desired from a given device, the variation of iron permeability μ_i with magnetization must be weighed in selecting the air gap. This information is available from manufacturers of core materials. See Sec. 6.4. The introduction of an air gap in an inductor carrying direct current is usually essential. The iron permeability and the ratio of air gap to length of iron path both contribute to effective permeability in accordance with Eq. (1.19). As the air gap is increased, the vertical intercept B_o decreases, and the coordinate H of the operating point decreases. This decrease in H of the operating point increases the iron permeability μ_i as the contribution of the air gap tends to make the effective permeability decrease. There is an optimum air gap determined by trial and error for which the effective permeability is a maximum. This air gap will give the maximum inductance for a given direct current and given core and coil geometry.

The larger the air gap, the less effect μ_i has on the effective permeability. The variation in μ_i with magnetizing conditions causes ferromagnetic inductors to be nonlinear. The larger the air gap, the closer the inductor will approach linear performance. For linear inductance the operating point is placed well down on the magnetization curve. This reduces the effective permeability and increases size and weight. What is referred to as the air gap is, for construction reasons, some type of solid insulating material which has no effect on the magnetic operation.

Illustrative problem

For the inductor in the illustrative problem on Faraday's law above, find the air gap that will give the maximum effective permeability and determine the value of the maximum inductance. The following data are repeated for reference:

> *Direct current*: 1.0 A
> *Number of turns*: 1000
> *Length of magnetic path*: 20.96 cm
> *Alternating current flux density*: 805 G
> *Effective core area*: 10.9 cm²

Solution

The magnetization curve used for this problem is shown in Fig. 1.3. The incremental permeability curves used are given in Fig. 1.4. Calculate the magnetomotive force using Eq. (1.12).

$$\text{mmf} = 0.4\pi N I_{dc}$$

$$= 0.4\pi \times 1000 \times 1$$

$$= 1257 \text{ Gb}$$

Determine the horizontal intercept of Eq. (1.18) by letting B equal zero.

$$H_o = \frac{\text{mmf}}{l_i}$$

$$= \frac{1257}{20.96}$$

$$= 60 \text{ Oe}$$

The horizontal intercept is beyond the range of the plot. This difficulty is overcome by calculating the slope of Eq. (1.18) from the horizontal and vertical intercepts and using that slope to draw the straight line through the vertical intercept. Make a trial selection of the vertical intercept of 12.0 kG. Determine the slope and draw the

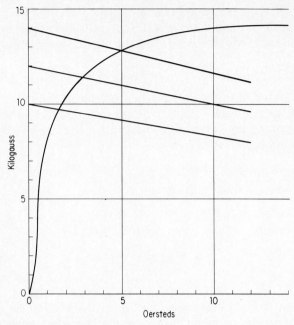

FIG. 1.3 Typical dc magnetization curve for silicon steel with air gap lines drawn.

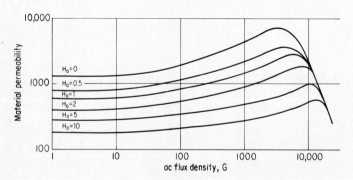

FIG. 1.4 Typical incremental permeability curves for silicon steel.

line for Eq. (1.18) as shown in Fig. 1.3. Note the value of H_{dc} where the straight line intersects the magnetization curve. With this value of H_{dc}, 3.0 Oe, and the value of ac flux density above, the value of μ_i from Fig. 1.4 is estimated at 700. Use Eq. (1.18) and the vertical intercept to determine the air gap.

$$l_a = \frac{\text{mmf}}{B_o}$$

$$= \frac{1257}{12,000}$$

$$= 0.105 \text{ cm}$$

Use Eq. (1.19) to calculate the effective permeability.

$$\frac{1}{\mu_{\text{eff}}} = \frac{l_a}{l_i} + \frac{1}{\mu_i}$$

$$= \frac{0.105}{20.96} + \frac{1}{700}$$

$$= 156$$

Repeat this process for a vertical intercept of 14 kG.

$$H_{\text{dc}} = 5 \qquad \mu_i = 460$$

$$l_a = \frac{1257}{14,000}$$

$$= 0.0898$$

$$\frac{1}{\mu_{\text{eff}}} = \frac{0.0898}{20.96} + \frac{1}{460}$$

$$\mu_{\text{eff}} = 155$$

Repeat again for a vertical intercept of 10 kG.

$$H_{\text{dc}} = 1.6 \qquad \mu_i = 1000$$

$$l_a = \frac{1257}{10,000}$$

$$= 0.126$$

$$\frac{1}{\mu_{\text{eff}}} = \frac{0.126}{20.96} + \frac{1}{1000}$$

$$\mu_{\text{eff}} = 143$$

The maximum effective permeability obtainable is about 156. Use Eq. (1.20) to calculate the inductance for that permeability.

$$L = \frac{4\pi N^2 \mu_{eff} A}{l_i} \times 10^{-9}$$

$$= \frac{4\pi (1000)^2 \times 156 \times 10.9 \times 10^{-9}}{20.96}$$

$$= 1.02 \text{ H}$$

1.7 RMS CURRENT

The current-carrying capacity of a conductor is determined by the average power dissipated in the conductor. The average power dissipated is equal to the square of the rms current flowing through the conductor multiplied by the resistance of the conductor. A complete description of the variation of a current with time must be available in order to determine its rms value. This description includes the duty cycle as well as variations during the time that current is flowing. If the current consists of two or more components, the rms value of the total current can be determined by taking the square root of the sum of the squares of the rms values of every component.

The duty cycle of a current is the ratio of time on to total time. This will be a number equal to or less than 1. If a component of current is constant during the time that it flows, the average power that it dissipates is the square of the current multiplied by the resistance of the conductor and by the duty cycle. The rms value of that component of current will be the current multiplied by the square root of the duty cycle. If a component of current having a duty cycle less than 1 varies during the time that it is on, then the rms value the component would have if its duty cycle were 1 must be determined. That value multiplied by the square root of the duty cycle is the component's rms value.

If the duty cycle is very small, the rms value of current will be dramatically reduced below its peak value. Conductor current-carrying capacity under this condition must consider, in addition to the rms value, the relationship between actual time during which the current flows and the thermal time constant of the conductor. The thermal time constant of that conductor will determine the temperature rise during the time that current is flowing. See Sec. 9.2 for a discussion of thermal time constant. The power dissipated in the conductor during the time that current flows must not be great enough to cause the insulation temperature to exceed its allowable limit.

Illustrative problem

A pulse modulation transformer has a winding which carries both a filament current and a pulse current. The winding consists of two parallel conductors. The windings are in parallel for the pulse current and in series for the filament current. The pulse current will divide equally between the two conductors, while each conductor will carry the full filament current. The filament current is a sine wave and flows continuously. The pulse current is a rectangular wave with a high peak value and a small duty cycle. Typical values for this device are:

> *Filament current.*   5 A
> *Peak pulse current.*   100 A
> *Duty cycle.*   0.002

Solution

The rms current in each conductor will be

$$I_{rms} = \sqrt{I^2_{fil} + (\text{duty cycle})I^2_{pk}}$$
$$= \sqrt{5^2 + (0.002)50^2}$$
$$= 5.48 \text{ A}$$

CHAPTER 2

Circuit Analysis

2.1 THE IDEAL TRANSFORMER

Faraday's law in integral form, Eq. (1.13), states that the flux produced in a coil is proportional to the time integral of voltage. The same law in differential form, Eq. (1.9), states that a voltage is developed across a coil proportional to the turns and the time rate of change of flux. Consider a coil to which a varying voltage is applied which generates a varying flux. If in close proximity to this coil a second coil is placed so that the flux generated by the first coil passes through the second coil, then in accordance with Eq. (1.9) a voltage will be developed across the second coil proportional to the rate of change of flux and the number of turns in the second coil. If all the flux generated by the first coil links all the turns of the second coil, the voltage in the second coil by Faraday's law will be proportional to the voltage applied to the first coil and the ratio of the turns in the second coil over the turns in the first coil. If the ends of the second coil are connected through a resistance, a current will flow in such a direction as to oppose any change in flux. The direction of current is dictated by energy considerations. The second coil cannot supply energy to the magnetic field but can only remove energy from it. The current that flows in the second coil tends to reduce the net flux linking both coils. To continue to satisfy Faraday's law, an additional current must flow in the first coil which will exactly offset the reduction in flux caused by the current flow in the second coil. If the resistance is replaced by a reactance, a similar situation exists. To supply energy to the reactance, the current in the second coil will flow in such a direction as to reduce the flux, as with a resistance. At some time during the voltage cycle the direction of energy transfer is reversed. The secondary current will tend to increase the flux. A lower instantaneous primary current will flow so that Faraday's law for the voltage applied to the primary is satisfied. This will cause a primary-current phase shift from the applied voltage, resulting in a transfer of energy back to the source during a portion of the cycle. Since the flux produced by either current is proportional to the product of current and turns, the current in the first coil will be equal to the current in the second coil multiplied by the ratio of the turns in the second coil over the turns in the first coil. Thus the current is transferred by the reciprocal of the ratio by which the voltage is transformed. The first coil is called the *primary winding,* and the second coil is called the *secondary winding.* By use of the suffixes *p* and *s* to designate primary and secondary, the following relationships can be stated:

$$E_s = \frac{N_s}{N_p} E_p \tag{2.1}$$

$$I_s = \frac{N_p}{N_s} I_p \tag{2.2}$$

By multiplying Eq. (2.1) by Eq. (2.2):

$$E_s I_s = E_p I_p \tag{2.3}$$

By dividing Eq. (2.1) by Eq. (2.2):

$$\frac{E_s}{I_s} = Z_s$$

$$= \left(\frac{N_s}{N_p}\right)^2 Z_p \tag{2.4}$$

Equation (2.3) shows that the volt-amperes in the primary and secondary are equal. Equation (2.4) shows that impedances are transformed as the square of the turns ratio. Equations (2.1) through (2.4) are the basic relationships for the ideal transformer.

2.2   THE TRANSFORMER EQUIVALENT CIRCUIT

Transformer utilization is concerned with the departure of realizable transformers from the ideal transformer. Analysis of this departure is usually made by means of equivalent circuits. Equivalent circuits are not exact replicas of real transformers; they are only convenient approximations. Judgment is required to determine which equivalent circuit to use and to evaluate the results of analysis by that circuit.

Transformer equivalent circuits are three-terminal networks in cascade with an ideal transformer. What is sometimes referred to as the complete equivalent circuit is shown in Fig. 2.1. The elements of this circuit are explained below.

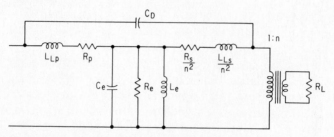

FIG. 2.1 Complete transformer equivalent circuit.

The current transformation for the ideal transformer neglects the current which flows in the primary independent of the load current. This current is required to produce the flux as explained in Sec. 1.3. Called the *magnetizing current,* this current is added to the transformed load current. It does not represent a power loss. The magnetizing current can, therefore, be accounted for by an inductance in parallel with the ideal transformer. This is shown in Fig. 2.1 as L_e. Any nonlinear effects due to ferromagnetic materials are represented by changes in this inductance.

Because ferromagnetic materials are themselves conductors, the changing flux in the core sets up small current loops in the core called *eddy currents.* These currents represent a power loss which is independent of load current. The continuous alignment and reversal of particles in the core which give the core its magnetic properties is another source of power loss by the transformer core. This latter loss is called *hysteresis loss.* The combined hysteresis and eddy current losses are known as *core loss.* Since the current associated with the core loss is in addition to the transformed load current, the core loss can be represented by a resistance in parallel with the primary of the ideal transformer. This is shown in Fig. 2.1 as R_e. The value of this resistance will vary with the operating conditions of the transformer.

Since both primary and secondary windings are made of conductors that have finite conductivity, each winding will have associated with it a resistance. Current in the secondary winding will flow through the secondary resistance only in response to an applied load. Therefore, the secondary resistance may be represented as a resistance in series with the load. Viewed from the primary, the secondary winding resistance is, in fact, indistinguishable from the load and transforms to the primary side by the square of the turns ratio the same as the load impedance. The secondary resistance is shown in Fig. 2.1 as R_s/n^2 on the primary side of the ideal transformer which has a turns ratio of $1:n$. The primary resistance carries not only the transformed load current but also the no-load current which is the sum of the magnetizing current and the core loss current. The primary resistance in the equivalent circuit is, therefore, placed before the parallel elements representing no-load current. The primary resistance is shown as R_p in Fig. 2.1.

Because transformers are composed of closely spaced turns with the primary and secondary often very close together and both close to the core and ground, there is an appreciable amount of stray capacitance present. The turns are at varying potentials to each other and ground, and the geometric capacitance of each turn varies, making the true effective capacitance hopelessly complex. This situation can sometimes be handled by representing the stray capacitance by two lumped capacitors, one in series and one in parallel, shown as C_e and C_D in Fig. 2.1. The positions of these capacitors in the equivalent circuit are subject to some adjustment depending upon the nature of the analysis.

The voltage transformation in the ideal transformer, Eq. (2.1), and the expression for magnetomotive force in a closely spaced coil, Eq. (1.12), both assume that all the flux links all the turns of all windings. Actually a small amount of leakage flux exists. The presence of this leakage flux is evidenced by the failure of the inductance of the device to be exactly proportional to the square of the number of turns, the failure of the voltage transformation ratio to be exactly equal to the turns ratio, and the existence of a small inductance detectable in the primary when the secondary is shorted, at which time there would otherwise be zero change in flux. These factors suggest for the equivalent circuit two series inductances called *leakage inductance,* represented in Fig. 2.1 as L_{Lp} and L_{Ls}/n^2. As with the secondary resistance, the secondary leakage inductance transforms to the primary by the square of the turns ratio. The leakage inductance is a distributed parameter similar to stray capacitance. Its representation as a lumped element is only an approximation. Even with its imprecision, the complete equivalent circuit is too unwieldy for most analysis. Usually it is further simplified into one or more circuits of restricted validity. Each circuit is then analyzed for those conditions for which the simplification is valid.

2.3 SIMPLIFICATIONS OF THE EQUIVALENT CIRCUIT

The simplest transformer is the power transformer operating from a single-frequency sine wave voltage source of zero impedance into a resistive load. The magnetizing inductance and stray capacitance can usually be neglected. The voltage drop across the leakage inductance subtracts in quadrature from the applied voltage, and so the effect on the magnitude of the output voltage is small and may be often neglected. Under these conditions, the equivalent circuit of Fig. 2.2 is applicable. The turns ratio is easily adjusted to compensate for the voltage drop through the winding resistance. The power loss due to the current flow through the winding resistances, called *copper loss,* and the core loss determine the efficiency of the transformer.

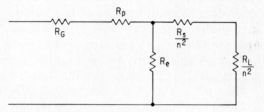

FIG. 2.2 Equivalent circuit for single-frequency power transformer with resistive load.

Under some conditions the leakage inductance must be considered in a single-frequency transformer. The leakage inductance may be unusually large as in the case in a high-voltage isolation transformer. The load impedance may be reactive. In those cases the equivalent circuit of Fig. 2.3 is sometimes used. All the leakage inductance may be considered to be one lumped value. The voltage drop across both the leakage reactance and the winding resistances must be considered. Particularly trou-

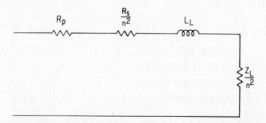

FIG. 2.3 Equivalent circuit for power transformers in which leakage inductance must be considered.

blesome is the case of a load having capacitive reactance. The voltage across the load may have resonant peaking. Since the parameters of the equivalent circuit and the load are known to only a limited accuracy, the use of taps to adjust the output voltage may be required.

When the source frequency varies over a wide range and when the source impedance cannot be neglected, the problem becomes more difficult. The complete equivalent circuit has the configuration of a bandpass filter. The output will be attenuated at both the highest and lowest frequencies. At the midband there will be an insertion loss due to the winding resistances and the core loss. Separate circuits are used for each of three response regions. Figure 2.4 shows an equivalent circuit for the low-frequency response region. The shunt inductance is the significant parameter. The leakage inductance and distributed capacitance can be neglected. The midband response will be approximated by Fig. 2.2. The high-frequency response may be analyzed by the circuit of Fig. 2.5 in which the open-circuit inductance is now neglected, but the leakage inductance and shunt capacitance must be considered. Square wave and pulse transformers are generally analyzed by three circuits shown in Figs. 2.6 to 2.8 for the leading edge, the top, and the trailing edge, respectively. Detailed information on the analysis of these equivalent circuits is given below.

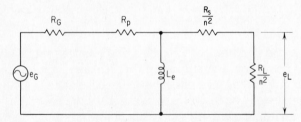

FIG. 2.4 Transformer equivalent circuit for evaluating low-frequency response with a generator having significant internal resistance.

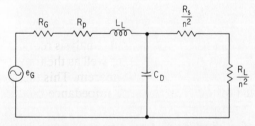

FIG. 2.5 Transformer equivalent circuit for evaluating high-frequency response with a generator having significant internal resistance.

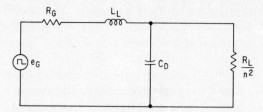

FIG. 2.6 Transformer equivalent circuit used for evaluating leading edge response to a rectangular voltage pulse.

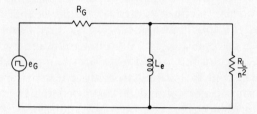

FIG. 2.7 Transformer equivalent circuit used to evaluate top response to a rectangular voltage pulse.

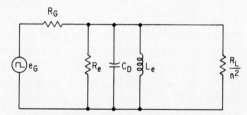

FIG. 2.8 Transformer equivalent circuit used to evaluate trailing edge response to a rectangular voltage pulse.

2.4 POWER SOURCES AND LOADS

Transformer response is a circuit problem. Analysis requires a knowledge of the characteristics of source and load as well as the transformer. Transformer circuit analysis uses Thévenin's theorem. This theorem states that the current which flows through a load impedance connected between two terminals of a network is the same as that which would flow if the load impedance were connected to a voltage source equal to the open-circuit voltage at the two terminals with the internal impedance of the network in series. The transformer primary resistance and leakage inductance will appear in series with the Thévenin generator impedance. The existence of nonlinear generator impedances complicates the analysis, a common problem in pulse and switching circuits. The impedance presented by the generator is not the same during the rise of the pulse as it is for the top of the pulse, and the impedance during the fall of the pulse is different from that during either the rise or the top.

Transformers are sometimes driven from current sources. Norton's theorem is useful in those cases. It states that the current which flows in a load impedance joining two terminals of a network is the same as that which would flow from a constant-current generator whose current is equal to the current that would flow if the two terminals were shorted and with an impedance equal to the internal impedance of the network shunted across the load impedance. The voltage developed across a loaded transformer connected to a Norton current generator will depend upon the impedance of the loaded transformer in parallel with the generator impedance and the current from the constant-current generator. The Norton current generator may be converted to a Thévenin generator whose series impedance is the generator impedance of the Norton current generator and whose open-circuit voltage is equal to that same impedance multiplied by the generator current. The circuit may then be analyzed in terms of voltage, a more convenient procedure for some calculations.

The most common representation of a load for transformer networks is a constant resistance. It is important to know when this simplest representation is valid and when a more complex representation is required. A common variant to the constant resistance is a resistance in series with a switch which opens and closes at specified times during the operating cycle. Sometimes two or more resistances and switches in parallel are necessary to approximate load characteristics. Pulse transformer loads are often so represented with different resistances for the rise, top, and trailing edges of the pulse.

Reactive elements in the load must generally be considered. The loads on distribution transformers are regularly given in terms of power factor. In a series representation of the load, the power factor is the ratio of the resistance to the total impedance. Power and distribution transformers are rated by volt-ampere capacity, i.e., by their ability to deliver a specified current and voltage without regard to the nature of the load. If the load is heavily reactive, little power is delivered for heavy utilization of the transformer and other current-rated equipment. Power factor is important for economical use of distribution facilities. A similar situation often exists in equipment transformers. Distorted current waveforms and large currents with low-duty cycles have disproportionately high rms values for power delivered. This is an inevitable consideration in the design of transformers with rectified loads. When reactive elements appear in the loads of wideband and nonsinusoidal transformers, those elements are usually lumped with the elements of the equivalent circuit. The simplest applicable form of the equivalent circuit is then chosen to analyze each region of operation.

2.5   EFFECT OF TRANSFORMERS ON CIRCUIT PERFORMANCE

The needed functions that transformers perform exact a toll on equipment performance in the form of reduced efficiency and fidelity. The methods used to examine these effects vary with circuit application. A simplification sometimes used is to replace a loaded transformer with a resistor whose value is the transformed load impedance. This is usually insufficient. More adequate methods used for analysis relate to the various equivalent circuits. In addition to analyzing transformer circuit performance on the basis of equivalent circuits, certain significant circuit characteristics require special consideration. These characteristics may appear in various transformer applications which may employ several different simplifications of the transformer equivalent circuit. For example, a transformer subjected to dc magnetization could be a power transformer driven from a commercial power source, an audio transformer, or an inverter transformer. Such special circuit characteristics are discussed separately.

CHAPTER 3

Circuit Performance of Power Transformers

In common parlance a power transformer operates over a narrow frequency band from a generator whose internal impedance is small compared with the series impedance of the transformer winding resistance and leakage inductance. Transformer applications which do not conform to this definition may involve the transformation of small or large amounts of power, but the methods used to analyze performance differ. Typical applications for power transformers include distribution systems for electric utilities, rectifiers providing dc electric power from commercial ac power, filament voltage for electron tubes, and miscellaneous utility voltages required in electronic equipment.

3.1 EFFICIENCY

Among the first considerations in power transformer applications is efficiency. Power transformer losses contribute to the economics of the system in which they operate. The heat from these losses must be considered in the heat transfer problems of the transformers and the equipment in which they are installed. The power loss in a transformer is due to the resistance of the windings and the losses in the core. The loss in each winding is equal to the resistance of that winding multiplied by the square of the current which flows through it. The core loss, which is determined by the core material and the transformer design, varies with the amplitude and frequency of the applied voltage. It is approximately true that the core or iron loss represents the no-load loss and that the winding resistance or copper loss represents the load loss.

The common definition of efficiency of a machine is the power output divided by the power input, the latter being the output power plus the machine losses. The phenomenon of power factor makes some explanation necessary when applying this definition to transformers. The losses in a power transformer remain constant for a given load current and applied voltage when the power factor of the load varies from 1 to zero. By use of the common definition, the efficiency of such a transformer would vary from a maximum to zero. It is customary, therefore, to include in the definition of transformer efficiency the stipulation that the power factor be unity. A corollary to this stipulation is that power transformers are rated by volt-ampere capacity, not wattage. Transformer efficiency then becomes volt-ampere output divided by volt-ampere output plus total losses in the transformer. This definition does not give weight to reactive no-load current flow since no significant power loss in the transformer is involved. There is no theoretical prohibition, with this definition, to the efficiency of a transformer reaching 100 percent. The practical limits to efficiency are size, weight, and cost.

3.2 REGULATION

In a power transformer, regulation is the difference between secondary no-load and secondary full-load voltages divided by the secondary full-load voltage with the primary voltage held constant. Often expressed as a percentage, regulation is part of the unavoidable price paid for transformer use. It is closely related to efficiency, and, like efficiency, there is no theoretical limit to perfection—only the practical limits of size, weight, and cost. If the leakage inductance can be neglected, and if the power factor of the load is unity, the regulation of a power transformer is approximately equal to the ratio of copper losses to load power. Leakage inductance becomes a factor in regulation when the size of the transformer is above a few hundred volt-amperes. It can be a factor in smaller transformers which devote an unusually large percentage of their coil volume to insulation. In many cases it is necessary to consider the reactance of the load in determining regulation. The equivalent circuit of Fig. 2.3 is used to compute regulation. The relative magnitudes of winding resistance, leakage reactance, load resistance, and load reactance determine which elements may be neglected and which elements must be considered in making this computation.

3.3   PHASE SHIFT

Occasionally it is necessary to know and limit the phase shift between input and output of a power transformer. The problem is likely to occur in such applications as a transformer supplying the voltage to the fixed phase of an ac servomotor or a transformer supplying a reference voltage for a phase detector. The series elements of winding resistance and leakage reactance that cause regulation also cause phase shift. The out-of-phase component of voltage drop across these series elements is usually caused by in-phase load current flowing through the leakage inductance or out-of-phase load current flowing through the winding resistance or a combination of both. In critical applications it may be necessary to consider the voltage drop caused by the exciting current flowing through the series elements. In such cases the resistance representing the core loss and the inductance representing the magnetizing current are considered either alone or as part of the load when determining the drop across the series elements. See the illustrative problem below for a sample calculation of regulation and phase shift.

Illustrative problem

Determine the regulation and phase shift in a transformer with equivalent circuit parameters and load given below. Use the equivalent circuit of Fig. 3.1.

Total winding resistance referred to primary:

$$R_w = 0.2 \ \Omega$$

Total leakage inductance referred to primary:

$$L_L = 0.002 \ \text{H}$$

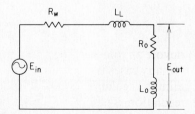

FIG. 3.1 Transformer equivalent circuit used for calculating regulation and phase shift in illustrative problem.

Load resistance referred to primary:

$$R_o = 10.0 \ \Omega$$

Input frequency:

$$f = 60 \text{ Hz}$$

Load inductance referred to primary:

$$L_o = 0.01 \text{ H}$$

Solution

$$\frac{E_o}{E_{in}} = \frac{R_o + j\omega L_o}{R_w + R_o + j\omega(L_o + L_L)}$$

$$= \frac{10 + j3.77}{10.2 + j4.524}$$

$$= \frac{10.687 \ /20.66°}{11.158 \ /23.92°}$$

$$= 0.9578 /{-3.26°}$$

Regulation

$$\frac{E_{in} - E_o}{E_o} = \frac{1}{0.9578} - 1$$

$$= 4.4\%$$

Phase shift

$$-3.26°$$

3.4 FREQUENCY RANGE

Power transformers are designed to operate near the maximum allowable flux density under steady-state conditions at the lowest rated frequency and highest rated voltage. See Secs. 1.2 and 1.5 for a discussion of flux density. The saturation threshold places an upper limit to the ratio of voltage over frequency. Operation above this limit can be destructive. Power transformers will operate satisfactorily at frequencies considerably above the minimum up to the region where the leakage inductance becomes a governing factor. Operation above this region results in reduced output voltage but is not destructive. This question comes up in connection with equipment used overseas but designed for use in the United States and equipment designed for ground use operated in air-craft. See the introductory remarks in Chap. 12 for a further discussion. Power transformers may operate without saturating at voltages higher than rated, provided the frequency is increased above the minimum by at least the same ratio as the voltage is increased above rated. This pro-cedure, while often done for test purposes, can be hazardous. Care must be taken to avoid operating the insulation system above safe limits or passing excessive currents through the windings.

3.5 STARTING TRANSIENTS

Transient conditions, particularly starting transients, can cause higher flux densities than exist in the steady-state condition. Power trans-formers, for reasons of economy, are not designed to absorb these tran-sients without saturating. This condition contributes to what is usually called the *inrush current*. The magnitude of the inrush current depends upon many factors including circuit characteristics. Some types of loads have starting currents much larger than operating currents. Motors and the filaments of large transmitting tubes are common examples. The point on the sine wave at which switch contact is made determines the volt-time area during the first half-cycle. If contact is made when the volt-age is zero, the volt-time area during the first half-cycle will be twice that developed when switch contact is made when the voltage is at its peak value. In the former case, saturation on the first half-cycle is assured. Sat-uration of the core reduces the value of the shunt inductance (L_e in Fig. 2.1) to zero. Current is then limited only by the leakage inductance and primary winding resistance. These two parameters combined constitute the short-circuit impedance of the transformer. Core saturation effects last only a few cycles of power frequency. The inrush current will be greater if the transformer is operating at the threshold of saturation in steady state. The results of this are sometimes observed in the form of occasional blown fuses or popped circuit breakers at turn on. A small reduction in initial cost is achieved if this situation is tolerated.

It is sometimes advantageous to deliberately build a large short-cir-cuit impedance into the transformer to limit high starting currents in the load. Winding resistances can be increased for this purpose up to the point where the limitation is the insulation temperature. Increasing the leakage inductance is a better way to increase the short-circuit impedance since no additional energy is consumed or dissipated in the form of heat in the transformer. Increasing the short-circuit impedance increases the regulation. If the steady-state current is predictable and constant, the poorer regulation can be tolerated. Deviations from nominal full-load voltage will be greater when the short-circuit impedance is made larger because of the limitation on the precision with which regulation can be predicted in design and controlled during manufacture.

Transformers contribute to the worsening of an inductive power factor of the load on a distribution system. Consideration of this effect is seldom given priority in equipment transformer design. The power factor will be improved when transformers are fully loaded at unity power factor and when the transformers are of minimum size for the application, since the inductive exciting current under these conditions is a smaller percentage of the total primary current. Power factor is a consideration not only in large distribution systems but also in smaller systems in vehicles and aircraft where keeping the size and weight of equipment to a minimum is a primary requirement.

3.6 HARMONIC DISTORTION

Harmonic distortion introduced by power transformers is sometimes of concern. Distortion is caused by a variation in the permeability of the transformer core as the flux density varies over its excursion from zero to maximum each cycle. A nonsinusoidal magnetizing current results from a sinusoidal applied voltage. The voltage drop across the transformer series impedance due to the magnetizing current will be nonsinusoidal. This voltage drop subtracted from the sinusoidal input voltage will give a nonsinusoidal output voltage. This effect is normally less than about 2 or 3 percent harmonic distortion. If the magnetization is symmetrical, only odd harmonics will appear in the magnetizing current. Unsymmetrical magnetization will occur when unbalanced direct currents flow in the windings, causing a dc magnetizing force to be applied to the core. When this occurs, both odd and even harmonics can appear in the magnetizing current. The usual method for reducing the generation of harmonics is to operate the core at a lower flux density. The magnetizing current experiences a much greater than proportional reduction by doing this, and the operating range of the flux density then excludes the region of greatest nonlinearity.

Although harmonic distortion is of greater interest in wideband transformers than in power transformers, there are power applications where distortion is a factor. Sometimes an accurate rms output voltage is required. Much of the instrumentation available is average sensing, but calibrated in rms values. This is satisfactory for measuring a distortion-less input, but if the transformer introduces harmonic distortion, output measurements with an average sensing instrument will be inaccurate. Bridging and balancing circuits frequently involve the cancellation of the fundamental component of an ac voltage which originates from a power transformer. If the transformer voltage is rich in harmonics, the residual null signal will be seriously distorted, and reduced sensitivity will result.

3.7   SPURIOUS COUPLING EFFECTS

Power transformers are intended to couple power sources to loads. Performing well in this role, they also provide unwanted coupling among circuits associated with them. This coupling occurs in a variety of ways. Power transformers often have several secondary windings instead of one as depicted in the equivalent circuits. The secondary circuits appear in parallel, each with its own secondary winding resistance and leakage inductance. Each current transformed to the primary contributes to the total primary current. Often transformer secondaries consist of one winding supplying a large percentage of the total volt-ampere load and one or more windings drawing very few volt-amperes. In this case the change in output voltage of all secondaries due to a change in primary current passing through the primary series impedance results mostly from a change in the current from the secondary winding supplying the preponderance of the load. Thus a change in load current in the latter winding has a noticeable effect on the other output voltages. If some of the light loads consist of sensitive circuits, malfunctioning can result. Sometimes separate transformers for the light loads are advisable.

Low-frequency signals pass through transformers by inductive coupling. The upper cutoff frequency in power transformers is far above the rated power frequency. The capacitive coupling between windings and from windings to ground allows the passage of signals above this cutoff frequency. Thus a power transformer can be an effective conduit for unwanted signals between circuits and circuits to ground. It is a fundamental characteristic of power transformers that they are capable of transforming frequencies far above the minimum rated. External preventive circuitry is needed when this cannot be tolerated. Capacitive coupling can be controlled to some extent in transformers. Electrostatic shields can be placed between adjacent windings. These shields may then be grounded or placed at a guard potential. The capacitance between windings can be reduced by several orders of magnitude with the use of electrostatic shields. The capacitance between winding and ground, if the shield is connected to ground, is usually increased by the use of an electrostatic shield.

Transformers are notorious avenues for introducing noise on power lines. The use of electrostatic shields between primary and secondary windings is routine in such applications as aircraft where noise levels on power lines must be kept low. These shields supplement but do not substitute for the customary power line filters. The electrostatic shield sometimes consists of a single layer of wire similar to the winding immediately above or below it. The shield can also consist of a layer of copper foil. The foil must overlap itself to be an effective shield. To prevent a shorted turn, the overlap must be insulated. Failure of this insulation and its inadvertent omission are classic pitfalls.

Transformers generate radiated as well as conducted interference. The magnetic field is not confined to the core and can be a source of troublesome hum. Since the field is highly directional, this problem can sometimes be minimized by orienting the transformer in the direction that produces the least interference, or the transformer may be relocated away from sensitive components. Magnetic shielding is an expensive last resort. The mild-steel cases commonly used to house transformers reduce the stray external magnetic field far below that of open-type transformers. A more exotic construction uses high-nickel-alloy magnetic steel as the case material. The benefit obtained from the high permeability of this material is partially lost in the fabrication and assembly techniques used by manufacturers. A greater attenuation of the stray field is achieved if a double-case construction is used. An inner case of magnetic steel is surrounded by a second case of mild steel.

3.8 LOW-CAPACITANCE TRANSFORMERS

Circuits requiring low capacitance to ground present a special challenge to the transformer designer. In such circuits the primary winding is most often effectively at ground potential. The task is to minimize the capacitance of the secondary winding to ground and the primary winding considered as a single electrode. The closest actual ground region is usually the core. The capacitance of the secondary to ground, like any other capacitance, is directly proportional to the area of the electrodes and the dielectric constant of the insulation between the electrodes and inversely proportional to the distance between the electrodes. Available solid and liquid insulating materials have dielectric constants 2 to 5 times that of air. Low-capacitance transformers are consequently often air-insulated with secondary windings of large diameter but small projected electrode area. The core windows are made large to increase the distance to the secondary winding. Thus low-capacitance transformers rated at more than a few volt-amperes are often cumbersome devices. In spite of the sacrifice made in size, it is usually difficult to reduce the capacitance to ground below about 25 pF. There is an unavoidable increase in leakage inductance when the capacitance between windings is reduced.

3.9   HIGH-VOLTAGE ISOLATION TRANSFORMERS

It is often necessary to apply an external voltage to a transformer winding, requiring electrical isolation of that winding. This is readily accomplished through the use of adequate insulation between the winding and other electrodes in the transformer. This additional insulation decreases the capacitance of the winding to ground, often a desirable effect, but increases the size, winding resistance, and leakage inductance, usually unwanted effects. The output voltage of a high-voltage isolation transformer will vary with power factor of the load much more than an otherwise similar transformer without high-voltage isolation. In fact, a high-voltage isolation transformer designed for use with unity power factor loads may be unusable with low power factor inductive loads. Capacitive loads can cause the output load voltage to rise above the output no-load voltage.

3.10   CURRENT TRANSFORMERS

Current transformers enjoy an undeserved reputation for precision because of their frequent use in instrumentation. This promise of precision is fulfilled only when the transformers are properly designed and constructed and then utilized within their ratings. A paragraph above discusses a method of analyzing transformer circuits driven from current sources. In the ideal current transformer the ampere-turns in the primary will equal the ampere-turns in the secondary. The existence of core loss and a finite shunt inductance render the transformer less than ideal. The primary current includes the no-load current. The transformer must be designed so that the no-load current is small compared with the transformed load current. Since the no-load current is mostly in quadrature with the load current, the effect of the no-load current is further reduced. Use of an impedance greater than rated will cause the voltage on the primary and the no-load current to increase. The load current will then be a smaller multiple of the no-load current, and the accuracy of the current transformation will be reduced. The smaller the no-load current is, the higher the impedance rating of the secondary can be made. The core configuration that provides the highest permeability is the toroid, which is the usual choice for instrument current transformers. The existence of a convenient hole in the center of the core through which a single turn may be passed is fortuitous.

There are many applications for current transformers other than instrumentation. Some of these applications involve the use of high-impedance circuitry, in which case it becomes necessary to load the current transformer with a relatively low resistance. When this is done, the accuracy of the resulting circuit is no longer a function of the current transformer alone, but is equally dependent upon the accuracy of the resistor. In fact, the resistor may be trimmed as a means of calibrating the output, in which case the accuracy requirement on the current transformer is very modest. Regardless of accuracy or type of construction used, it is important to keep the secondary of a current transformer loaded or shorted. Destructive voltages can develop if the transformer is operated with the secondary open-circuited.

3.11 AUTOTRANSFORMERS

Autotransformers are three-terminal devices connected as shown in Fig. 3.2a and b. The advantage of an autotransformer is a reduction in size, weight, and cost compared with an equivalent isolation transformer, achieved by lowering the required current-carrying capacity in that part of the winding common to both primary and secondary mesh currents. The advantage of the autotransformer improves as the voltage transformation ratio approaches 1. The autotransformer has little advantage when the voltage transformation is very large. Given these limitations, the autotransformer can often be used effectively when electrical isolation between primary and secondary is not required.

The voltage-current relationships in an autotransformer are the same as in an isolation transformer if the definitions of primary and secondary

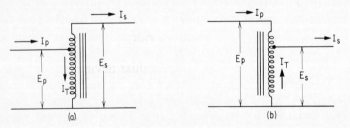

FIG. 3.2 (a) Step-up autotransformer schematic; (b) step-down autotransformer schematic.

in Fig. 3.2 are used. Following those definitions, N_p is the number of turns across which the primary voltage is applied, and N_s is the number of turns across which the secondary voltage is developed. As in other transformers, the primary volt-amperes will equal the secondary volt-amperes neglecting losses. Then for a step-up auto transformer

$$E_s = \frac{N_s}{N_p} E_p \qquad (3.1)$$

$$E_p I_p = E_s I_s \qquad (3.2)$$

$$I_T = I_p - I_s \qquad (3.3)$$

$$I_T N_p = I_s(N_s - N_p) \qquad (3.4)$$

Care is required in specifying the ratings of an autotransformer. Although the volt-ampere load will be $E_s I_s$, this value does not give the net volt-amperes transformed nor does it give an accurate picture of the rating when compared with an isolation transformer. A more meaningful volt-ampere rating for a step-up autotransformer is

$$E_p I_T = (E_s - E_p)I_s \qquad (3.5)$$

This equivalent volt-ampere rating corresponds to the rating of an isolation transformer of the same size as the autotransformer. Specified in this manner, the volt-ampere rating approaches zero as the secondary voltage approaches the primary voltage, reflecting more correctly the actual situation.

For the step-down autotransformer the current voltage relationships are Eqs. (3.1), (3.2), and the following:

$$I_T = I_s - I_p \qquad (3.6)$$

$$I_T N_s = I_p(N_p - N_s) \qquad (3.7)$$

The equivalent volt-ampere rating for the step-down autotransformer is

$$E_s I_T = (E_p - E_s)I_p \qquad (3.8)$$

The variable autotransformer is a familiar device in most laboratories. Capable of functioning as either a step-up or a step-down transformer, this device follows the current-voltage relationships given above. The output current capability of the variable autotransformer is determined by the carbon brush contact arrangement of the variable output, a more restrictive mechanism than the current-carrying capacity of the winding in a fixed transformer. The variable output has a finite and for some applications a too-coarse voltage resolution. The brush contact is made wide enough to engage more than one turn simultaneously. This improves the resolution beyond one turn, by permitting single or dual contact, and prevents circuit interruption at every turn. A partial shorted turn with current limited by the brush resistance exists when two turns are contacted simultaneously, a condition that further limits the current rating of the device. The variable autotransformer is one of the few dissipationless devices that can provide controlled ac voltage output without distortion.

CHAPTER 4

Circuit Performance of Audio and Wideband Transformers

An *audio transformer* is obviously a transformer which is used over the audio-frequency range. The more general term, *wideband transformer,* is often applied to transformers that operate in the megahertz range. The lowest operating frequency for such transformers may be well above the audio range. The theory of operation of wideband and audio transformers is the same. The term wideband transformer is used here to describe both types of transformers in which the operating frequency band is sufficiently wide to include regions at both the upper and lower ends of the band where the transformer response differs from the response at the center. The power and impedance levels achievable vary with the frequency range, as does construction.

Wideband performance is needed where complex waveforms with high-order harmonics must be transformed with low distortion. Typical applications include high-fidelity audio and video transmission, sonar search and communications equipment, and feedback circuits.

4.1 INSERTION LOSS

In the midband region of wideband transformers there will be a power loss due to the resistance of the windings and the core. Known as *insertion loss,* this loss of power is approximately constant over the operating range. The equivalent circuit of Fig. 2.2 can be used to evaluate insertion loss. The insertion loss of the transformer is given by

$$dB = 10 \log \frac{P_L}{P_L + P_T} \tag{4.1}$$

where P_L is the power delivered to the load and P_T is the sum of the winding and core losses.

The insertion loss can also be expressed in the following manner:

$$dB = 20 \log \frac{E}{E'} \tag{4.2}$$

where E is the voltage across the load with the actual transformer in place and E' is the voltage that would appear across the load if the transformer were ideal.

In wideband applications the transformer is commonly used to match the load to the source. This is done not only to affect maximum power transfer but also to prevent reflections from transmission line effects and establish optimum damping conditions for transients. The turns ratio may be more critical for these latter reasons than for maximum power transfer. A common approximation is to set the turns ratio so that R_G is exactly equal to R_L/n^2. This assumes the ideal case in which the transformer losses are zero. Half of the power is delivered to the load and half is dissipated in the source resistance. When there is a need for precision matching, the resistances of the windings must be considered. When this is done, the following relationship exists:

$$R_G = R_p + \frac{R_s}{n^2} + \frac{R_L}{n^2} \tag{4.3}$$

The correct turns ratio for this condition will also be correct for providing a match on the secondary. The effect of core loss on the match can usually be neglected. When the condition of Eq. (4.3) is met, half of the power will still be dissipated in the generator resistance. The other half will be dissipated in the transformer load combination.

4.2 LOW-FREQUENCY RESPONSE

The equivalent circuit of Fig. 2.4 is used to analyze the low-frequency response of wideband transformers. The generator and load impedances are assumed to be constant resistances of known magnitude. In the analysis the primary and secondary winding resistances are included in the generator and load resistances, respectively. The ratio of output to generator voltage is

$$\frac{e_L}{e_G} = \frac{\dfrac{j\omega L_e R_L}{RL + j\omega L_e}}{R_G + \dfrac{j\omega L_e R_L}{R_L + j\omega L_e}} = \frac{R_L}{R_L + R_G + \dfrac{R_G R_L}{j\omega L_e}} \qquad (4.4)$$

Both the magnitude and the phase of this ratio are of interest. Equation (4.4) may be rearranged to

$$\frac{e_L}{e_G} = \frac{R_L}{R_L + R_G} \frac{1}{\sqrt{1 + \left(\dfrac{R_T}{\omega L_e}\right)^2}} \angle \theta \qquad (4.5)$$

in which

$$R_T = \frac{R_L R_G}{R_L + R_G}$$

$$\theta = \tan^{-1} \frac{R_T}{\omega L_e}$$

The cutoff frequency is defined as

$$\omega_L = \frac{R_L R_G}{L_e(R_L + R_G)} = \frac{R_T}{L_e}$$

in which R_T is in ohms and L_e is in henrys. Equation (4.5) then becomes

$$\frac{e_L}{e_G} = \frac{R_L}{R_L + R_G} \frac{1}{\sqrt{1 + \left(\dfrac{\omega_L}{\omega}\right)^2}} \angle \theta \qquad (4.6)$$

in which

$$\theta = \tan^{-1} \frac{\omega_L}{\omega} \qquad (4.7)$$

In Eq. (4.7) ω is angular frequency in radians per second.

The magnitude of the ratio in Eq. (4.6) may be expressed in decibels:

$$dB = 20 \log \left(\frac{R_L}{R_L + R_G} \frac{1}{\sqrt{1 + \left(\frac{\omega_L}{\omega}\right)^2}} \right)$$

$$= 10 \log \left(\frac{R_L}{R_L + R_G} \right)^2 + 10 \log \frac{1}{1 + \left(\frac{\omega_L}{\omega}\right)^2} \tag{4.8}$$

The first term in Eq. (4.8) represents the midband loss. This equation may be simplified by considering the low-frequency response as the ratio of output at low frequency to output at midband:

$$dB = 10 \log \frac{1}{1 + \left(\frac{\omega_L}{\omega}\right)^2} \tag{4.9}$$

Equations (4.7) and (4.9) are normalized expressions for the phase and magnitude of the transformer circuit low-frequency transfer characteristic. These functions are plotted in Fig. 4.1. The reciprocal of the frequency term in Eqs. (4.7) and (4.9) is plotted on the abscissa in order to have the frequency increase from left to right. By calculating the cutoff frequency ω_L, the voltage output magnitude and phase at any frequency ω may be determined from Fig. 4.1.

In Eqs. (4.7) and (4.9) R_G and R_L contain the winding resistances, and so the contribution they make in determining performance is

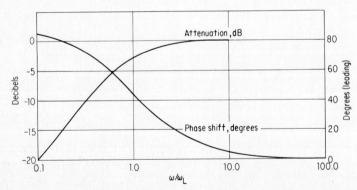

FIG. 4.1 Low-frequency transformer circuit response magnitude and phase vs. normalized frequency.

included. The contribution core loss makes to the insertion loss has been neglected. The core loss as a portion of load power in decibels is

$$dB = 10 \log \frac{P_L}{P_L + P_{fe}} \tag{4.10}$$

The core loss in decibels from Eq. (4.10) may be added to the first term in Eq. (4.8) to obtain the total insertion loss.

4.3 HIGH-FREQUENCY RESPONSE

The high-frequency response of wideband transformers is usually analyzed by means of the simplified circuits of Figs. 4.2 and 4.3. Although neither circuit is a precise representation of an actual transformer, Fig. 4.2 approximates a step-up transformer and Fig. 4.3 approximates a step-down transformer. The stray capacitances in the transformer have a greater effect on the high-impedance side. Therefore, in a single lumped constant approximation of distributed parameters, it is more nearly correct to place the shunt capacitance on the high-impedance side of the leakage inductance. As will be seen, this choice has little significance if optimum design criteria are met. In the analysis the primary winding

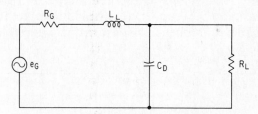

FIG. 4.2 Equivalent circuit used to analyze high-frequency response of step-up transformers.

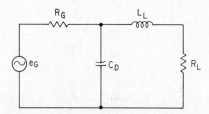

FIG. 4.3 Equivalent circuit used to analyze high-frequency response of step-down transformers.

resistance is lumped with the generator resistance, and the secondary winding resistance is lumped with the load resistance. The core loss is neglected. This is usually a valid simplification since the core loss decreases more rapidly with decreasing flux density than it increases with frequency. There is an additional loss at high frequencies; this is the dielectric loss in the insulating materials. Dielectric loss is usually more significant in insulation temperature considerations than in circuit analysis.

In Fig. 4.2 for a step-up transformer the ratio of output to input voltage is equal to the ratio of the impedance of the parallel combination of distributed capacitance and load resistance to the total input impedance:

$$\frac{e_L}{e_G} = \frac{\dfrac{R_L}{j\omega C_D \left(R_L + \dfrac{1}{j\omega C_D} \right)}}{R_G + j\omega L_L + \dfrac{R_L}{j\omega C_D \left(R_L + \dfrac{1}{j\omega C_D} \right)}} \tag{4.11}$$

In Fig. 4.2 the low-frequency ratio of e_L to e_G, which actually is the midband response, will be equal to the ratio of the load resistance to the sum of the load and generator resistances. If the reciprocal of this ratio is multiplied by the ratio of load to generator voltages, the resulting term is the ratio of the output voltage at any frequency ω to the output voltage at midband, that ratio being unity at midband frequency. Equation (4.11) may be normalized to midband response and rearranged to yield

$$\frac{e_L}{e_G}\left(\frac{R_L + R_G}{R_L} \right) = \frac{1}{1 - \dfrac{R_L\omega^2 L_L C_D}{R_L + R_G} + j\omega\left(\dfrac{L_L + C_D R_G R_L}{R_L + R_G} \right)} \tag{4.12}$$

Equation (4.12) may be normalized for both frequency and resistance by defining the following parameters:

$$\omega_H = \sqrt{\frac{1}{L_L C_D}} \tag{4.13}$$

$$R_G = K_1 \sqrt{\frac{L_L}{C_D}} \tag{4.14}$$

$$R_L = K_2 \sqrt{\frac{L_L}{C_D}} \tag{4.15}$$

Equation (4.13) defines the high-frequency cutoff. Equations (4.14) and (4.15) define the characteristic impedance of the transformer when the constant is 1. Frequency is in radians per second and impedance is in ohms. Substituting Eqs. (4.13) through (4.15) in Eq. (4.12) yields

$$\frac{e_L}{e_G}\left(\frac{R_L + R_G}{R_L}\right) = \frac{1}{1 - \left(\dfrac{\omega}{\omega_H}\right)^2 \dfrac{K_2}{K_1 + K_2} + j\dfrac{\omega}{\omega_H}\left(\dfrac{1 + K_1 K_2}{K_1 + K_2}\right)} \tag{4.16}$$

Equation (4.16) may be placed in polar form:

$$\frac{e_L}{e_G}\left(\frac{R_L + R_G}{R_L}\right) = \frac{\angle\,\theta}{\sqrt{\left[1 - \left(\dfrac{\omega}{\omega_H}\right)^2 \dfrac{K_2}{K_1 + K_2}\right]^2 + \left[\dfrac{\omega}{\omega_H}\left(\dfrac{1 + K_1 K_2}{K_1 + K_2}\right)\right]^2}} \tag{4.17}$$

in which

$$\theta = \tan^{-1}\frac{\dfrac{\omega}{\omega_H}\left(\dfrac{1 + K_1 K_2}{K_1 + K_2}\right)}{1 - \left(\dfrac{\omega}{\omega_H}\right)^2 \dfrac{K_2}{K_1 + K_2}} \tag{4.18}$$

Equations (4.17) and (4.18) are the normalized equations for the magnitude and phase of the output voltage in the equivalent circuit of Fig. 4.2.

The high-frequency response of a step-down transformer using the equivalent circuit of Fig. 4.3 can be determined from the following mesh equations:

$$e_G = i_1\left(R_G + \frac{1}{j\omega C_D}\right) - i_2\left(\frac{1}{j\omega C_D}\right) \tag{4.19}$$

$$i_1\left(\frac{1}{j\omega C_D}\right) = i_2\left(R_L + j\omega L_L + \frac{1}{j\omega C_D}\right) \tag{4.20}$$

The simultaneous solution of Eqs. (4.19) and (4.20) produces

$$\frac{e_L}{e_G}\left(\frac{R_L + R_G}{R_L}\right) = \frac{1}{1 - R_G\dfrac{\omega^2 L_L C_D}{R_L + R_G} + j\omega\left(\dfrac{L_L + C_D R_G R_L}{R_G + R_L}\right)} \tag{4.21}$$

Equation (4.21) is similar to Eq. (4.12) for the step-up transformer, being the ratio of the output voltage at any frequency ω to the output voltage at midband. By use of the parameters defined in Eq. (4.13) to (4.15), Eq. (4.21) may be normalized and placed in polar form:

$$\frac{e_L}{e_G}\left(\frac{R_L + R_G}{R_L}\right) = \frac{\angle\theta}{\sqrt{\left[1 - \left(\frac{\omega}{\omega_H}\right)^2 \frac{K_1}{K_1 + K_2}\right]^2 + \left[\frac{\omega}{\omega_H}\left(\frac{1 + K_1 K_2}{K_1 + K_2}\right)\right]^2}}$$

$$(4.22)$$

where

$$\theta = \tan^{-1}\left[\frac{\frac{\omega}{\omega_H}\left(\frac{1 + K_1 + K_2}{K_1 + K_2}\right)}{1 - \left(\frac{\omega}{\omega_H}\right)^2 \frac{K_1}{K_1 + K_2}}\right]$$

$$(4.23)$$

Equations (4.22) and (4.23) are the normalized equations for the magnitude and phase of the output voltage in the equivalent circuit of Fig. 4.3 for the high-frequency response of a step-down transformer.

Equations (4.17) and (4.18) for step-up transformers are identical to Eqs. (4.22) and (4.23) for step-down transformers except that parameters K_1 and K_2 are transposed. If the source and load resistances are matched, K_1 and K_2 are equal, and the distinction between step-up and step-down equivalent circuits vanishes. Equation (4.17) expressed in decibels is plotted in Fig. 4.4. Equation (4.18) is plotted in Fig. 4.5. In the parametric equations, Eqs. (4.13) to (4.15) used with Figs. 4.4 and 4.5, the leakage inductance L_L is in henrys, the distributed capacitance C_D is in farads, and angular frequency ω is in radians per second. R_G and R_L are in ohms, and the parameters K_1 and K_2 are dimensionless. By transposing K_1 and K_2, the curves in Figs. 4.4 and 4.5 are applicable to Eqs. (4.22) and (4.23) for step-down transformers. The higher the cutoff frequency ω_H, the better will be the high-frequency response. The most favorable matching condition is usually when K_1 and K_2 are both equal to 1. This means matching the source and load resistances by means of the transformer turns ratio and, if necessary, using Eq. (4.3). In addition, the characteristic impedance of the transformer is matched to the source-load resistance.

Illustrative problem for wideband transformers

Determine the turns ratio to match the generator to the load, the midband output, the insertion loss, and the frequencies at which

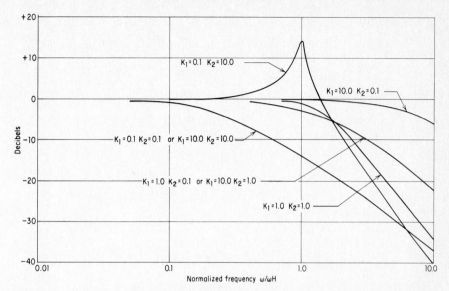

FIG. 4.4 High-frequency amplitude response of wideband transformers in decibels below midband vs. normalized frequency with matching constants as parameters.

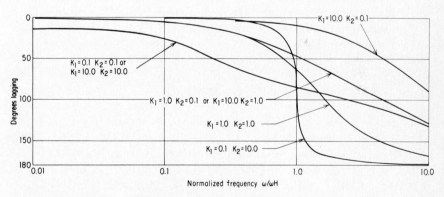

FIG. 4.5 High-frequency phase shift response of wideband transformers in degrees vs. normalized frequency with matching constants as parameters.

the output is 3 dB below midband output for a sonar matching transformer with the following characteristics: $R_G = 100\ \Omega$, $R_L = 1000\ \Omega$, $R_p = 2\ \Omega$, $R_s = 20\ \Omega$, $C_D = 4000$ pF, $L_L = 0.0001$ H, $L_e = 0.01$ H, and core loss is negligible. The inductances and capacitance are referred to the primary.

Solution

The turns ratio is calculated for midband operation with Eq. (4.3):

$$n^2 = \frac{R_L + R_s}{R_G - R_p}$$

$$= \frac{1000 + 20}{100 - 2}$$

$$n = 3.226$$

By reference to Fig. 2.2 and neglecting core loss R_c, the ratio of output to input is

$$\frac{e_L}{e_G} = \frac{R_L/n^2}{R_G + R_p + (R_L + R_s)/n^2}$$

$$= \frac{1000/(3.226)^2}{100 + 2 + (1000 + 20)/(3.226)^2}$$

$$= 0.4804$$

The output at midband for an ideal transformer in which the generator impedance is matched to the load will be half the no-load generator voltage. The loss below that value will be the insertion loss of the transformer:

$$\text{dB} = 20 \log \frac{0.4804}{0.50}$$

$$= -0.35$$

The low cutoff frequency [see Eq. (4.6)] is

$$\omega_L = \frac{R_L R_G}{L_e(R_L + R_G)}$$

$$= \frac{102 \times 98}{0.01(98 + 102)}$$

$$= 4998$$

From the attenuation curve in Fig. 4.1 it is seen that the normalized frequency for the 3-dB point is 1. From the normalized frequency at the 3-dB point and the low cutoff frequency, the actual 3-dB low frequency may be calculated:

$$\frac{\omega}{\omega_L} = 1 \qquad \omega = \omega_L$$

$$f = \frac{\omega_L}{2\pi}$$

$$= \frac{4998}{2\pi}$$

$$= 795$$

From Eqs. (4.13) to (4.15) the high cutoff frequency, the characteristic impedance, and the matching constants are determined:

$$\omega_H = \frac{1}{\sqrt{L_L C_D}}$$

$$= \frac{1}{\sqrt{0.0001 \times 4000 \times 10^{-12}}}$$

$$= 1.58 \times 10^6$$

$$Z = \sqrt{\frac{L_L}{C_D}}$$

$$= \sqrt{\frac{0.001}{4000 \times 10^{-12}}}$$

$$= 158$$

$$K_1 = \frac{Z}{R_G} = \frac{158}{100} = 1.58$$

$$K_2 = \frac{Z}{R_L} = \frac{158}{98} = 1.61$$

From Fig. 4.4 the normalized frequency at 3 dB down for the above matching constants may be estimated. That number is approximately 1. Substituting that value and the above matching constants in Eq. (4.16) gives a response of −1.7 dB. By successive approximation the normalized frequency for 3 dB down is found to be 1.26. Substitution of 1.26 for the normalized frequency and the above matching constants in Eq. (4.16) expressed in decibels yields

$$dB = 20 \log \frac{1}{\sqrt{\left[1 - \left(\dfrac{\omega}{\omega_H}\right)^2 \dfrac{K_2}{K_1 + K_2}\right]^2 + \left(\dfrac{\omega}{\omega_H}\right)^2 \left(\dfrac{1 + K_1 K_2}{K_1 + K_2}\right)^2}}$$

$$= 20 \log \frac{1}{\sqrt{\left[1 - (1.26)^2 \dfrac{1.61}{1.58 + 1.61}\right]^2 + (1.26)^2 \left(\dfrac{1 + 1.58 \times 1.61}{1.58 + 1.61}\right)^2}}$$

$$= 3 dB$$

Using the value of 1.26 for the normalized frequency, the actual 3-dB high frequency is determined as follows:

$$\frac{\omega}{\omega_H} = 1.26$$

$$\omega = 1.26 \omega_H$$

$$= 1.26 \times 1.58 \times 10^6$$

$$f = \frac{1.26 \times 1.58 \times 10^6}{2\pi}$$

$$= 31.7 \text{ kHz}$$

CHAPTER 5

Circuit Performance of Pulse Transformers

Pulse transformers are those which transform video pulses. Resembling wideband transformers, they differ primarily in the ways by which they are specified and designed. While a so-called wideband transformer of proper bandwidth may function as a pulse transformer, convenience and custom have placed pulse transformers in a separate category. The notation used here follows industry practice. Pulse transformer technique was developed for use in radar systems, and radar has continued to be its major patron. However, the increased use of square wave voltages has led to extending pulse transformer theory to other circuits such as inverters. The transition is straightforward.

Pulse transformers may be classified by the following types: blocking oscillator, isolation, interstage, and output. In all these applications, the quality of the pulse transformer is measured in terms of its deteriorating effect on a rectangular pulse. Figure 5.1 illustrates the output of a pulse transformer for a rectangular input pulse. Some commonly used terms describing features of the pulse are shown in this figure. Universal agreement on the finer details of these terms is lacking. For example, the pulse width is illustrated as being the value at half amplitude. This is a logical choice. The product of this value and the peak voltage is equal to the volt-time area used to calculate flux density for an equivalent trapezoidal wave. However, the pulse width of the input pulse is the time between the start of the rise and the start of the fall of the pulse. These two intervals are not necessarily of equal value. The rise and fall times of the pulse

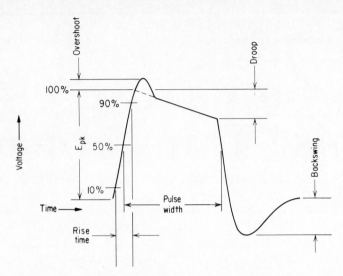

FIG. 5.1 Pulse shape nomenclature.

are frequently defined as the time between 10 and 90 percent of the pulse amplitude. This is done as a convenience in measurement since the corners of the pulse are often indeterminate. In specifying pulse shapes, it is necessary to define the terms carefully.

The utilization of pulse transformers in equipment design requires a common ground for communication between the circuit designer and the transformer designer. The principal design tool available to the pulse transformer designer is linear analysis of simplified equivalent circuits. Three different simplified circuits are used corresponding to the rise of the pulse, the top of the pulse, and the fall of the pulse. Linear circuit analysis requires a knowledge of the source and load impedances for each of these conditions. These values are frequently not known with precision. In fact they may not even be constants. Under this condition, approximate values for the impedances must be selected. A prototype transformer designed to these values must then be evaluated in the circuit and changes made if necessary. The linear circuit theory is based on resistive source and load impedances. The most common departure from this in practice is for the load to have some capacitance in parallel with the load resistance. This can be accommodated by adding this capacitance to the shunt capacitance in the transformer. Series inductance can be accommodated by adding to the leakage inductance. Shunt inductance may be included by reducing the shunt inductance of the transformer, L_e. More difficult source and load impedances are usually not included in the design analysis.

5.1 CHARACTERISTICS OF VARIOUS TYPES OF PULSE TRANSFORMERS

Pulse output transformers have perhaps the most readily defined circuit functions. The source and load impedances, at least during the top of the pulse, are usually known approximately, and there are usually reasonably defined limits to the rise and fall times and the droop of the top. Pulse output transformers are often subject to extraneous complicating requirements. Typical of these requirements are the need to feed filament current through the pulse transformer, the presence of very high voltages, and the need for current and voltage monitoring of pulse and filament power. Provisions for such necessary functions inevitably result in an increase in leakage inductance and distributed capacitance.

Pulse interstage transformers, since they handle only small amounts of power, have higher source and load impedances than pulse output transformers. High impedances place more stringent requirements for shunt inductance and distributed capacitance. Pulse transformers do not perform well in circuits where the output is essentially open-circuited. The relatively low shunt inductance then becomes the load presented to the source with usually unsatisfactory results.

Pulse transformers are sometimes needed to provide isolation between primary and secondary from an externally applied voltage. A typical example would be transforming a grid signal to a high-voltage tube whose plate is near ground and whose cathode is at a high negative potential. The requirement for high-voltage isolation means additional insulation between the windings. This increases the leakage inductance. The transformer operates with a handicap, and pulse fidelity suffers. Source and load impedances have the same impact on design and performance of pulse isolation transformers as they do on pulse output and interstage transformers.

The operation of blocking oscillator transformers is the most complex of the three general types of pulse transformers. A blocking oscillator is a regenerative amplifier which is driven to full output. Having reached full output, the amplifier is caused to reverse its regenerative action by one of several mechanisms, and the output is reduced to zero. Figure 5.2 shows a simplified blocking oscillator circuit. The primary winding of a blocking oscillator transformer is placed in the collector circuit of the transistor. A secondary winding is connected between the base and emitter of the transistor with polarity such that any increase in collector current results in an increase in emitter-to-base current. Initially biased so that the collector current is zero, the transistor starts to conduct by the application of a triggering signal at e_1. The regenerative effect of the transformer causes a rapid rise of both base and collector currents. The point is quickly reached when no further increase in collector current is possi-

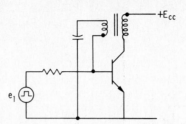

FIG. 5.2 Simplified blocking oscillator circuit.

ble. The current limitation can be caused by saturation of either the transistor or the transformer. If the transistor saturates, that is, for a further increase in base current there is no further increase in collector current, then the voltage across the transformer which had been developed by the increasing collector current begins to drop. This initiates decreasing base and collector currents. The degenerative effect drives the transistor back to cutoff. Alternatively the transistor may remain in the linear region where the collector current is proportional to the base current during the entire operating cycle. The transformer will have a voltage impressed upon it which is proportional to the time rate of change of current. The current will tend to increase at a constant rate, maintaining a constant voltage across the transformer. This will continue until the transformer reaches saturation, at which time the voltage across the transformer will abruptly drop. The base current will decrease rapidly, and the degenerative effect will, as before, drive the transistor to cutoff.

The rise and fall of the pulse are controlled by the pulse transformer characteristics in a manner similar to other pulse transformer applications. In addition, the blocking oscillator transformer determines the pulse width. If the transformer is operating in the linear mode, the shunt impedance controls the collector current. That current is equal to the time integral of the voltage across the transformer divided by the shunt inductance. The shunt inductance determines the time when the collector current reaches saturation. That time interval is the pulse width. If the transformer operates in the saturating mode, the pulse width is established by the volt-time integral needed to reach saturation. Rigorous analysis of blocking oscillator operation is tedious. Experimental evaluation of circuit and transformer design is usually necessary.

5.2 ANALYSIS OF PULSE RISE IN TRANSFORMERS

Transformer output voltage rise in response to a step function constant voltage input may be analyzed by means of the equivalent circuit of Fig.

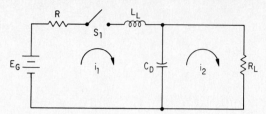

FIG. 5.3 Equivalent circuit of step-up transformer for analyzing pulse rise.

5.3. This circuit is applicable to step-up transformers. The shunt inductance has been omitted in this equivalent circuit because the current that would flow through it during the rapid pulse rise would be negligible compared with the currents through the capacitance and the load. The mesh equations for the circuit of Fig. 5.3 are

$$e_G = i_1 R_G + L_L \frac{di_1}{dt} + \frac{1}{C_D} \int i_1 \, dt - \frac{1}{C_D} \int i_2 \, dt \qquad (5.1)$$

$$\frac{1}{C_D} \int i_1 \, dt = i_2 R_L + \frac{1}{C_D} \int i_2 \, dt \qquad (5.2)$$

Switch S_1 closes at time $t = 0$. This input voltage function is used in the Laplace transform expression for i_2 obtained from the simultaneous solution of Eqs. (5.1) and (5.2).

$$I_2[S] = \frac{E_G}{L_L C_D R_L S \left[S^2 + S \left(\dfrac{R_G}{L_L} + \dfrac{1}{R_L C_D} \right) + \dfrac{R_L + R_G}{L_L C_D R_L} \right]} \qquad (5.3)$$

From Eq. (5.3) the normalized load voltage is obtained:

$$\frac{E_L[S]}{E_G} = \frac{1}{L_L C_D S \left[S^2 + S \left(\dfrac{R_G}{L_L} + \dfrac{1}{R_L C_D} \right) + \dfrac{R_L + R_G}{L_L C_D R_L} \right]} \qquad (5.4)$$

In Eq. (5.4) the roots of the denominator are

$$S = \frac{-b \pm \sqrt{b^2 - 4c}}{2}$$

where $b = \dfrac{R_G}{L_L} + \dfrac{1}{R_L C_D}$

$\quad\ c = \dfrac{R_L + R_G}{L_L C_D R_L}$

The inverse transform of Eq. (5.4) is

$$\frac{e_L(t)}{E_G} = \frac{R_L}{R_L + R_G}\left[1 - \exp\left(-\frac{b}{2}t\right)\left(\frac{b}{\sqrt{b^2 - 4c}}\sinh\frac{\sqrt{b^2 - 4c}}{2}t\right.\right.$$
$$\left.\left. + \cosh\frac{\sqrt{b^2 - 4c}}{2}t\right)\right] \tag{5.5}$$

If in Eq. (5.5) $4c < b^2$, the discriminant remains real and the circuit is overdamped. If $4c > b^2$, the discriminant becomes imaginary, and Eq. (5.5) becomes

$$\frac{e_L(t)}{E_G} = \frac{R_L}{R_L + R_G}\left[1 - \exp\left(-\frac{b}{2}t\right)\left(\frac{b}{\sqrt{4c - b^2}}\sin\frac{\sqrt{4c - b^2}}{2}t\right.\right.$$
$$\left.\left. + \cos\frac{\sqrt{4c - b^2}}{2}t\right)\right] \tag{5.6}$$

Normalized curves of Eq. (5.5) and (5.6) may be obtained by plotting

$$\frac{e_L}{E_G}\frac{R_L + R_G}{R_L} \quad \text{vs.} \quad \frac{\sqrt{c}}{2\pi}t$$

for various values of

$$\sqrt{\frac{b^2}{4c}} = \frac{b}{2\sqrt{c}} = \frac{R_G R_L C_D + L_L}{2\sqrt{R_L L_L C_D(R_G + R_L)}}$$

which is the damping coefficient. These curves are shown in Fig. 5.4. An examination of this figure shows that the best compromise between rise time and overshoot occurs when the circuit is slightly underdamped. When $R_L = R_G$ and when $R_L = \sqrt{L_L/C_D}$, the damping factor is 0.707. This condition is the usual objective when the generator and load impedances are matched. If R_G is made equal to zero, a condition approximated in some circuits, and the damping factor is made equal to 0.707, then

$$R_L = 0.707 \sqrt{\frac{L_L}{C_D}}$$

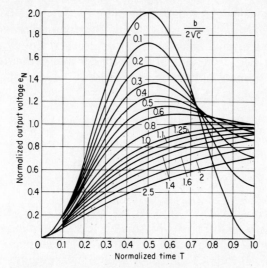

FIG. 5.4 Normalized output voltage rise vs. normalized time with damping factor as parameter. *(G. N. Glasoe and J. V. Lebacqz, Pulse Generators, vol. 5, Radiation Laboratory Series, McGraw-Hill, New York, 1948. Reproduced by permission of McGraw-Hill.)*

For a step-up transformer, equivalent circuit Fig. 5.3:

$$T = \left(\frac{1}{2\pi} \sqrt{\frac{R_L + R_G}{L_L C_D R_L}} \right) t$$

$$\frac{b}{2\sqrt{C}} = \frac{R_G R_L C_D + L_L}{2\sqrt{R_L L_L C_D (R_G + R_L)}}$$

For a step-down transformer, equivalent circuit Fig. 5.5:

$$T = \left(\frac{1}{2\pi} \sqrt{\frac{R_G + R_L}{L_L C_D R_G}} \right) t$$

$$\frac{b}{2\sqrt{C}} = \frac{R_L R_G C_D + L_L}{2\sqrt{R_G L_L C_D (R_L + R_G)}}$$

For both step-up and step-down transformers:

$$e_N = \frac{e_L}{E_G} \left(\frac{R_L + R_G}{R_L} \right)$$

R is in ohms, L is in henrys, C is in farads, and t is in seconds.

The rise time and general shape of the pulse rise for any other set of conditions can be estimated by substituting the appropriate values in the expression for damping factor and using the curves of Fig. 5.4. These curves may be used also to obtain the absolute value of the rise time from the normalized time T plotted on the abscissa. The absolute time will be

$$t = \frac{2\pi}{\sqrt{c}} T$$

$$= 2\pi T \sqrt{\frac{L_L C_D R_L}{R_L + R_G}}$$

From this expression it is seen that the quantity $\sqrt{L_L C_D}$ is a figure of merit for pulse transformers. The smaller this value, the faster the rise time of the output will be in response to a rectangular pulse.

The circuit of Fig. 5.3 serves for analyzing step-up transformers. In a step-down transformer stray and load capacitance in the secondary is less significant than stray and source capacitance in the primary. When the capacitance in the primary predominates, Fig. 5.5 is a better equivalent circuit for the rise of pulse analysis. The mesh equations for the circuit of Fig. 5.3 are

$$E_G = i_1 R_G + \frac{1}{C_D} \int i_1 \, dt - \frac{1}{C_D} \int i_2 \, dt \tag{5.7}$$

$$\frac{1}{C_D} \int i_1 \, dt = \frac{1}{C_D} \int i_2 \, dt + L_L \frac{di_2}{dt} + i_2 R_L \tag{5.8}$$

The Laplace transform for i_2, obtained from the simultaneous solution of Eqs. (5.7) and (5.8), is

$$I_2[S] = \frac{E_G}{R_G L_L C_D S \left[S^2 + S \left(\dfrac{R_L}{L_L} + \dfrac{1}{R_G C_D} \right) + \dfrac{R_G + R_L}{L_L C_D R_G} \right]} \tag{5.9}$$

The Laplace transform of the normalized voltage from Eq. (5.9) is

$$\frac{E_L[S]}{E_G} = \frac{R_L}{R_G L_L C_D S \left[S^2 + S \left(\dfrac{R_L}{L_L} + \dfrac{1}{R_G C_D} \right) + \dfrac{R_G + R_L}{L_L C_D R_G} \right]} \tag{5.10}$$

Equation (5.10) differs from Eq. (5.4) only by the ratio R_L/R_G and by the transposition of R_L and R_G. The ratio maintains the correct final value

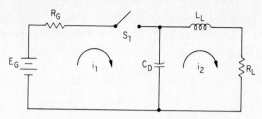

FIG. 5.5 Equivalent circuit of step-down transformer for analyzing pulse rise.

of the normalized voltage after the transposition of R_L and R_G. Equation (5.5) is therefore the transformation of Eq. (5.10) with the following values of b and c:

$$b = \frac{R_L}{L_L} + \frac{1}{R_G C_D} \qquad c = \frac{R_G + R_L}{L_L C_D R_G}$$

Equations (5.5) and (5.6) may be used for both step-up transformers, equivalent circuit of Fig. 5.3, and for step-down transformers, equivalent circuit of Fig. 5.5, provided that R_L and R_G are properly positioned. Figure 5.4 is a family of curves from Eqs. (5.5) and (5.6) for various values of damping factor. The notes accompanying this figure give the appropriate positions for R_L and R_G with both step-up and step-down transformers. If R_L and R_G are matched, the distinction between step-up and step-down response disappears.

5.3   ANALYSIS OF TOP OF PULSE IN TRANSFORMERS

The top of a rectangular pulse, when applied to a transformer, is analyzed by means of the equivalent circuit of Fig. 5.6. The leakage inductance and distributed capacitance have little effect on the top of the pulse and may be neglected. This equivalent circuit will be recognized as the same as that

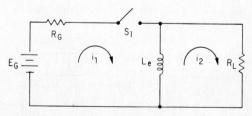

FIG. 5.6 Transformer equivalent circuit for analyzing top of pulse response.

used to analyze the low-frequency response of wide-band transformers. The source and load resistance effective at the top of the pulse must be known and used for the analysis to be accurate. Switch S_1 closes at time $t = 0$. The mesh equations for Fig. 5.6 are

$$E_G = i_1 R_G + L_e \frac{di_1}{dt} - L_e \frac{di_2}{dt} \qquad (5.11)$$

$$L_e \frac{di_1}{dt} = L_e \frac{di_2}{dt} - i_2 R_L \qquad (5.12)$$

The simultaneous solution of these equations yields for e_L

$$e_L = E_G \frac{R_L}{R_L + R_G} \exp\left[-\frac{R_L R_G}{L_e(R_L + R_G)} t \right] \qquad (5.13)$$

Equation (5.13) may be normalized by the use of the parameter

$$K = \frac{R_L}{R_L + R_G} \qquad (5.14)$$

Equation (5.14) gives the ratio of the initial voltage across the load to the no-load source voltage. Using the parameter in this equation in Eq. (5.13) yields

$$\frac{e_L}{KE_G} = \exp\left(-\frac{KR_G}{L_e} t \right] \qquad (5.15)$$

Equation (5.15) is plotted in Fig. 5.7; with $e_L/E_G K$ as ordinate and $R_G/L_e t$ as abscissa, K is a parameter. In the most frequent case the generator resistance is matched to the load resistance, and $K = 0.5$. When $K = 1$, the load is open-circuited; when $K = 0$, the load is short-circuited. The value of the ordinate must be multiplied by the parameter K and the generator voltage E_G to obtain the initial output voltage. This initial voltage may be multiplied by the value of the ordinate for the appropriate K to find the output voltage at any normalized time shown on the abscissa. The final value of output voltage will occur at actual time t equal to the pulse width. In this analysis resistance is in ohms, inductance in henrys, and time in seconds.

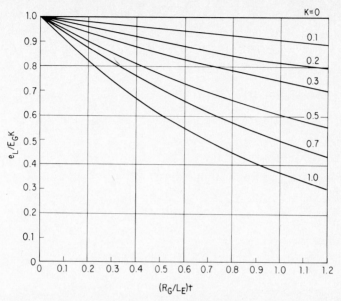

FIG. 5.7 Top of pulse response, normalized output voltage vs. normalized time with output voltage ratio as parameter. Resistance is in ohms, inductance in henrys, and time in seconds. Equivalent circuit is Fig. 5.6.

5.4 ANALYSIS OF TRAILING EDGE RESPONSE OF PULSE TRANSFORMERS

The trailing edge of the pulse developed at the output of a transformer is analyzed by means of the equivalent circuit in Fig. 5.8. The voltage source is not shown since it has been turned off and does not enter into the analysis. The resistance R_T represents the total shunt resistance during the trailing edge interval. It includes core loss and source and/or load resistance if connected. In this analysis there are two initial conditions that must be considered: the initial charge on the capacitor and the initial current in the shunt inductance. The initial charge on the capacitor is

$$Q_o = C_D E_o \tag{5.16}$$

where E_o is the amplitude of the input-voltage pulse, assumed to be constant during the pulse. The initial current is the exciting current flowing at the end of the pulse. This current may be determined by integrating

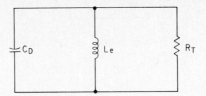

FIG. 5.8 Equivalent circuit used to analyze trailing edge response of pulse transformers.

the defining relationship for inductance:

$$e = L\frac{di}{dt}$$

$$I_o = L_e \int_0^\tau e\, dt$$

$$= \frac{E_o \tau}{L_e} \tag{5.17}$$

Equation (5.17) is based on a rectangular pulse of width τ.
The node equation for the circuit of Fig. 5.8 is

$$C_D \frac{de}{dt} + \frac{1}{L_e}\int e\, dt + \frac{e}{R_T} + I_o = 0 \tag{5.18}$$

With the initial conditions of current and charge, the Laplace transform of Eq. (5.18) is

$$C_D S E[S] - C_D E_o + \frac{E[S]}{L_e S} + \frac{E[S]}{R_T} + \frac{I_o}{S} = 0 \tag{5.19}$$

Substituting the expression for initial current from Eq. (5.17) in Eq. (5.19) and solving for $E[S]$ yields

$$E[S] = E_o \left(S - \frac{\tau}{L_e C_D}\right) \bigg/ \left(S + \frac{1}{2R_T C_D} + \sqrt{\left(\frac{1}{2R_T C_D}\right)^2 - \frac{1}{L_e C_D}}\right)$$

$$\times \left(S + \frac{1}{2R_T C_D} - \sqrt{\left(\frac{1}{2R_T C_D}\right)^2 - \frac{1}{L_e C_D}}\right) \tag{5.20}$$

The inverse transform of Eq. (5.20) is

$$
\begin{aligned}
e(t) = E_o \exp\left(-\frac{t}{R_T C_D}\right) &\left[\cosh \sqrt{\left(\frac{1}{2R_T C_D}\right)^2 - \frac{1}{L_e C_D}}\, t \right. \\
&\left. - \frac{\dfrac{\tau}{L_e C_D} + \dfrac{1}{2R_T C_D}}{\sqrt{\left(\dfrac{1}{2R_T C_D}\right)^2 - \dfrac{1}{L_e C_D}}} \sinh \sqrt{\left(\frac{1}{2R_T C_D}\right)^2 - \frac{1}{L_e C_D}}\, t \right] \quad (5.21)
\end{aligned}
$$

Equation (5.21) may be normalized for graphical representation by defining three parameters:

Damping factor $D = 1/2R_T \sqrt{L_e/C_D}$
Initial current factor $F = \tau/\sqrt{L_e C_D}$
Normalized time $T = t/2\pi\sqrt{L_e C_D}$

By use of these parameters Eq. (5.21) becomes

$$
\begin{aligned}
\frac{e(t)}{E_o} = \exp(-2\pi DT) &\left(\cosh 2\pi\sqrt{D^2 - 1}\,T \right. \\
&\left. - \frac{D + F}{\sqrt{D^2 - 1}} \sinh 2\pi\sqrt{D^2 - 1}\,T \right) \quad (5.22)
\end{aligned}
$$

If the damping factor D is less than 1, Eq. (5.22) becomes oscillatory:

$$
\begin{aligned}
\frac{e(t)}{E_o} = \exp(-2\pi T) &\left(\cos 2\pi\sqrt{1 - D^2}\,T \right. \\
&\left. - \frac{D + F}{\sqrt{1 - D^2}} \sin 2\pi\sqrt{1 - D^2}\,T \right) \quad (5.23)
\end{aligned}
$$

Equations (5.22) and (5.23) have been plotted in Figs. 5.9 to 5.12 for four values of the exciting-current factor from zero, for which the exciting current is zero, to 5, for which the exciting current is in excess of most operating conditions. In each figure the damping factor is plotted as a parameter. These curves show that satisfactory trailing edge response requires control over both the damping factor and the exciting current. For any given damping factor, performance is improved by decreasing the exciting current. This can be done by increasing the shunt inductance.

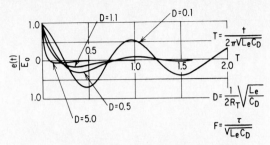

FIG. 5.9 Normalized pulse transformer trailing edge output voltage vs. normalized time for an initial current factor F of 0.0. Damping factor D is a parameter. Inductance L is in henrys; capacitance C is in farads; resistance R is in ohms; time t and pulse width τ are in seconds. Equivalent circuit is Fig. 5.8.

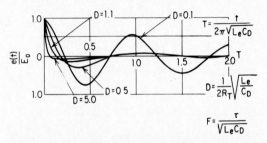

FIG. 5.10 Normalized pulse transformer trailing edge output voltage vs. normalized time for an initial current factor F of 0.1. Damping factor D is a parameter. Inductance L is in henrys; capacitance C is in farads; resistance R is in ohms; time t and pulse width τ are in seconds. Equivalent circuit is Fig. 5.8.

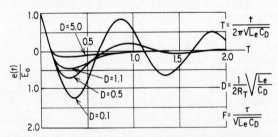

FIG. 5.11 Normalized pulse transformer trailing edge output voltage vs. normalized time for an initial current factor F of 1.0. Damping factor D is a parameter. Inductance L is in henrys; capacitance C is in farads; resistance R is in ohms; time t and pulse width τ are in seconds. Equivalent circuit is Fig. 5.8.

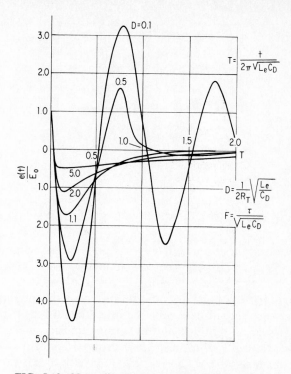

FIG. 5.12 Normalized pulse transformer trailing edge output voltage vs. normalized time for an initial current factor F of 5.0. Damping factor D is a parameter. Inductance L is in henrys; capacitance C is in farads; resistance R is in ohms; time t and pulse width τ are in seconds. Equivalent circuit is Fig. 5.8.

This also increases the damping factor, a generally good effect, but unfortunately it also increases the response time. To obtain fast recovery, not only must the shunt inductance be made large, but the distributed capacitance must be made small.

It often occurs that R_T is large because the load and source resistances are disconnected during the trailing edge response interval. There are often restrictions on increasing the shunt inductance and decreasing the distributed capacitance. This situation leads to an unsatisfactory trailing edge response. It is sometimes necessary to use external circuitry to dissipate the reactive energy in the transformer. A clipping diode across the primary in series with a resistor, as shown in Fig. 5.13, will allow current to flow in the reverse direction without interfering with operation during the time when the input voltage pulse is applied. This clipping diode circuit reduces the amplitude of the backswing and either elim-

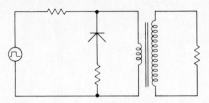

FIG. 5.13 Clipping diode circuit used to improve trailing edge response of pulse transformer.

inates or greatly attenuates subsequent voltage of the same polarity as the applied pulse.

Rapid recovery is of particular importance where a train of closely spaced pulses is transformed. A certain amount of backswing is inevitable. If the succeeding pulse is applied during the backswing of the first pulse, the charge on the capacitor and the current in the inductance at the start of the succeeding pulse become the initial conditions for that pulse, reducing its forward amplitude and increasing its backswing. This effect is cumulative, resulting in a pulse train droop analogous to the top of pulse droop in a single pulse. The general approach to this problem is to reduce the reactive energy stored in the transformer during each pulse and to dissipate that energy as rapidly as possible after each pulse with auxiliary circuitry.

Illustrative problem for pulse transformer response

A pulse transformer is used to match a generator to a magnetron load. The match is achieved during the top of pulse interval. During the pulse rise interval, the resistance of the load is 10 times the value during the top of the pulse. The generator resistance is the same for the rise and top of pulse intervals. Both generator and load are disconnected during the trailing edge interval. The pulse widths will be 0.25 and 1.0 μs. The transformer equivalent circuit elements referred to the secondary are:

$$R_G = R_L = 850 \ \Omega$$
$$L_L = 52 \ \mu\text{H}$$
$$L_e = 14 \ \text{mH}$$
$$R_T = 33,000 \ \Omega$$
$$C_D = 60 \ \text{pF}$$

Determine the approximate shape of the output pulse.

Solution

PULSE RISE

Refer to Fig. 5.4. The value of the load resistance during the pulse rise will be $10 \times 850 = 8500 \, \Omega$. Determine the damping factor using the formula for a step-up transformer.

$$\frac{b}{2\sqrt{c}} = \frac{R_G R_L C_D + L_L}{2\sqrt{R_L L_L C_D (R_G + R_L)}}$$

$$= \frac{850 \times 8500 \times 60 \times 10^{-12} + 52 \times 10^{-6}}{2\sqrt{8500 \times 52 \times 10^{-6} \times 60 \times 10^{-12}(850 + 8500)}}$$

$$= 0.5$$

Calculate the pulse rise normalized voltage factor:

$$\frac{R_G + R_L}{R_L} = \frac{850 + 8500}{8500}$$

$$= 1.1$$

The pulse rise interval will last until the output pulse reaches the value which will be maintained during the top of pulse interval, at which moment the load resistance will switch to its lower value. The normalized voltage factor during the top of pulse interval is

$$\frac{R_G + R_L}{R_G} = \frac{850 + 850}{850}$$

$$= 2$$

The normalized rise time is determined from the pulse rise normalized voltage. This will be the pulse rise normalized voltage factor divided by the normalized voltage factor for the top of the pulse:

$$\frac{1.1}{2} = 0.55$$

This value is used with the curve in Fig. 5.4 for which the damping factor is 0.5 to determine the normalized rise time of approximately 0.22. The actual rise time is calculated from the normalized rise time using the formula for step-up transformers:

$$t = 2\pi \left(\sqrt{\frac{L_L C_D R_L}{R_L + R_G}} \right) T$$

$$= 2\pi \left(\frac{52 \times 10^{-6} \times 60 \times 10^{-12} \times 8500}{8500 + 850} \right) \times 0.22$$

$$= 0.07 \times 10^{-6} \text{ s}$$

TOP OF PULSE

Refer to Fig. 5.7 and the section above on top of pulse response. Calculate the value for K.

$$K = \frac{R_L}{R_G + R_L}$$

$$= \frac{850}{850 + 850}$$

$$= \frac{1}{2}$$

Determine the normalized time for the longer pulse width.

$$T = \frac{R_G}{L_e} t$$

$$= \frac{850}{0.014} \times 10^{-6}$$

$$= 0.06$$

By use of the above values of K and T, the normalized voltage at the end of the longer pulse is determined from Fig. 5.7:

$$\frac{e_L}{E_G K} = 0.97$$

The actual voltage at the end of the pulse is

$$\frac{e_L}{E_G} = 0.97K$$

$$= \frac{0.97}{2}$$

$$= 0.485$$

The droop during the top of pulse interval is

$$\frac{K - \dfrac{e_L}{E_G}}{K} = \frac{0.5 - 0.485}{0.5}$$

$$= 0.03$$

TRAILING EDGE OF PULSE

Refer to the section above on trailing edge response and to Figs. 5.9 through 5.12. Calculate the damping factor.

$$D = \frac{1}{2R_T} \sqrt{\frac{L_e}{C_D}}$$

$$= \frac{1}{2 \times 33{,}000} \sqrt{\frac{0.014}{60 \times 10^{-12}}}$$

$$= 0.23$$

Calculate the initial current factor.

$$F = \frac{\tau}{\sqrt{L_e C_D}}$$

$$= \frac{1 \times 10^{-6}}{\sqrt{0.014 \times 60 \times 10^{-12}}}$$

$$= 1.09$$

Figure 5.11 has the initial current factor closest to the above value. The trailing edge will be intermediate between the curves for $D = 0.1$ and $D = 0.5$. The ringing period in normalized time is approximately 1. The ringing period in actual time is

$$t = (2\pi \sqrt{L_c C_D})T$$
$$= (2\pi \sqrt{0.014 \times 60 \times 10^{-6}}) \times 1$$
$$= 5.8 \times 10^{-6} \, \text{s}$$

The trailing edge will have a backswing with an amplitude approximately equal to the initial pulse for the longer pulse width. The composite pulse will have an appearance similar to that shown in Fig. 5.14. In this figure the rise and top of the pulse are satisfactory. The trailing edge is not. The trailing edge for the longer pulse suffers to obtain a satisfactory rise time for the shorter pulse. The transformer could probably function in this application if a clipping diode circuit were used.

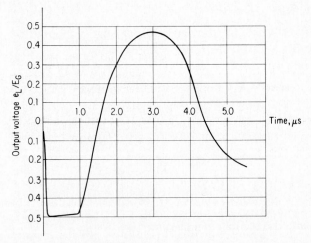

FIG. 5.14 Composite pulse from data in illustrative problem for pulse transformer response.

CHAPTER 6

The Use of Ferromagnetic Materials in Transformers and Inductors

The features of the ideal transformer may be translated into requirements for the ideal core material. The ideal transformer has infinite shunt inductance requiring a core material with infinite permeability. The ideal transformer has zero core loss. In the ideal transformer leakage inductance, winding resistances, and stray capacitance are all zero, achievable with a core material with infinite saturation flux density. The departure of practical transformers from the ideal reflects in large measure the departure of the core materials from these ideal properties.

The magnetic designer has available a number of magnetic materials from which to choose, none of which are fully adequate. Cost, schedules, and performance requirements influence the choice. Economics dictates the least expensive option that will ensure performance to specification on schedule. Since in general the materials with the best properties are the most expensive, it is an economic necessity to exploit the properties of the lowest-performance core material that will do the job. This task requires a knowledge of the properties of the core material and an understanding of the impact of these properties on the device in which it is to be used. In Sec. 6.1 magnetic properties and materials are discussed. In Sec. 7.2 core geometry is discussed.

6.1 THE MAGNETIZATION CURVE

Ferromagnetic materials have properties which are conveniently described by a magnetization curve. A magnetization curve is a plot of flux density vs. magnetizing force, terms defined in Sec. 1.2. Magnetizing force and flux density are related by Eq. (1.4), repeated here for reference:

$$B = \mu H \qquad (1.4)$$

In this equation μ is the permeability, a property of the medium. In the practical system of units used in magnetics, the permeability of a vacuum is 1. The permeability of air and most other nonferrous materials is also approximately 1. The permeability of most nonferrous materials is independent of the value of flux density. This is not true of ferromagnetic materials. In ferromagnetic materials the initial permeability at zero flux density is many times that of air. As the flux density increases, the permeability increases above its initial value, then decreases, approaching the value of air in the region called saturation. If the magnetizing force is decreased after an increase, the magnetization curve is found not to be retraced. The flux density assumes another set of values higher than those obtained when the magnetizing force was increasing. When the magnetizing force is decreased to zero, the flux density remains at a finite level. To reduce the flux density to zero, it is necessary to reverse the direction of the magnetizing force. If the magnetizing force is increased, then decreased in the new direction, symmetrical effects in that direction will occur. This sequence is called *hysteresis*. If an alternating current is used to develop the magnetizing force, a closed double-valued magnetization curve is traced called a *hysteresis curve,* illustrated in Fig. 6.1. The shape

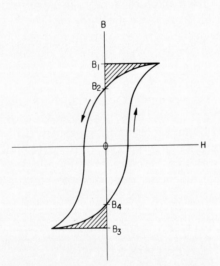

FIG. 6.1 Hysteresis loop illustrating areas of stored and dissipated energy. Shaded areas represent stored energy.

of the hysteresis curve, a matter of concern in magnetics, is a function of the material and geometry of the magnetic circuit. It is also a function of the amplitude and frequency of the magnetizing force.

6.2   CORE LOSS

Core loss is the power dissipated in the core. There are two types of core losses, hysteresis and eddy current.

Hysteresis loss is the energy used to align and rotate the elementary magnetic particles of the core material. The energy delivered to a coil containing a ferromagnetic core during an interval τ, the time spent in one traverse of the hysteresis loop, is

$$J = \int_0^\tau ei \, dt \tag{6.1}$$

In Eq. (6.1) e is in volts, i is in amperes, t is in seconds, and J is in joules. To describe J in magnetic quantities, use is made of two equations given in another section, repeated below for reference:

$$e = N\frac{d\phi}{dt} \times 10^{-8} \tag{1.9}$$

$$H = \frac{0.4\pi Ni}{l_i} \tag{1.16}$$

The flux in a ferromagnetic circuit is approximately uniform across the cross-sectional area of the core, so that

$$\phi = BA \tag{6.2}$$

By rearrangement Eq. (1.16) becomes

$$i = \frac{Hl_i}{0.4\pi N} \tag{6.3}$$

Substituting Eqs. (1.9), (6.2), and (6.3) in Eq. (6.1) yields

$$J = \int_0^\tau NA\frac{dB}{dt} \times 10^{-8} \times \frac{Hl_i}{0.4\pi N} \, dt$$

$$= \frac{Al_i \times 10^{-8}}{0.4\pi} \int_0^B H \, dB \tag{6.4}$$

The integral portion of Eq. (6.4) may be separated into intervals with the aid of Fig. 6.1 as follows:

$$\int_0^B H\, dB = \int_0^{B_1} H\, dB - \int_{B_1}^{B_2} H\, dB + \int_{B_2}^0 H\, dB$$
$$+ \int_0^{B_3} H\, dB - \int_{B_3}^{B_4} H\, dB + \int_{B_4}^0 H\, dB \quad (6.5)$$

In Eq. (6.5) the positive terms indicate energy supplied to the coil, and the negative terms indicate energy returned to the circuit. Thus the net energy absorbed by the core due to hysteresis is proportional to the area enclosed by the hysteresis loop. The product Al_i in Eq. (6.4) is the volume of the core, called V below. If the core magnetization is repeated with a frequency f, then Eq. (6.4) may be written as a power loss:

$$P = \frac{fV \times 10^{-8}}{0.4\pi} \int H\, dB \quad (6.6)$$

In Eq. (6.6) P is in watts, f is in hertz, V is in cubic centimeters, H is in oersteds, and B is in gauss.

Resistive losses in the core and winding are not considered in Eq. (6.6). The assumption was made that there is no voltage drop due to current flowing through the winding resistance and that all the current produces flux. Errors from neglecting the winding resistance are usually negligible. The assumption that all the current is flux-producing neglects the frequently significant parallel current path representing eddy currents. It is possible to view the dynamic hysteresis loop on an oscilloscope. The same voltage which is applied to the coil is applied to an integrating circuit to obtain a voltage which is proportional to flux density in accordance with the integral form of Faraday's law. This voltage is applied to the vertical plates of the oscilloscope. The current flowing through the coil is sampled by means of a low resistance in series with the coil. The voltage across this resistor is applied to the horizontal plates of the oscilloscope. The voltage on the horizontal plates is proportional to the magnetizing force. The scope may be calibrated and the display used for test purposes. The circuit is illustrated in Fig. 6.2. To prevent errors due to eddy currents, the frequency should be kept low.

Eddy current loss is caused by circulating currents in a conductive magnetic core. The currents are caused by voltages induced by the changing flux in the core in the same manner that voltages are induced in the secondary windings of transformers. (See Sec. 2.1.) Eddy currents tend to follow circular paths normal to the direction of the magnetic flux. To

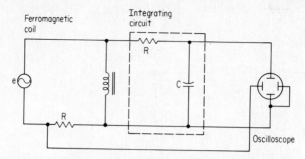

FIG. 6.2 Circuit for viewing hysteresis loop on an oscilloscope.

evaluate eddy current losses, consider Fig. 6.3. Dimensions W and T represent the width and thickness of a section of a magnetic core. In this section of unit length the flux is normal to the paper and is assumed to be uniform over the area $W \times T$. Consider an elementary path, $2x$, $2y$, in this section. This path constitutes a shorted turn to the flux contained in the area $4xy$. By Faraday's law the voltage induced in this path is

$$e = \frac{d\phi}{dt} \times 10^{-8}$$

$$= 4xy \frac{dB}{dt} \times 10^{-8} \qquad (6.7)$$

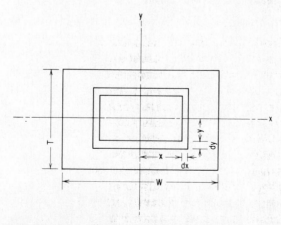

FIG. 6.3 Representation of a magnetic core cross section used for analyzing eddy currents.

where ϕ is the flux included in the area $4xy$ and B is the flux density over the area.

The impedance of the circuit represented by the path $2x$, $2y$ is mostly resistive and will be assumed to be completely so. Then the power dissipated in the path is

$$p = \frac{e^2}{R} \tag{6.8}$$

where p and e are the instantaneous power dissipated and voltage induced, respectively. The resistance of the path is

$$R = \rho\left(\frac{4y}{dx} + \frac{4x}{dy}\right) \tag{6.9}$$

in which ρ is the resistance per unit area per unit length. If the proportions of the cross section are known, the following relationship may be used to reduce the number of variables:

$$y = Kx$$

Then the area $4xy$ becomes $4Kx^2$, and Eq. (6.9) becomes

$$R = \rho\left(\frac{4Kx}{dx} + \frac{4x}{K\,dx}\right) \tag{6.10}$$

Equation (6.7) becomes

$$e = 4Kx^2 \frac{dB}{dt} \times 10^{-8} \tag{6.11}$$

Substituting Eqs. (6.10) and (6.11) in Eq. (6.8) gives an expression for the instantaneous differential power:

$$dp = \frac{\left(4Kx^2 \dfrac{dB}{dt} \times 10^{-8}\right)^2}{\rho\left(\dfrac{4Kx}{dx} + \dfrac{4x}{K\,dx}\right)} = \frac{4K^2 \times 10^{-16}\left(\dfrac{dB}{dt}\right)^2 x^3\,dx}{\rho\left(K + \dfrac{1}{K}\right)} \tag{6.12}$$

Equation (6.12) may be integrated over the area $W \times T = KW^2$:

$$p = \frac{4K^2 \times 10^{-16} \left(\dfrac{dB}{dt}\right)^2}{\rho \left(K + \dfrac{1}{K}\right)} \int_0^{W/2} x^3 \, dx = \frac{K^2 \times 10^{-16} \left(\dfrac{dB}{dt}\right)^2 W^4}{16\rho \left(K + \dfrac{1}{K}\right)} \qquad (6.13)$$

The volume of the core section per unit length is KW^2. The instantaneous power dissipated by the circulating eddy currents per unit volume is

$$p = \frac{K^2 W^2 \times 10^{-16} \left(\dfrac{dB}{dt}\right)^2}{16\rho(K^2 + 1)} \qquad (6.14)$$

The average power dissipated per unit volume will be

$$P_{av} = \frac{1}{T} \int_0^T p \, dt \qquad (6.15)$$

Three input-voltage cases are of particular interest, a sine wave voltage, a symmetrical square wave voltage, and a pulse voltage of single polarity. For each of these cases the rate of change of flux density must be determined for use in Eq. (6.14). For a sine wave voltage that determination can be made from Faraday's law in integral form:

$$B = \frac{E}{A} \times 10^8 \int \sin \omega t \, dt$$

from which:
$$B = -B_{max} \cos \omega t$$

then:
$$\frac{dB}{dt} = \omega B_{max} \sin \omega t$$

and:
$$\left(\frac{dB}{dt}\right)^2 = \omega^2 B_{max}^2 \sin^2 \omega t \qquad (6.16)$$

Substituting Eq. (6.16) in Eq. (6.14) yields

$$p = \frac{K^2 W^2 \omega^2 B_{max}^2 \sin^2 \omega t}{16\rho(K^2 + 1)} \times 10^{-16} \qquad (6.17)$$

Equation (6.17) is the instantaneous power. The average power is obtained by integrating between the limits of 0 and $1/2f$:

$$P_{av} = \frac{2fK^2W^2B_{max}^2(\omega^2 \times 10^{-16})}{16\rho(K^2 + 1)} \int_0^{1/2f} \sin^2 \omega t\, dt$$

$$= \frac{\pi^2K^2W^2B_{max}^2f^2}{8\rho(K^2 + 1)} \times 10^{-16} \tag{6.18}$$

In a symmetrical square wave the voltage of one polarity is constant for the first half of the period of the wave, then switches to the opposite polarity at the same amplitude for the second half of the period. By Faraday's law the time rate of change of flux density is constant when the voltage is constant. The flux density is a symmetrical triangular wave whose peaks are reached at the instant the polarity of the voltage reverses. The flux density changes from negative maximum to positive maximum during one half-period interval. Since the rate of change is constant:

$$\frac{dB}{dt} = \frac{2B_{max}}{\frac{1}{2}T} = 4B_{max}f \tag{6.19}$$

in which f is the frequency of the square wave. Since the voltage is constant at both polarities of a symmetrical square wave, the instantaneous power is equal to the average power. The average power dissipation per unit volume due to eddy currents from this waveform can be obtained by substituting Eq. (6.19) in Eq. (6.14):

$$P_{av} = \frac{K^2W^2B_{max}^2f^2}{\rho(K^2 + 1)} \times 10^{-16} \tag{6.20}$$

For a pulse voltage of single polarity, the peak power dissipated per unit volume due to eddy currents will be

$$p = \frac{K^2W^2B_{max}^2}{16\rho(K^2 + 1)\tau^2} \times 10^{-16} \tag{6.21}$$

To find the average power [Eq. (6.21)], p must be multiplied by the duty cycle which is equal to the pulse width multiplied by the repetition frequency f:

$$P_{av} = \frac{K^2 W^2 B_{max}^2}{16\rho(K^2 + 1)\tau^2} f\tau \times 10^{-16}$$

$$= \frac{K^2 W^2 B_{max}^2 f}{16\rho(K^2 + 1)\tau} \times 10^{-16} \qquad (6.22)$$

In Eqs. (6.18), (6.20), and (6.22) P_{av} is in watts per cubic centimeter, W is in centimeters, K is dimensionless, B is in gauss, f is in hertz, and ρ is in ohms per square centimeter per centimeter. In Eq. (6.22) τ is the pulse width in seconds and f represents the pulse repetition frequency. These equations neglect magnetic skin effects. They are of value in showing the factors which affect losses rather than for obtaining quantitative information.

The term $K^2/(1 + K^2)$ appears in all the equations for eddy current loss. The smaller the value of K, the less will be the eddy current loss. To reduce losses, cores are made of thin strips or laminations stacked together to obtain the required core area. The laminations are separated from each other by means of insulating films on the lamination surface. Some core materials are formulated to have high resistivities for reducing core losses. See Sec. 6.4.

6.3 THE EFFECT OF CORE MATERIALS ON BANDWIDTH

The properties of core materials determine the achievable bandwidth of transformers. Refer to Chaps. 4 and 5 concerning the effect of transformers on circuit performance and to Chap. 10 on the determination of equivalent circuit parameters. The low cutoff frequency is determined by the shunt inductance. The high cutoff frequency is determined by the leakage inductance and distributed capacitance. The shunt inductance is determined by the permeability of the magnetic core, the number of turns in the primary, and the core geometry. The higher the permeability and the greater the number of turns, the greater will be the open-circuit inductance. The leakage inductance and distributed capacitance are determined by the number of turns and the coil geometry. The leakage inductance and distributed capacitance are both roughly proportional to coil volume. To achieve wide bandwidth, a core material of high permeability is needed to obtain high shunt inductance. The smaller the number of turns used to achieve the shunt inductance needed for the low-frequency response, the less will be the leakage inductance and distributed capacitance, the product of which determines the high-frequency cutoff.

As the frequency decreases, the flux density increases for a fixed voltage. The permeability of the core decreases with the onset of saturation. This places the additional requirement on the core material that it have a high saturation flux density for good low-frequency response.

6.4 MAGNETIC MATERIALS FOR TRANSFORMERS AND INDUCTORS

The properties of magnetic materials are affected by the composition of the material, by the way the material is fabricated, and by the heat treatment of fabricated parts. Iron, nickel, and cobalt are the ferromagnetic elements of commercial interest. Iron, by far the most plentiful, is alloyed to produce magnetic steels with enhanced properties. Nickel alloys are used in high-performance devices where the need for their special properties justifies the cost. Cobalt alloys possess valuable properties, but cobalt is so scarce and expensive that it is seldom used. Supplementing the metallic alloys are ferrites and powdered iron alloys.

Figure 6.4 defines some of the significant features of a hysteresis curve. A material for use as a permanent magnet should have high remanent flux and high coercive force. Materials with these properties are

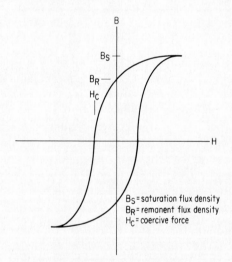

FIG. 6.4 Hysteresis curve defining points used to describe properties of magnetic materials.

called *hard*. Devices which develop alternating fields require materials with a low coercive force. These materials are called *soft*. Devices with alternating fields in addition require high permeability and low core losses. Magnetic switching devices require high remanence. In other alternating-field devices high remanence may be a disadvantage.

Metallic alloys are milled in thin sheets. The thickness of the sheet and the rolling process used to form the sheet affect magnetic properties. Sheet stock is wound, punched, and sheared. These processes impair the properties of the alloys. Heat treating after fabrication helps to restore and enhance the magnetic properties. Silicon steel sheet is the most widely used soft magnetic material. It contains between 0.5 and 3.25 percent silicon. When used as an alloying element in iron, silicon affects the crystalline structure so the anisotropy of the alloy is increased. This may be exploited in the rolling, fabricating, and heat-treating processes to increase permeability and decrease core losses in the direction of rolling. Silicon steel alloys are made in a number of different grades, thicknesses, and surface treatments accompanied by various rolling and annealing processes. Some silicon steels have been given industry classifications through the American Iron and Steel Institute. Identified by a number preceded by "M-", this classification is based on core loss at a single frequency and flux density of standard test samples in which grain orientation is observed and optimum annealing practice is followed. The user is interested not only in core loss but other properties, each under various conditions other than the standard test conditions. Both rolling mills supplying sheet and fabricators supplying finished parts assign proprietary designations to the grades of silicon steel for which advantageous characteristics are claimed. The user must depend upon these suppliers for performance data, usually supplied in the form of curves showing typical properties as functions of such variables as frequency, flux density, and temperature. Information on maximum excursions from these curves is seldom available. Users' assembly and processing operations further modify performance, making establishment of useful performance minima difficult. Silicon steel alloys are universally used for power frequencies. In thin gauges their use extends into the audio range. The thinnest gauges are employed in pulse transformers using C core geometry. Asymmetrical magnetization and the usual associated air gap in the magnetic path cause a smaller percentage change in the effective permeability of silicon steel cores than in the cores made from more exotic materials. For applications in which dc magnetization occurs, silicon steel is frequently used when higher-performance materials would otherwise be indicated.

Nickel-iron alloys have greatly increased permeability over silicon steel but saturate at lower flux densities. The 50 percent nickel alloy saturates at approximately 13 kG. The 80 percent nickel alloy saturates at about 8 kG. The 80 percent nickel alloy has the highest initial permeability of any of the commercially available materials. Special heat treating and use of favorable core geometry impart other properties to nickel-iron alloys. Square hysteresis loops are obtainable in toroidal cores of nickel alloys which are useful in saturating and switching devices. High permeability makes wideband devices and operation with high-circuit impedances possible. Nickel-iron alloys are usually associated with high-performance circuits in the audio and video frequencies.

The cobalt-iron alloy Supermendur[1] saturates at a higher flux density than any other commercially available soft magnetic alloy. It is available in wound toroidal and cut cores. The scarcity and cost of cobalt limit the commercial significance of this alloy.

Magnetic metallic glasses are alloys of iron, boron, and silicon which are quenched so rapidly upon casting that crystal formation typical of other magnetic alloys is prevented. Metallic glasses have higher resistivity than crystalline magnetic alloys and lower core losses as a result. The saturation flux density is high. Presently available only in 1-mil ribbon, metallic glasses show promise for use in high-frequency applications where a higher saturation flux density would be an advantage over ferrites. Metallic glasses have a limited availability in wound toroidal and cut cores.

Powdered magnetic alloys are useful in high-frequency circuits where high Q's are required. These materials are available in toroidal cores and slugs. Saturation flux density is similar to the 80 percent nickel alloy. Permeability is lower than either solid metallic alloys or ferrites. The large effective air gap in powdered alloy cores makes them tolerant of dc magnetization. This material finds applications into the 200-kHz range.

Ferrites are an important class of magnetic materials for use at frequencies from the high audio range well into the megahertz range. Ferrites consist of ferric oxide in loose chemical union with one or more oxides of manganese, zinc, and nickel. Manganese-zinc ferrites have higher permeabilities at low frequencies and saturate at higher flux densities than nickel ferrites. Nickel ferrites maintain their permeability over a wider frequency range and have low losses at high frequencies. The importance of ferrites in magnetics technology is largely due to their high resistivity, which is in the semiconductor range. This is a decisive advantage over metallic alloys at high frequencies. Ferrite materials are not standardized. Each supplier of ferrites has proprietary formulas for mate-

---

[1] Registered trade name of Arnold Engineering Co.

TABLE 6.1 Typical Properties of Important Soft Magnetic Materials

Material	Approximate Composition, %	B_s, kG	B_R, kG	H_c, Oe	Core Loss, W/lb	Permeability		Application Notes
						Initial	Maximum	
Silicon Steels								
0.014 in thick AISI M-6	Fe 97 Si 3	19	14	0.1	0.66 at 15 kG 60 Hz	350	50,000	Widely used for laminations and I bars at power and audio frequencies. Most effective when flux path is in direction of rolling.
0.012 in thick AISI M-5	Fe 97 Si 3	19	14	0.1	0.58 at 15 kG 60 Hz	350	50,000	Used in wound C cores mostly for operation at 50/60 Hz. Has low core losses at high flux densities.
0.0185 in thick AISI M-19	Fe 97 Si 3	19	—	0.5	0.80 at 10 kG 60 Hz	300	10,000	Used for laminations when higher core loss is acceptable. Less costly.
0.025 in thick AISI M-22	Fe 97 Si 3	19	—	0.6	0.9 at 10 kG 60 Hz	300	10,000	Higher losses and less costly than M-19.
0.004 in thick Grain-oriented	Fe 97 Si 3	19	14	0.4	10.0 at 15 kG 400 Hz	350	50,000	Widely used in wound cores cut and uncut at 400 Hz and higher frequencies.
0.002 in thick Grain-oriented	Fe 97 Si 3	19	14	0.5	14.0 at 10 kG 1.0 kHz	350	50,000	Used in wound cores cut and uncut for high-frequency and pulse use.

TABLE 6.1 (continued)

Material	Approximate Composition, %		Properties						Application Notes
			B_s, kG	B_R, kG	H_c Oe	Core Loss, W/lb	Permeability Initial	Permeability Maximum	
			Nickel-Iron Alloys						
0.014 in thick 50% Ni	Fe 50 Ni 50		13	11	0.15	3.0 at 10 kG 400 Hz	5,000	40,000	Used in laminations to provide high permeability at high flux densities at audio frequencies.
0.006 in thick 50% Ni	Fe 50 Ni 50		13	11	0.15	2.0 at 10 kG 400 Hz	5,000	100,000	Applications similar to 0.014-in-thick material but lower losses at high frequencies. Labor cost of stacking laminations very high.
0.014 in thick 80% Ni	Fe 20 Ni 80		7.5	6	0.05	0.8 at 6 kG 400 Hz	30,000	100,000	Used in laminations to provide very high initial permeability. Losses high at high frequencies. Saturates at low flux density.
0.006 in thick 80% Ni	Fe 20 Ni 80		7.5	6	0.05	0.4 at 6 kG 400 Hz	30,000	60,000	Applications similar to 0.014-in-thick material with lower losses at high frequencies. Labor cost of stacking laminations very high.
0.004 in thick 80% Ni	Fe 20 Ni 80		7.5	6	0.05	8.0 at 6 kG 5 kHz	30,000	60,000	Used in wound cores cut and uncut to provide high permeability and low losses at high frequencies.
0.004 in thick Square loop Ni-Fe	Fe 50 Ni 50		15	14.5	0.11	1.3 at 10 kG 400 Hz	—	—	Used in saturating and switching devices. Square hysteresis loop realizable only in toroidal cores.

	Composition							Remarks
*Supermendur**								
0.004 in thick	Fe 51 Co 49	22	21	0.2	14.0 at 20 kG 400 Hz	800	27,000	Used in wound cores providing highest saturation flux density. Very costly.
Metallic Glass								
0.001 in thick	Fe 81 B 13 Si 3.5	16	11	0.06	10.0 at 6 kG 10 kHz	2,500	100,000	Has low losses at high frequencies and high flux densities. Available in cut cores and toroids. Very costly. Under development.
Powdered Alloys								
	Variable Fe Ni Mo	8	—	—	9.0 at 1 kG 20 kHz	200	210	Used in high-frequency high-Q applications. Available in toroids and slugs.
Ferrites								
Mn-Zn	$Fe_2O_3 \cdot MnO$ $Fe_2O_3 \cdot ZnO$	4.5	1.0	0.2	0.05 W/cm^3	2,700	4,800	Used in high-frequency and relatively high flux density applications. Available in pot cores and proprietary shapes.
Ni	$Fe_2O_3 \cdot NiO$	3.2	2.6	4.0	—	120	150	Available in pot cores for use at frequencies into the megahertz range.

*Registered trade name of Arnold Engineering Co.

rials whose properties are under the control of the individual supplier. Minimum performance data for these materials are not likely to exist for practical operating conditions. Ferrites are ceramics possessing many of the physical properties of more familiar ceramics. They are formed by molding and firing, which results in hard, brittle materials. The need for a cavity for each individual shape limits the versatility of ferrites. The nature of the material and the forming process restricts the maximum size of ferrite cores. A unique limiting feature of ferrites is their low Curie temperature. In most electromagnetic devices, high temperature is of first concern because of the operating temperature limits of coil insulation. When ferrite cores are used, the Curie temperature is often the limiting consideration. Since magnetic properties deteriorate gradually as the Curie temperature is approached, the effect of temperature on ferrite performance in the operating range of common insulating materials must be given careful attention.

Table 6.1 is a compilation of typical properties of the most important soft magnetic materials. This table is intended as a guide in the selection of materials. Detailed performance data required for design should be obtained from the individual supplier.

CHAPTER 7

Mechanical Considerations

Magnetic devices receive first consideration for their electrical performance. With satisfactory electrical performance established, the cost and reliable operation are largely determined by mechanical requirements. There are many options available in mechanical construction, each with advantages and limitations. The objective is to obtain effective execution of the most appropriate option.

To choose from among the various mechanical options available, it is first necessary to select a construction that will perform under the actual operating conditions. The construction must enable the device to survive the rigors of transportation and installation and meet the user's specifications. In actual operation, the device must function in the circuit with rated electrical conditions even in the most disadvantageous environment.

Transportation is an important but often neglected consideration in mechanical construction. Equipment for fixed ground installation may, upon being uncrated at the site, be found to be a shambles. Consideration must be given to the shipping problem regardless of how modest the functional mechanical needs of the device are.

Although the user ultimately wants simply to have the equipment work satisfactorily under the most adverse conditions likely to be found, the means used to achieve this can be torturous. The user of magnetic components is confronted with a plethora of general requirements. The users are faced with the standards of various groups, such as military organizations, the insurance industry, state and municipal governments, the federal government, environmental protection agencies, industry associations, and even the user's own organization. It is not surprising, therefore, that in this maze of specifications, the user will ask for features that are not needed and sometimes omit features that are needed. Fear of the latter leads to overspecifying. When overspecified requirements are not negotiable, they must be weighed irrespective of rationality.

7.1   ENCLOSURES

Untreated core and coil assemblies are fragile with generally inadequate insulation structures. Enclosures and/or supplementary treatments are needed to achieve reliable performance under most environmental conditions. Moisture is the number one enemy. Many coil materials are hygroscopic. This property is intensified by the interstices of the coil. Moisture reduces insulation resistance and dielectric strength by serving as the solvent for airborne solutes such as carbon dioxide, a natural constituent of the atmosphere, and other ionic gases and solids, both natural and anomalous. Moisture also accelerates the corrosion of metals used in core and coil assemblies. These assemblies are subject to the mechanical abuse of shock, vibration, and being struck by other objects. Enclosures reduce vulnerability to mechanical damage. Many core and coil assemblies must be submerged in insulating media other than air to complete the insulating system or to improve heat transfer. If so, an enclosure is needed to contain the insulating medium. The insulating properties of air decrease with altitude. Open cores and coils that work satisfactorily at sea level may fail at high altitudes. Enclosures are needed for these high-altitude applications.

Hermetically Sealed Cases

The hermetically sealed metal case is the most effective protection available for magnetic components. This protection is enhanced through the use of potting materials for which the case provides the container. Hermetic sealing is not an absolute condition. Leaks in the most carefully constructed hermetically sealed containers can be detected with sensitive instruments. The rate of leaking is a function of the pressure differential between the inside and the outside of the container. A nominal seal at negligible pressure differential may leak intolerably at a moderate pressure differential. There is no quantitative answer to the question of how good is good enough, but some qualitative tests are in general use. The bubble test consists of stabilizing the unit to be tested at room temperature, then submerging it in a tank of hot water. The heat increases the pressure of entrapped air in the unit. A sustained stream of bubbles from a seam or orifice is an indication of a leak. Oil-filled units are tested by heating as described in Sec. 17.2. This is a realistic test. The unit is tested to determine that it does not leak oil, a reasonable requirement for oil-filled units. If it does not leak oil, there is little likelihood that atmospheric contaminants will enter the case.

The construction used in hermetically sealed cases must provide a means for installing the core and coil assembly, attaching and sealing covers, bringing out properly rated electrical terminations, and offering suitable mounting means, all while maintaining the integrity of the hermetic seal. Other required features of the case may substantially affect construction. Typical of such features are lifting facilities, high-voltage terminations, and access requirements. Cases are made of aluminum, brass, or steel, with steel being the most common. Brass cases are used with some high-frequency devices to reduce losses from eddy currents, induced by stray magnetic fields, flowing in the case. A higher coefficient of conductivity causes the losses in brass to be less than those in steel. Sheet-metal aluminum is rarely used. Cast aluminum construction is used for very special applications. Steel sheet can meet most requirements at an overall cost less than that of other materials.

All constructions have seams that must be sealed. The methods available for sealing seams are gasketing, welding, brazing, and soft soldering. The best gaskets are O rings. They are most effective when used with grooves and surfaces machined to close tolerances, obtainable with cast-aluminum construction. O-ring gaskets are not suitable for most sheet-metal construction. Flat gaskets are generally preferred because they do not require precision surfaces. Welding is a familiar process for joining metals. It requires fusing the base metal and usually alloying with a filler metal. Brazing joins metal by fusing copper or silver alloys which form the filler material. The base metal is not fused in brazing. Soft soldering is the joining of metals with fused lead-tin alloys at relatively low temperatures. Weldments are strong and are preferred for large enclosures. Hermetically sealed weldments are more difficult to make than weldments made solely for their strength. Very strong welded seams can be porous or have gaps. It is more difficult to obtain a good weldment with aluminum than with steel. Welding quality control is a perennial problem in all materials. Servicing of welded enclosures is facilitated when the cover seam is made on the edge of an outside lip. This permits removal of the cover by grinding off the weld bead and later rewelding. Brazing is used where the higher temperature required in welding is objectionable. Brazing is not as strong as welding but much stronger than soft solder. Soft soldering is a universal process for making electrical connections. Soft soldering for hermetic sealing is less well known and more difficult. To make a satisfactory soft-solder seal, the contact surfaces must be prepared. The correct pretreatment is hot-solder dip plating, a process that is sometimes incorrectly called hot-tin dipping. Tin plating is usually applied electrolytically and is unsatisfactory for hermetic sealing. Tin changes from its white crystalline state to amorphous gray at a temperature of about 13°C (55°F). In the amorphous state tin will not alloy with solder. The necessity for surface preparation limits the size of cases that can be sealed with soft solder. It is difficult to obtain the uniform heating needed for good solder flow on cases with large thermal masses. Small cases with good surface preparation are readily soft-solder-sealed. The hot-solder plating used with soft-solder-sealed cases is an excellent protection for steel. Cases that are sealed by welding or brazing cannot be preplated because the plating interferes with and is destroyed by the heat from those processes. Welded seams are particularly subject to oxidation. Welded steel cases are generally painted, first with a rust-inhibiting primer, then with a finish coat.

Types of Construction

The case illustrated in Fig. 7.1 is a drawn shell open at one end. The open end is closed with a drawn cover which fits inside the open end of the shell. In drawn construction good control can be maintained over the clearance between case and cover. This aids the soft-solder-sealing process. The existence of only one seam in this construction reduces assembly labor and the probability of leaks. The procurement of drawing tools is expensive and time-consuming. Most applications use cases for which tooling already exists. This practice frequently precludes optimum case dimensions. Drawn cases have generous corner radii which subtract from the available inside volume. The practicalities of fabrication and assembly restrict the location of mounting facilities and terminations to the single end cover. The drawn case yields a product that is rugged and of high quality. The fully fabricated case illustrated in Fig. 7.2 is widely used because of its dimensional versatility. The corners are bent to minimum radii. Folding leaves butt seams at several places, making solder bridges necessary. This objection can be overcome by welding the folded seams. The welded seams may require grinding to restore dimensional precision. Hot-solder plating is done after welding and grinding. Both ends are open in this construction, sometimes an advantage in assembly. Without welded seams, this construction has a high probability of leaking. Consequently, it is seldom used with oil-filled units. Figure 7.3 shows a compromise construction in which the case wall is fabricated and the outside

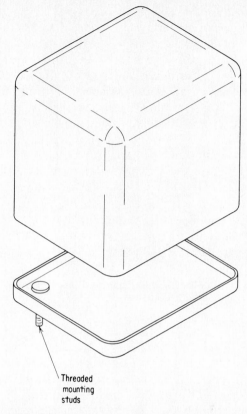

FIG. 7.1 Drawn steel case with inside-fitting end cover.

fitting end covers are drawn. Drawing tooling is required only for the end covers. This type of construction is readily soft-solder-sealed. The seams have good integrity. The case wall usually has a lock seam which is easily sealed with soft solder. Welded construction is used in large cases and oil-filled tanks. A welded steel tank with a gasketed top cover is illustrated in Fig. 7.4. Reinforcing sections are frequently added to large welded enclosures. Irregular shapes and special features can be provided with welded construction.

Mounting Facilities

Small cases usually have threaded studs on the cover as illustrated in Fig. 7.1. Internal assembly is facilitated when the terminals are on the same

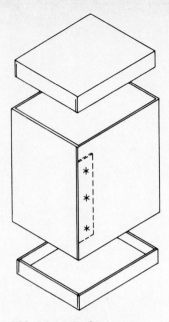

FIG. 7.2 Fully fabricated case with inside-fitting end covers.

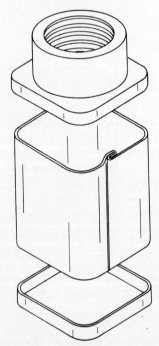

FIG. 7.3 Case with fabricated wall, outside-fitting drawn end covers, and bellows for oil expansion.

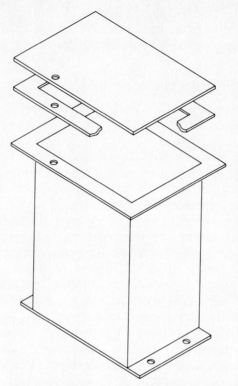

FIG. 7.4 Flanged and welded steel tank with gasketed and bolted top cover.

surface. Threaded studs may be single-ended or double-ended. Double-ended studs provide a means for internal mounting of the core and coil, thus transferring the load directly to the mounting without its passing through seams. For strength the stud passes through a hole in the cover and is held in place by a shoulder on the stud which bears against the interior side of the cover. Most studs are projection-welded. Projection welding adds strength to the case assembly, but it is not a very reliable seal. A fillet of solder may be added around the stud shoulder to ensure a seal. This is a simple operation if the case is hot-solder-plated after projection welding. Threaded inserts are also used for mounting. Inserts are projection-welded or rolled over. As with studs, threaded inserts may be sealed with solder fillets. Inserts do not increase the external outline as do studs, but they decrease the available volume inside. Mounting flanges, as illustrated in Fig. 7.4, are used on both small and large units. On small units mounting flanges are less preferable because the additional mounting area they require is a large percentage of the total area needed by the entire case. Flanges are common on large units which are too heavy to be secured by studs or inserts. Casters permit large units to be rolled into position in confined spaces. A convenience for later servicing, the casters are chocked with the unit in position.

Oil-Filled Hermetically Sealed Enclosures

Oil has a high thermal coefficient of expansion, changing in volume as much as 10 percent over the range from minimum to maximum oil temperature. This change in volume must be accommodated in the mechanical design of the enclosure since oil is an incompressible liquid. Making this accommodation is especially difficult when oil is needed as part of a high-voltage insulation system. A change in oil volume causes a change in a fluid level or in the position of a mechanical device. If the provision for volume change reaches its limit before the minimum oil temperature is reached, the oil volume will decrease as the temperature is decreased below the lower limiting temperature. A low-pressure void will form in the enclosure which will contain volatile constituents from the oil and air if the unit was imperfectly filled. The dielectric strength of this void is very poor. If the void reaches a critical high-voltage region, breakdown will occur. Movement of the unit causes the void to migrate. High-voltage units mounted in equipment subject to motion during operation cannot tolerate low-temperature voids. If the mechanical provision for expansion reaches its limit below the maximum oil temperature, the oil will expand as the oil temperature increases above the limiting temperature until excessive pressure forces a leak. Oil will flow from the leak until the internal and external pressures equalize. When the oil temperature is reduced, the internal pressure is lowered. The leak draws air, increasing the probability of later failure. Thus any mechanical device providing for oil expansion must adequately cover the entire oil temperature range.

The simplest method of providing for oil expansion is an expansion volume at the top of the enclosure. This arrangement works well for units that operate in a fixed position. The pressure developed will depend upon the size of the expansion volume, the change in volume of the oil, the temperature and pressure at which the unit was sealed, and the temperature limits of the oil. A refinement of this technique provides a separate expansion chamber leaving the main enclosure full of oil. To be effective, the separate expansion chamber must positively exclude air from the main enclosure. Void-free construction can be accomplished with flexible case sides or metal bellows. The latter are mounted in a cylindrical well exterior to the main enclosure. By exposing the open end of the bellows to the exterior, the contribution of the bellows assembly to the oil volume is kept small. It is advantageous to limit the total oil volume, thereby reducing the demand on the oil expansion device. Near optimum enclosure dimensions are needed to meet this objective. Oil volume can be further reduced with solid blocks or filler material such as fine gravel.

Unsealed Enclosures

Some of the problems and expense of hermetically sealed cases may be avoided by using unsealed enclosures. This is particularly appropriate for oil-filled cases. If the expansion volume is vented to the atmosphere, the internal and external pressures are equalized, and the dual problems of low pressure voids and excessive high pressure are avoided. Venting is usually done on units which will operate in one position, the vent being placed on the top. The vent may be filtered and restricted, but it must allow the passage of air and therefore moisture. Oil is hygroscopic. Its dielectric properties decline with increased moisture content but remain good, and its mechanical properties are unimpaired. Exposure to atmospheric moisture is not a prohibitive deterrent to venting oil-filled enclosures. Venting increases the exposure of oil to oxygen. When so exposed, oil oxidizes slowly, forming sludge. Most distribution transformers are oil-filled and vented. Oil-filled distribution transformers used by utilities have remarkable longevity. Maintenance procedures for the oil contribute to this achievement.

The protection provided by polymerizing materials is adequate for some applications without the need for hermetically sealed cases. Under certain conditions there is an advantage to using a metal case as a container for the core and coil and a polymerizing potting material. The case is filled with the resin which hardens. The exposed surface takes the place of a sealed cover. Since this surface is nonconducting, it is useful for terminations. Some of the advantages of casting may be realized with this construction without the expense of molding cavities.

Plastic Embedment

Plastic embedment is an attractive enclosure option for applications requiring good environmental protection as well as minimum size and weight. Embedment processes surround the device with plastic, providing environmental protection. Supplemental processes also improve the insulation structure. The extent of environmental protection and insulation enhancement depends on the type of embedment used and the quality of the process. The best do not provide as good environmental protection as do hermetically sealed cases. The insulation structure of embedments can be made the equal of any other solid insulation system. The nonconductive surface of embedments is an advantage for small high-voltage devices. While fabricating problems generally limit hermetically sealed cases to either rectangular or cylindrical shapes, plastic embedment permits convoluted shapes with minimum enclosed volumes. Plastics are available that permit operation at high temperatures and allow temporary overload conditions. Embedded devices have good heat transfer characteristics because they permit the ambient air to be in close proximity to the windings. The greatest heat dissipation densities and minimum sizes are generally obtained with plastic embedments. At low temperatures, plastic embedments do not perform as well. Available plastics have higher coefficients of thermal expansion than the devices embedded in them. The plastic becomes brittle and contracts around the embedded object when the temperature is lowered. The combination of contraction and embrittlement at low temperatures makes the embedding plastic subject to cracking. Maximum stress is developed when the device suddenly dissipates maximum losses at minimum ambient temperature after being stabilized in the off condition at that temperature. The embedded device begins to expand while the plastic coating is still cold and brittle. There is a high probability that magnetic devices embedded in plastic will crack if subjected to repeated cycling of this nature. Cracks occur in regions where stress concentrations develop. Large units crack more readily than small units. The cracks that develop from thermal stresses do not necessarily lead to immediate electrical failure. They allow the entrance of moisture and compromise insulation barriers. Cracks may go undetected until a later electrical failure occurs, at which time it may be difficult to identify the source of the failure. With its limitations recognized, plastic embedment can be a satisfactory compromise solution to many enclosure problems.

Conformal Coating

Magnetic devices are conformal-coated by dipping them in viscous thermosetting resins. The resin hardens by polymerization usually accelerated by elevated temperature. The general shape of the device is maintained through the dipping process. The resin must be removed manually from the terminal and mounting surfaces, which is more easily done when the resin is incompletely hardened and hot. The resins used in conformal coating are thixotropic. These resins undergo an apparent reduction in viscosity when stirred or agitated. The viscosity gradually increases to its original condition if the resin is left standing. This useful property permits reduction of the viscosity by stirring immediately before dipping. The low viscosity contributes to a thin, uniform coating. An increase in viscosity after dipping prevents excessive runoff. Varying between 1 and 3 mm, the thickness of the coating is determined by the properties of the resin, the temperature of the device, the temperature of the resin, the length of time the device stands at room temperature, and the elevated cure temperature. Control of runoff is complicated by an initial reduction in viscosity of the resin with increased temperature followed by an increase in viscosity as the hardening process advances, a characteristic of thermosetting resins. Runoff during the initial reduction in viscosity can be reduced by increasing the standing time at room temperature and by reducing the elevated cure temperature. If the coating is too thick, the device will be oversized. If the coating is too thin, the protection against moisture will be reduced.

Dip coating does not fill the interstices of the coil. Additional processing is required. Air contained in trapped voids will expand when heated, causing blow holes in a completely dipped device while the resin is still fluid. A void-free structure, in addition to preventing blow holes, improves moisture resistance and insulation strength. To fill the voids, the device is first partially dipped in the coating resin, leaving a portion of the coil, usually an end, uncoated. The initial coating is hardened, forming a reservoir. With the uncoated portion facing up, the unit is vacuum-impregnated. A low-viscosity polymerizing resin is used, filling the reservoir created by the initial coating. The filled unit is carefully transferred to a baking oven without spilling the liquid impregnant from the reservoir. The impregnant is hardened in the oven. A second dip coat seals the open end through which the impregnation was performed.

The enclosed volume of the conformally coated finished product is the smallest of any of the various embedment processes. The dielectric properties are somewhat indeterminate because the obstruction of impregnant flow by the coil can cause voids. The quality of this process is more dependent upon the skill of the operator than are other more mechanized processes. The process has a relatively high labor content. There is no tooling cost, making it a suitable choice for many short production run items.

Transfer Molding

Transfer molding is a production method used for embedding small magnetic devices. A high-quantity process, it provides good quality at low unit cost. Transfer molding requires a molding machine, a mold cavity, and a chamber for holding the mold charge from which the molding material is transferred to the cavity. The chamber may be an integral part of the mold tool. The transfer is accomplished by a hydraulically operated plunger which forces the material from the transfer chamber through a narrow passage, called a *runner,* into the mold cavity. Transfer molding uses thermosetting plastics. In the state used for molding, these plastics undergo an initial softening phase with increased temperature followed by a rapid irreversible hardening. In the molding operation, the molding charge is heated. While soft, the charge is forced by the plunger under high pressure into the heated mold cavity. The plastic hardens in the cavity. To permit removal of the molded article, the cavity is made in two pieces. The mold halves are opened and closed with powerful hydraulic rams needed to resist the high pressure developed. Phenolic resins are commonly used in transfer molding. They have good physical properties for molding and good electrical properties for embedment of electrical components. Since they require lower pressures, epoxy resins compounded for molding are also used. Lower pressures are less likely to damage delicate electrical assemblies. When used for embedment of electrical assemblies, transfer molding is essentially the same process as that used to mold plastic articles without embedment except for the need to hold the assembly to be embedded in position in the cavity and the need for lower molding pressures.

Like many other processes, transfer molding embedment is most likely to be successful when the basic design of the article provides for the molding process. Means for holding the article in the cavity must be provided. The article must be rugged and should not require complete resin penetration for proper electrical performance. Its geometry should favor the needs of the mold cavity.

The use of transfer molding embedment is governed by economic considerations. There is a substantial investment in the mold cavity. Most molds have several cavities, permitting simultaneous molding of several pieces. This reduces unit cost and increases mold cost. In this process, the tooling cost is high, and the unit cost is low in comparison with alternative processes. Full control of electrical and mechanical properties of the finished product is best obtained by molding the article in the shop where the electrical assembly is made. This requires investment in a molding machine. Such an investment needs the support of adequate sales of production quantities. Having the embedment done by a molder avoids the investment in the molding machine but not the mold cavity. Paid for separately by the purchaser, the mold generally remains in the custody of the molder. Purchase agreements and construction often limit molds to the original molder so that multiple sourcing usually means multiple tooling costs. The molder is not an electrical manufacturer, and so the part must be designed and built to reduce the probability of the work affecting electrical function. These considerations lead to transfer molding being infrequently used for embedment of magnetic devices.

Plastic Casting

Embedment by casting is a process by which a cavity containing the device to be embedded is filled with liquid resin by pouring. After the resin is hardened, the casting is removed from the cavity. The process differs from potting, in which the container remains as part of the finished article. Casting differs from molding in that no pressure is used in the pouring operation, although the pouring is sometimes done in a vacuum, atmospheric pressure being applied to the poured surface after the pour is completed. Casting is similar to transfer molding in that the geometry of the completed part is determined by the cavity. The time required for the resins used in casting to harden is hours compared with a few minutes or seconds in molding. The resin volume is frequently much larger in castings than in transfer molding where the resin volume per cycle is limited by the capacity of the molding machine. Casting is not a refined production process like transfer molding, but with proper techniques reasonable production efficiency can be attained. The mechanics of casting are less difficult than those of transfer molding. Most casting embedment is done by the same shop that manufactures the coils. The cavities used in casting can be simple or intricate, as needed. A casting cavity differs from a transfer mold cavity in that the resin flows into the cavity through a large opening rather than a small orifice. The larger opening is needed because gravity, not pressure, forces the resin into the cavity. The large opening makes cleanup after molding more of a task and restricts to some extent the cavity geometry. The cavity must be constructed so that the embedded part is held in position during the pouring and resin-hardening operations. Mounting facilities and terminals must be left available with a minimum amount of cleanup required. The most obvious way to construct a casting cavity is to assemble it from machined detail parts. This is known as a *permanent cavity*. The production rate from a single permanent cavity is one or two pieces a day. If sufficient permanent cavities were available, a reasonable production rate could be achieved, but the tooling cost would be excessive. Some means of reproducing the cavity from a single tool is needed for efficient production. To reproduce the cavity, a pattern is constructed from which castings are made. Castings can be made from plaster of paris or silicone rubber. With care silicone rubber cavities may be used several times. Lacking rigidity, rubber cavities do not have the dimensional precision of other materials, but they have the advantage of permitting removal around minor undercuts. The cavity may be reproduced by slush casting. In slush casting an aluminum pattern is dipped into a pot of molten low-melting-point alloy. The metal adjacent to the pattern freezes. The pattern is withdrawn with a shell of metal around it which is the negative of the pattern. The shell is slipped off the pattern after cooling briefly. The device to be embedded is mounted in the slush casting, which is then filled with resin. The alloy material may be salvaged after the slush casting is removed from the hardened resin casting. Slush castings may be made in two parts and fitted together to permit undercuts. The design of patterns, the accommodations of needed drafts for withdrawing the pattern, and making provision for mounting the object to be embedded are challenging mechanical design problems. The metal used for slush casting is a eutectic alloy of 42 percent tin and 58 percent bismuth. It has a melting point of 138°C (281°F). This alloy has the unusual property of expanding slightly upon freezing, aiding the removal of the shell from the pattern. It also makes the shell jam between protrusions, but generous draft angles help relieve this problem. The metal is sold as Cerrotru, a trade name of the Cerro Corporation, and as AsarcoLo 281, a trade name of the Asarco Corp. Figure 7.5 is a photograph of a pattern used to make a slush casting. Figure 7.6 shows the pattern being dipped into a melt pot, and Fig. 7.7 shows a core and coil being mounted in a slush casting. Some resins require baking at temperatures above the melting point of the slush casting alloy in order to harden completely. In such cases the resin is given a partial cure

FIG. 7.5 Aluminum slush casting pattern. *(Reproduced through the courtesy of Cerro Metal Products Corp.)*

FIG. 7.6 Slush casting pattern being dipped in melt pot. *(Reproduced through the courtesy of Cerro Metal Products Corp.)*

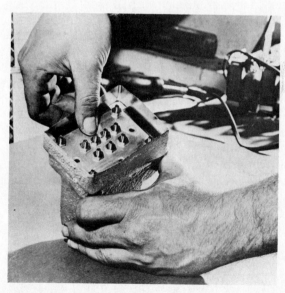

FIG. 7.7 Core and coil being mounted in cavity made by slush molding. *(Reproduced through the courtesy of Cerro Metal Products Corp.)*

at a temperature below the melting point of the alloy. The semicured embedment is then removed from the shell for postcuring outside the shell at a higher temperature. Some resins tend to adhere to the surface of the cavity. To prevent this, the surfaces are coated with a mold-release compound. Adhesion of the resin to the embedment is needed especially at the exit surfaces of terminals and mounting devices. Mold release must not be allowed to touch these surfaces. The slush casting alloy is expensive. There is some consumption of material with use due to dross formation on the surface of the melt, and this is a factor in the production cost. The process is useful because the production rate and production lot size are not restricted by the number of patterns or mold cavities available. Efficiency can be high and setup costs low.

Open Construction

The lowest-cost magnetic components use open construction. In open construction varnish impregnation provides the sole protection for the core and coil. Some manufacturers use various coil-wrapping techniques of mostly cosmetic value. Wrapping actually impedes the flow of varnish to the portions of the coil where it is most needed. Varnish impregnation strengthens the sheet insulation, making the coil less fragile and less susceptible to moisture. As installed, the steel cores are unprotected against corrosion. Any of the various plating and painting processes for protecting steel would damage the magnetic properties of the core. The varnish impregnation is reasonably effective in protecting the core in a room environment, but it will not protect the core against a wet or corrosive atmosphere. Open construction does not provide protection against dust and mechanical abuse, and so units using this construction should be located in protected areas. Although it does not meet military environmental requirements, open construction is sometimes used in military equipment when the magnetic device is mounted in a sealed enclosure with other components. Here the cost saving can be realized because the enclosure provides satisfactory environmental protection. Open construction includes air as part of the insulation system, making it a poor choice for high-voltage devices regardless of how benign the environment may be. Because of its low cost, open construction is widely used in noncritical applications.

7.2 TYPES OF CORES

In achieving satisfactory properties of magnetic cores, both the material and the geometry of the core must be considered. The best materials can be crippled with inappropriate geometries or poor assemblies. The choice of geometry, a preoccupation of core users, must consider cost, winding method, core assembly method, grain orientation, lamination thickness, size, weight, and air gap usage.

Laminations

Metallic cores are laminated to reduce eddy current losses. The surface of the laminations is covered with a high-resistivity mill finish. Laminated cores are built up from shapes made of lamination sheet stock to obtain the required three-dimensional form. The technology of lamination production is well developed. Magnetic alloy strip is fed into high-speed punch presses. The presses are equipped with progressive carbide dies. The operation is fully automated. Dies are available for scores of different shapes, each die representing a substantial investment. Some shapes are industry standards. Similar shapes that are not industry standards are available from more than one source. These similar shapes may not be interchangeable because of minor differences. Single-source shapes may indicate low demand, requiring special setups and long procurement delays. Single suppliers are at liberty to discontinue the item at their convenience, and so it may be unavailable for future procurement.

The most common laminations used for small transformers are shown in Table 7.1. Most of these laminations are scrapless EI shapes. The name is derived from the shapes of the two pieces that form the lamination and from the fact that there is no scrap since the window portion of the E piece forms the I. Figure 7.8 shows two E and two I pieces as they are punched from sheet stock. Figure 7.9 shows how the pieces are alternately placed to form a core with a small effective air gap. The coil is placed at the center leg of the core. The flux passing through the center leg divides at the end of that leg, with one-half flowing through each of two paths around the outside of the coil. The cross-sectional area of the outside paths need be only one-half of the area of the center leg. The conditions of constant flux path area and scrapless punching fully define the proportions of the scrapless lamination. EI laminations are often known by the width of the center leg expressed in inches. Figure 7.8 shows the direction of rolling. It is apparent that the EI lamination cannot take full advantage of the low core loss and high permeability features of anisotropic materials. The capacity of a given lamination may be varied by varying the stack height.

The DU lamination, which is less extensively tooled, overcomes some of the disadvantages of the EI shape. Illustrated in Fig. 7.10, this shape, instead of maintaining a uniform flux path area, has enlarged areas in the regions where the flux path is perpendicular to the direction of

TABLE 7.1 Dimensions of Commonly Used Laminations

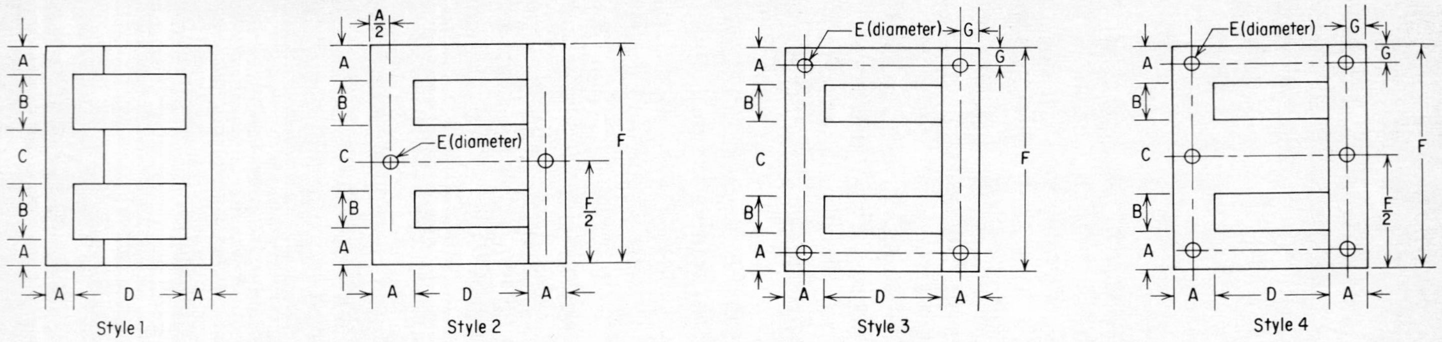

Lamination	Style	Units	A	B	C	D	E	F	G
			Dimensions						
EE 2425	1	in	1/8	1/4	1/4	1/2			
		mm	3.2	6.4	6.4	12.7			
EE 2627	1	in	3/16	1/4	3/8	11/16			
		mm	4.8	6.4	9.5	17.5			
EI—3/8	2	in	3/16	5/16	3/8	3/4	3/32	1 3/8	
		mm	4.8	7.9	9.5	19.1	2.4	34.9	
EI—1/2	2	in	1/4	5/16	1/2	13/16	1/8	1 5/8	
		mm	6.4	7.9	12.7	20.6	3.2	41.3	
EI—5/8	2	in	5/16	5/16	5/8	15/16	5/32	1 7/8	
		mm	7.9	7.9	15.8	23.8	4.0	47.6	
EI—3/4	2	in	3/8	3/8	3/4	1 1/8	0.132	2 1/4	
		mm	9.5	9.5	19.1	28.6	3.4	57.2	

EI—7/8	3	in	7/16	7/16	7/8	1 5/16	5/32	2 5/8	7/32
		mm	11.1	11.1	22.2	33.3	4.0	66.7	5.6
EI—1	3	in	1/2	1/2	1	1 1/2	7/32	3	1/4
		mm	12.7	12.7	25.4	38.1	5.6	76.2	6.4
EI—1 1/8	3	in	9/16	9/16	1 1/8	1 11/16	7/32	3 3/8	9/32
		mm	14.3	14.3	28.6	42.9	5.6	95.3	7.9
EI—1 1/4	3	in	5/8	5/8	1 1/4	1 11/16	7/32	3 3/8	9/32
		mm	15.9	15.9	31.8	47.6	5.6	85.7	7.1
EI—1 3/8	3	in	11/16	11/16	1 3/8	2 1/16	7/32	3 3/4	5/16
		mm	17.5	17.5	34.9	52.4	5.6	104.8	8.7
EI—1 1/2	3	in	3/4	3/4	1 1/2	2 1/4	7/32	4 1/2	3/8
		mm	19.1	19.1	38.1	57.2	5.6	114.3	9.5
EI—1 3/4	3	in	7/8	7/8	1 3/4	2 5/8	9/32	5 1/4	7/16
		mm	22.2	22.2	44.4	66.7	7.1	133.4	11.1
EI—19	3	in	7/8	1 3/4	1 3/4	3	17/64	7	7/16
		mm	22.2	44.4	44.4	76.2	6.7	177.8	11.1
EI—2	3	in	1	1	2	3	5/16	6	5/16
		mm	25.4	25.4	50.8	76.2	7.9	152.4	7.9
EI—2 1/4	4	in	1 1/8	1 1/8	2 1/4	3 3/8	5/16	6 3/4	5/16
		mm	28.6	28.6	57.2	85.7	7.9	171.5	7.9
EI—2 1/2	4	in	1 1/4	1 1/4	2 1/2	3 3/4	25/64	7 1/2	3/8
		mm	31.8	31.8	63.5	95.3	9.9	190.5	9.5
EI—3	3	in	1 1/2	1 1/2	3	4 1/2	3/8	9	3/8
		mm	38.1	38.1	76.2	114.3	9.5	228.6	9.5
EI—4	4	in	2	2	4	6	13/32	12	7/16
		mm	50.8	50.8	101.6	152.4	10.3	304.8	11.1
EI—5	4	in	2 1/2	2 1/2	5	7 1/2	35/64	15	9/16
		mm	63.5	63.5	127.	190.5	13.9	381.	14.3

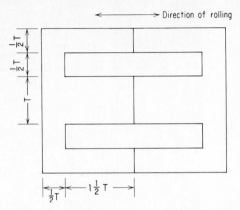

FIG. 7.8 Configuration of scrapless EI lamination.

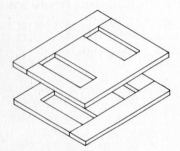

FIG. 7.9 Laminations showing alternating stack to reduce effective air gap.

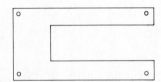

FIG. 7.10 Typical DU lamination for obtaining high effective permeability.

rolling and where the air gap introduces high reluctance. The increased width at the end legs makes alternate stacking of this single-piece lamination more effective in reducing the air gap reluctance. The DU lamination, as a result, ranks among the best shapes in effective permeability with values approaching those of toroidal cores. Economics and lack of versatility in available sizes have limited the use of the DU lamination.

The proportion of window area to core area is a matter of concern in many coil designs. The dimensions of punchings fixes this proportion which is often far from optimum. A method for overcoming this limitation, especially for large cores, is to build up the core from I-shaped pieces. I pieces may be sheared and punched without the need for special tooling. They can, therefore, be made economically to arbitrary dimensions. Methods for forming various core shapes from I pieces are shown in Fig. 7.11. The flux is in the direction of rolling in all legs of these cores.

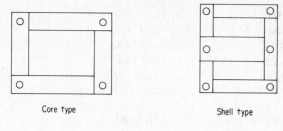

Core type   Shell type

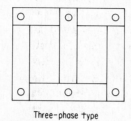

Three-phase type

FIG. 7.11 Various core types made with I pieces.

Wound Cores

Wound cores are made from preslit strip wound around a mandrel. Sufficient turns are made around the mandrel to obtain the required core cross section. The mandrel upon which the strip is wound determines the window size and shape. The strip width and the number of turns determine the core area. The turns are sometimes cemented together with adhesive. The advantage of the wound core is that the flux is at all times parallel with the direction of rolling. Various gauge thicknesses may be used for the strip. The thinner the gauge, the less are the eddy current losses, the better is the high-frequency performance, and the greater is the cost. Increasing cost and diminishing improvement in performance limit the minimum gauge of strip in practical cores to about 0.0005 in (0.013 mm). Wound cores have the further advantage of providing an economical means of obtaining thin-gauge stacks. The cost of manual stacking the thinner gauges is prohibitive.

Wound cores may be cut or uncut. Cut cores permit the use of preformed coils. Uncut cores require special winding machines. The cut wound core is called a "C" core because the two halves of a cut core wound on a rectangular mandrel form two shapes resembling the letter C. This type of core was developed by Westinghouse Electric Corporation, which coined the trade name *Hipersil,* an acronym for HIgh

PERmeability SILicon. The cut cores using Hipersil are made of a low-core-loss grade of magnetic silicon steel. Cutting the wound core causes a serious reduction in permeability. This is frequently an acceptable price to pay for the opportunity to use preformed coils since the core loss at power frequencies is unaffected. Cut cores are generally wound on rectangular mandrels since this geometry is most appropriate for preformed coils. Cut cores are also wound in a three-phase configuration called "E" cores. A C core is illustrated in Fig. 7.12 and an E core in Fig. 7.13.

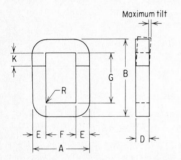

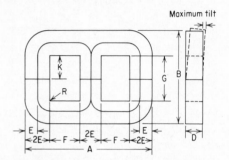

FIG. 7.12 C core showing industry standard letter designations for dimensions.

FIG. 7.13 E core showing industry standard letter designations for dimensions.

Cut cores have tempting magnetic properties. They may be operated at flux densities 20 percent higher than lamination structures, but serious mechanical constraints reduce this ostensible advantage. Dimensional control of cut cores is poor. Industry tolerances are shown in Tables 7.2 and 7.3. The outside corner radii are undefined, which, combined with poor dimensional control makes mechanical support of the core troublesome. E cores tend to be noisy because the inherent mechanical imperfections of the geometry permit vibration when magnetically excited. The halves of cut cores are generally held together by means of steel banding straps similar to those used in packaging for shipment. The strap is secured by means of a banding seal either soft-soldered or crimped to secure it in place. The seal often increases the overall dimensions of the core and coil assembly. The magnetic properties of the core are partly dependent upon the skill with which the banding operation is performed. In assembling the core halves into a preformed coil, a corner of the core may cut into the inside surface of the winding form. A small amount of insulating material may become dislodged and trapped between the core halves. A disastrous effect on permeability can result.

Cut cores wound from high-nickel alloys are available at substantially higher cost than silicon steel alloy cores. The high-nickel alloy cores have lower losses at high frequencies, but they saturate at lower flux densities than do the silicon steel cores. High-nickel cut cores degrade the effective permeability below the inherent material permeability to a greater extent than do interleaved cores.

The treatment of the mating faces of the cut core affects the permeability. If the surfaces are smooth and coplanar so that the core halves meet with a minimum air gap, the effective permeability will be high. Uneven mating surfaces cause a serious deterioration in permeability. In reducing the mating surfaces to a smooth finish, the laminations may become shorted together, increasing the core loss. To prevent this, the surfaces may be acid-etched. The acid will attack first the shorting burrs on the edges of the laminations with minimum effect on the surface finish.

The inherent permeability of high-nickel alloys is best realized in the form of wound uncut cores, the most common shape being the toroid. In the toroid are achieved the most exotic magnetic properties, notably square hysteresis loops, highest permeability, and highest saturating flux densities. Some of these properties are vulnerable to destruction from mechanical abuse. Hence many toroidal cores are mounted in protective boxes. The box increases the cross-sectional dimensions of the toroid so that the effective magnetic area is considerably less than the geometric area. The usable size range of toroidal cores is limited to the available winding-machine capacity. While astonishingly small toroids can be wound, the largest toroids are small compared with the largest sizes of cut wound cores.

Three-Phase EI Laminations

A typical three-phase EI lamination is shown in Fig. 7.14. As discussed in Sec. 11.5, the flux components in the coil legs of a three-phase core separate at the end of the legs and recombine in the yokes. During this transition there exists the possibility of higher flux densities than exists in the coil legs. The flux is perpendicular to the direction of rolling in the yoke of the E piece. These factors make the core loss higher in the yokes than in the coil legs. The yoke width is often made greater than the coil leg width, which reduces the flux density and core loss in the yoke. Some three-phase EI laminations have scrapless configurations. Those that have some scrap may still use the window area to form the I piece. In either case the flux in the I piece is in the direction of rolling. There is a fairly good inventory of three-phase EI laminations available from various suppliers. The shapes are not fully standardized. Data from each individual supplier should be used for design.

TABLE 7.2 Mechanical Tolerances for C Cores*

Dimension	Material	Allowable Tolerances—Finished Cores
A	All	$+ \frac{1}{32}$ max when $A \leq 1\frac{1}{2}$ $+ \frac{3}{64}$ max when $A > 1\frac{1}{2} \leq 2\frac{1}{2}$ $+ \frac{1}{16}$ max when $A > 2\frac{1}{2} \leq 3\frac{1}{2}$ $+ \frac{3}{32}$ max when $A > 3\frac{1}{2}$
B	0.001 and 0.002	$+ \frac{1}{16}$ max when $B \leq 2$ $+ \frac{3}{16}$ max when $B > 2 \leq 4$ $+ \frac{3}{8}$ max when $B > 4$
	0.004 and 0.012	$+ \frac{1}{16}$ max when $B < 3$ $+ \frac{5}{32}$ max when $B \geq 3 \leq 4$ $+ \frac{3}{16}$ max when $B > 4 \leq 6$ $+ \frac{3}{8}$ max when $B > 6 \leq 12$ $+ \frac{7}{16}$ max when $B > 12$
D	All	$+ \frac{1}{32}$—0 when $D \leq 1$ $+ \frac{3}{64}$—0 when $D > 1 \leq 2\frac{13}{16}$ $+ \frac{1}{16}$—0 when $D > 2\frac{13}{16}$ $+ \frac{3}{32}$—0 when $E > 2\frac{1}{2}$
E	0.001, 0.002, and 0.004	$\pm \frac{1}{64}$ when $E \leq \frac{1}{4}$ $+ \frac{1}{32}$—$\frac{1}{64}$ when $E > \frac{1}{4} \leq 1$ $\pm \frac{1}{32}$ when $E > 1$
	0.012	$\pm \frac{1}{64}$ when $E < \frac{1}{4}$ $+ \frac{1}{32}$—$\frac{1}{64}$ when $E \geq \frac{1}{4} < \frac{9}{16}$ $\pm \frac{1}{32}$ when $E \geq \frac{9}{16}$
F	All	$- \frac{1}{64}$ min
G	All	$- \frac{1}{64}$ min
Tilt	All	$\frac{1}{32}$ when $B < 3\frac{1}{2}$ $\frac{1}{16}$ when $B \geq 3\frac{1}{2}$

Ferrite Cores

Ferrite cores are molded ceramics. The materials are described in Sec. 6.4. A mold is required for each shape. Most new electrical designs use shapes for which molds already exist. The mold tools are expensive, and so the introduction of new shapes proceeds at a slow pace. The production yield decreases as the size increases, causing a rapid increase in cost with size. A resulting abundance of shapes in the smaller sizes contrasts with a paucity of available large sizes. The available shapes include an industry standard series of pot cores (Table 7.4) as well as proprietary toroids, UU, UI, and EE shapes. A pot core is illustrated in Fig. 7.15. Pot-core coils are usually wound on bobbins available from the core manufacturer. Bobbins for printed circuit board mounting in accordance with standard grid patterns are available for most pot cores. The quality of the mating surfaces of the pot cores determines the air gap and the effective permeability. This varies with suppliers. Pot cores are available with deliberately introduced air gaps by grinding the center post down by a specific amount. Ferrite cores require care in handling because they have low impact resistance.

Application Notes for Core Selection

Many applications can be served by several choices with almost equally satisfactory technical results. The objective in core selection is to choose the core that will do the job at the least cost. Operating frequency is the first clue to core selection because it is the determiner of core loss. Where a wide band of frequencies must be considered, the choice is usually made on the basis of the lowest frequency. Laminated structures are used

---

*Data are taken from EIA Standard RS-217. Letter designations refer to Fig. 7.12.

$$K = \ G/2 \text{ if } G < 3\tfrac{3}{4}$$

$$1\tfrac{11}{16} \text{ if } G \geq 3\tfrac{3}{4}$$

When $F \leq 2$ and $G \leq 2$

$$R = \tfrac{1}{32} \text{ for } 1, 2, \text{ and } 4 \text{ mils}$$

$$R = \tfrac{1}{16} \text{ for } 12 \text{ mils}$$

When F or $G > 2$ and F or $G \leq 5$

$$R = \tfrac{1}{8} \text{ for } 1, 2, 4, \text{ and } 12 \text{ mils}$$

When F or $G > 5$

$$R = \tfrac{5}{32} \text{ for } 1, 2, 4, \text{ and } 12 \text{ mils}$$

TABLE 7.3 Mechanical Tolerances for E Cores*

Dimension	Material	Allowable Tolerances—Finished Cores
A	0.004	$+ \frac{3}{32}$ max when $A \leq 5$ $+ \frac{3}{16}$ max when $A > 5 \leq 10$ $+ \frac{5}{16}$ max when $A > 10$
	0.012	$+ \frac{1}{8}$ max when $A \leq 5$ $+ \frac{1}{4}$ max when $A > 5 \leq 10$ $+ \frac{3}{8}$ max when $A > 10$
B	0.004	$+ \frac{3}{32}$ max when $B \leq 5$ $+ \frac{5}{32}$ max when $B > 5 \leq 10$ $+ \frac{1}{4}$ max when $B > 10$
	0.012	$+ \frac{1}{8}$ max when $B \leq 5$ $+ \frac{3}{16}$ max when $B > 5 \leq 10$ $+ \frac{5}{16}$ max when $B > 10$
D	0.004 and 0.012	$+ \frac{1}{32}$—0 when $D < 1$ $+ \frac{3}{64}$—0 when $D \geq 1 < 2$ $+ \frac{1}{16}$—0 when $D \geq 2$ $+ \frac{3}{32}$—0 when $2E > 2\frac{1}{2}$
2E	0.004 and 0.012	$\pm \frac{1}{32}$ when $2E \leq 1$ $+ \frac{1}{16}$—$\frac{1}{32}$ when $2E > 1 \leq 2$ $\pm \frac{1}{16}$ when $2E > 2$
F	0.004 and 0.012	$-\frac{1}{64}$ min
G	0.004 and 0.012	$-\frac{1}{64}$ min
Tilt	0.004 and 0.012	$\frac{1}{32}$ max when $B < 3\frac{1}{2}$ $\frac{1}{16}$ max when $B \geq 3\frac{1}{2}$

*Data are taken from EIA Standard RS-217. Letter designations refer to Fig. 7.13.

$$K = G/2 \text{ if } G \leq 5$$
$$G/3 \text{ to nearest } \tfrac{1}{16} \text{ if } G > 5$$
$$\tfrac{1}{16} \text{ when } F \leq 2 \text{ and } G \leq 2$$
$$R = \tfrac{1}{8} \text{ when } F \text{ or } G > 2 \text{ and } F \text{ and } G \leq 5$$
$$\tfrac{5}{32} \text{ when } F \text{ or } G > 5$$

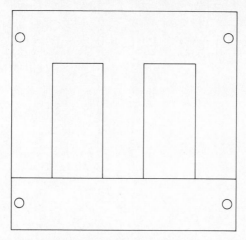

FIG. 7.14 Typical three-phase lamination with wide end yokes for reducing core loss.

for frequencies of 60 Hz or less. Wound cut cores can also be used economically in some specialized applications. The aircraft power frequency, 400 Hz, requires thinner-gauge laminations, usually 4 mils, making wound cores the economical choice in most cases. Cut wound cores are generally more economical than uncut cores. The latter are reserved for applications requiring the special properties available from those configurations, such as very high permeability and square hysteresis loops. For frequencies of 20 kHz and above, molded ferrite cores are usually the best choice.

Power-handling requirements bear upon the choice of core type. High-power and high-voltage requirements often exceed the capabilities of available lamination shapes, requiring the use of cores formed from I pieces, as illustrated in Fig. 7.11, or large C cores. High-power requirements at high frequencies present a particularly difficult problem. Thermal considerations dictate larger sizes, limiting the availability of suitable shapes. The use of lower-frequency configurations is sometimes forced in these high-frequency applications.

Some applications require the use of a deliberately introduced air gap. This is the case where direct currents cause asymmetrical magnetization and when adjustment of effective permeability is required to obtain a specific inductance value. EI laminations and cut cores are well adapted for this purpose, as well as the EE and UU ferrites. Uncut cores are unsuitable here. Cores formed of I pieces may be adapted to this application with some difficulty.

The thinner the lamination, the greater will be the cost. Configura-

TABLE 7.4 Dimensions of Standardized Pot Cores

Core Type	Units	Dimensions*											A_e	l_i
		A	B	C	D	E	F	G	H	I	J	K		
7 × 4	in	0.289	0.228	0.118	0.067	0.083	0.055	0.041			0.166	0.110	0.011	0.36
	mm	7.35	5.8	3.0	1.8	2.1	1.4	1.05			4.2	2.81	7.0	10.0
9 × 5	in	0.366	0.295	0.154	0.071	0.106	0.071	0.079	0.291		0.212	0.142	0.015	0.493
	mm	9.3	7.5	3.9	1.8	2.7	1.8	2.0	7.4		5.4	3.6	10.1	12.5
11 × 7	in	0.445	0.354	0.185	0.075	0.131	0.087	0.079	0.278		0.262	0.174	0.026	0.608
	mm	11.3	9.0	4.7	1.9	3.3	2.2	2.0	7.1		6.6	4.4	16.7	15.5
14 × 8	in	0.559	0.457	0.236	0.106	0.167	0.110	0.118	0.386	0.015	0.334	0.220	0.039	0.781
	mm	14.2	11.6	6.0	2.5	4.25	2.8	3.0	9.8	0.38	8.5	5.6	25.1	19.8
18 × 11	in	0.717	0.586	0.299	0.126	0.211	0.142	0.118	0.54	0.015	0.422	0.284	0.067	1.02
	mm	18.2	14.9	7.6	3.2	5.4	3.6	3.0	13.7	0.38	10.7	7.2	43.3	25.8
22 × 13	in	0.866	0.704	0.370	0.126	0.268	0.181	0.173	0.607	0.019	0.536	0.362	0.099	1.23
	mm	22.0	17.9	9.4	3.2	6.8	4.6	4.4	15.4	0.50	13.6	9.2	63.5	31.5
26 × 16	in	1.024	0.834	0.453	0.126	0.321	0.216	0.213	0.725	0.019	0.642	0.432	0.147	1.48
	mm	26.0	21.2	11.5	3.2	8.15	5.5	5.4	18.4	0.50	16.3	11.0	94.8	37.6
30 × 19	in	1.201	0.984	0.532	0.145	0.374	0.256	0.213	0.922	0.019	0.748	0.512	0.214	1.78
	mm	30.5	25.0	13.5	3.7	9.5	6.5	5.4	23.4	0.50	19.0	13.0	138.	45.2
36 × 22	in	1.418	1.177	0.638	0.157	0.433	0.287	0.213	1.055	0.019	0.866	0.574	0.313	2.1
	mm	36.0	29.9	16.2	4.0	11.0	7.3	5.4	26.8	0.50	22.0	14.6	202.	53.2

*Letters refer to dimensions in Fig. 7.15. A_e is the effective area in square units; l_i is the magnetic path length.

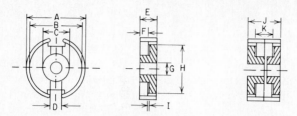

FIG. 7.15 Standard pot core with letters indicating dimensions given in Table 7.4.

tions forcing special winding techniques such as toroidal cores are more costly. Standard and existing shapes will cost less than special shapes requiring setups or tooling. Wound cores of thin-gauge material are expensive but are often the most economical choice because of the high labor cost of alternatives. Lamination economics have changed since the conception of scrapless laminations. Some laminations that are not scrapless are competitive with scrapless laminations. In this case the choice of lamination can be made on the basis of the best geometry for design purposes. Cost determinations on an individual basis are advisable.

The proportion of window to core area affects weight, cost, and performance. There is an optimum ratio for minimum weight determinable by successive approximation. The same is true of material costs. The impact of the ratio on labor cost is somewhat difficult to assess. The proportion of window to core area is often dictated by performance requirements such as the need for high-voltage spacing and the need to control the equivalent circuit parameters of leakage inductance and distributed capacitance in high-frequency applications.

7.3 TYPES OF WINDINGS

The concentration of magnetic fields needed for circuit functions requires the placement of large numbers of turns in small volumes. Various winding methods have been developed to achieve this. Each has its own advantages and limitations. For each application the selection of the most suitable winding method involves compromise in economics and performance.

Most winding methods involve rotating the winding in a winding machine. The magnet wire start lead of the winding is fixed to the coil structure before starting. As the machine rotates, turns are added to the winding. Counters, often with automatic stop capability, count the number of rotations made during the winding. Automatically applied brakes are a common feature of winding machines. They bring rapidly revolving machines to a stop quickly, preventing unwanted turns from being added to the winding.

Magnet wire is supplied on reels the size of which increases with the diameter of the magnet wire. This packaging is standardized to facilitate use in tension devices needed to control the payout of the wire as the coil rotates. A tension device is a mechanical servo which maintains a predetermined tension on the wire independent of its linear velocity through a system of pulleys, lever arms, and brakes. Efficient production is dependent upon properly functioning tension devices.

Layer-Wound Coils

Layer-wound coils have helical windings with sheet insulation between layers. The layer insulation is wider than the winding length, permitting a margin at each coil end. Margins and layer insulation are factors in both the mechanical and electrical design of the coil.

Layer-type winding machines are equipped with a wire guide which positions the wire on the coil. The wire guide advances at a rate slightly greater than the wire diameter per revolution of the coil, ensuring an even lay to the winding. The rate of travel of the wire guide is precise and adjustable for wire size. At the end of each layer of wire sheet, insulation is inserted. The direction of wire guide travel is reversed, and another layer of wire is wound over the layer insulation. With this construction the voltage stress of the layer insulation is a maximum at one end of the coil, while at the other end the stress is zero since the voltage induced in the winding divides in proportion to the turns.

Subsequent windings are readily added to layer-wound coils. Over the top of the last layer of the first winding is placed interwinding insulation of correct thickness for the voltage stress. Then a second winding is added in the same fashion as the first, usually with a different wire size and larger margins. The start and finish leads of layer-wound coils are anchored with pressure-sensitive tape or cord.

Layer-wound coils are wound on winding forms of mostly rectangular cross section with sufficient wall thickness to prevent collapse from the wire tension. The proportions of the winding-form rectangle help determine its ability to resist collapse. Its strength decreases along the longer side as the rectangle departs from a square.

Bobbin-Wound Coils

A bobbin is a receptacle for magnet wire turns into which a magnetic core is inserted after the required turns have been placed on the bobbin. A typical bobbin is shown in Fig. 7.16. Bobbins are dimensioned for specific

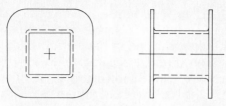

FIG. 7.16 Typical bobbin for use with cores having a square cross section.

cores and lamination stacks. They are available for square stacks of most of the smaller standard laminations and the standard series of ferrite pot cores. The availability of bobbins for other than square lamination stacks is limited. Bobbins are mostly made by injection molding. The selection of materials on the basis of molding requirements is frequently in conflict with the needs of the coil for electrical, thermal, and processing reasons. Bobbins are wound similarly to layer-wound coils except that the placement of turns is more random. When the wire is laid in place with a wire guide and a controlled number of turns per layer, the winding becomes scrambled after a few layers. Insulation insertion is generally not attempted in bobbins. Bobbins make more efficient use of the window area of a core than do layer-wound coils because thin walls replace the margins of layer-wound coils and there is no interlayer insulation. Terminations are a problem in bobbins. A common way to bring out the start lead is along the bobbin wall where it is insulated with tape. This method limits the induced voltage the coil can withstand because the start lead will be in close proximity to the turns near the finish of the winding. If the start lead is brought out through a hole in the bobbin wall, this problem is averted, but the likelihood of breakdown to the core is increased. The random placement of turns and the absence of layer insulation increase the likelihood of intrawinding failures from induced voltages. The insulation between winding and ground is provided by the bobbin. Voltages greater than that which the bobbin can safely withstand cannot be accommodated. Where bobbins are feasible, they are usually the least costly winding method.

Universal Winding

The turns of a universal winding, in contrast to those of the helical winding, do not lie adjacent to each other but follow an oscillating pattern extending for the length of the coil. The turns cross each other at an angle. There is no insulation introduced between turns other than the insulation on the magnet wire. The coil is self-supporting. A favorable voltage distribution is maintained between turns in close contact because of the systematic placement of turns. The finish turn is well removed from the start turn. The mechanical stability of the universal winding depends upon friction at wire surfaces. Film-insulated magnet wire, as used in bobbin and layer-wound coils, has too low a coefficient of friction to be suitable. Served wire is often used. The type of film-insulated magnet wire especially processed to increase its coefficient of friction is often used. Universal coils are wound on machines with cam-actuated wire guides. The throw of the cam determines the coil length. The ratio between cam and coil rotations determines the number of crossovers per turn. This winding machine function is adjustable by varying gear ratios.

The universal winding is a technique used for frequencies above 20 kHz where the advantages of lower distributed capacitance and higher Q become significant. Higher Q is obtained because of lower dielectric losses and the reduction of copper loss by eliminating the effect of current density concentration due to high-frequency current flow in parallel conductors in close proximity. A venerable institution whose origins go back to the early days of radio, the universal winding is enjoying renewed interest in combination with ferrite cores which have like serviceable frequency ranges.

Toroidal Winding

The toroidal core and coil is the *pièce de résistance* of magnetics technology. Ideal performance can be most closely approximated with this most expensive construction in general use. Toroidal winding is a challenging machine design problem. It is not possible to rotate the coil during winding as is done in helical coil winding, nor can multiple winding arrangements be had. Toroidal winding machines are available from several manufacturers. All the machines operate upon the same basic principle. The toroidal core is held in a seat composed of a drive wheel and idler wheels which rotate the core on its axis during winding, providing for winding advance. The machine is equipped with a shuttle of toroidal shape which is split radially. The cross section of the shuttle allows preloading with magnet wire. By separating at the split, the shuttle may be inserted through the center of the core. A second set of drive and idler wheels supports the shuttle, with the plane of the shuttle perpendicular to the plane of the core. The shuttle is then preloaded with magnet wire. The finish end of the preloaded wire is fixed to the core, and the shuttle is rotated in the opposite direction from that used to load the shuttle, placing turns on the core as the core rotates. Since the length of a turn on the core is much less than the circumference of the shuttle, it is necessary to equip the shuttle with a tensioning wire guide which rotates around the shuttle in a direction opposite to that of the shuttle rotation in a track on the side of the shuttle. The wire guide slips along the track with each rotation a sufficient amount to pay out the length of wire needed for one turn on the core. The wire guide rotation, which is not the same as the shuttle rotation, determines the number of turns placed on the core. This complicates accurate turns counting. After the winding is completed, the excess wire in the shuttle must be removed. A length counter measures the length of wire loaded on the shuttle, minimizing the loss of wire.

More than one winding may be placed on a toroid. Insulation may be placed between windings but not between turns of the same winding. As in a bobbin winding, turn-to-turn insulation is limited to the film on the magnet wire. Voltage between adjacent turns can be controlled better than in a bobbin but not as well as in layer-wound coils. Toroidal windings are generally considered a poor choice for high-voltage applications.

Winding Arrangements

Winding arrangements may be selected for specific characteristics. Instead of winding several layers in alternating directions, a series of single-layered windings all tending in the same direction may be connected in series after the coil is completed. This technique results in lower distributed capacitance. It is more costly, and the interconnections may be troublesome, particularly at high voltages.

Most devices are constructed with single coils in which the required number of windings is included in the single-coil structure. Both EI laminations and C cores regularly use single-coil construction. They are called *shell type* and *simple type,* respectively. The C core may also employ two coils which is known as *core-type construction.* Various series and parallel connections among the windings on the two coils allow the selection of favorable current-carrying capacities, leakage inductance, and distributed capacitance. Core-type construction has good space utilization and low leakage inductance. The two coils present a higher proportion of exposed coil surface for heat dissipation. Core-type construction is often favored when minimum size is the first consideration. The winding and assembly costs are higher than in single-coil constructions.

Saturable reactors require three coils on a shell-type core. The two outside legs of the core hold the ac windings, and the center leg holds the dc control winding. These various winding arrangements are illustrated in Fig. 7.17.

7.4   TERMINATIONS

Ranging from the trivially simple to the very complex, terminations are the connections and assemblies from the winding proper to the point

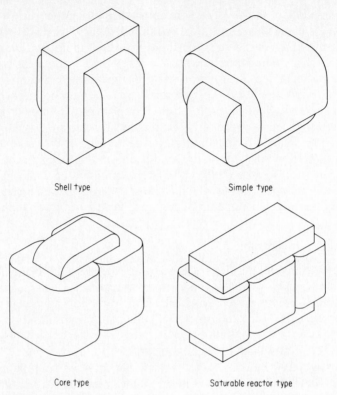

Shell type   Simple type

Core type   Saturable reactor type

FIG. 7.17 Various winding arrangements on conventional core shapes.

where the user makes the connections. Terminations are among the most troublesome features of magnetic devices. As a frequent source of opens and shorts, they play a crucial role in reliable performance. The termination must carry rated current with temporary overloads at rated voltage and transients of higher than rated voltage. Terminations are a major factor in the overall cost of the device. There are various types of terminations appropriate to the type of enclosure used. See Sec. 7.1.

Terminations for Open Construction

The simplest and least expensive termination for open-type core and coil construction is the flying lead. A flying lead is a flexible lead coming directly from the coil. The ends of the magnet wire forming the body of the coil are anchored in the coil. The larger sizes of magnet wire may be covered with insulating sleeving and brought out directly from the coil. Called an *extended winding lead,* this technique is illustrated in Fig. 7.18.

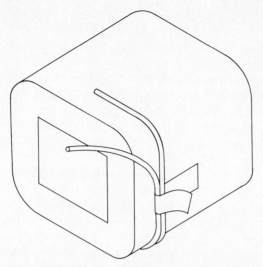

FIG. 7.18   Extended winding magnet wire lead on coil with wrapper removed to show tape anchor.

Two alternative flying lead arrangements in which the insulated stranded wire replaces the solid magnet wire are in common use. The first is to terminate the magnet wire in the winding proper by attaching the stranded lead to the magnet wire before the lead anchor and anchoring the stranded wire in the same manner as the extended winding lead. The second is to bring the magnet wire lead to the coil periphery across the coil margins. On the coil face the stranded wire is anchored and soldered to the magnet wire. The lead dress is covered with a wrapper around the entire coil. The use of stranded wire avoids the tendency solid wire has to fracture with repeated flexing. Anchoring the stranded wire in the coil proper avoids the risk of breakdown between the lead as it crosses the margin and the end turns of the windings in the coil. Dressing the magnet wire leads on the face of the coil to stranded wire leads or other terminations is a conventional and less costly low-voltage technique. A variation of the extended winding lead for heavy magnet wire is to form the flying lead into a terminal at the coil end. After tinning, the terminal formed in this manner makes a serviceable connection for the user's lead. When the last winding in a coil is at high voltage with respect to ground and the other windings, flying leads are often brought out through a hole in the center of the outer wrap. This lead position increases the creepage distance to the other windings and ground.

Flying leads are fragile. Both the copper conductor and the insulation are subject to fracture with repeated flexing. The vulnerable exit from the coil is a region of mechanical stress concentration. Flying leads, whether stranded or extended winding, are put in place before the core and coil are impregnated with varnish or treated in another way. The treatments tend to make the point of exit of the wire rigid, increasing the probability of fracture. Some ageing and embrittlement of the lead insulation are common during baking, adding to the likelihood of lead failure. Lead lengths are a problem. Since leads that are too short cannot be lengthened, flying leads are almost inevitably made longer than needed. They must be cut off, stripped, and tinned manually. Leads cut to the correct length and tinned before treatment become covered with varnish, making retinning necessary.

A very popular and serviceable terminal arrangement for low voltages and moderate currents is to mount terminal lugs on the face of the coil to which the extended winding leads are dressed. A wide variety of terminals, both solder and screw type, may be used in this fashion. The lugs are equipped with eyelets which are peened over into tough sheet insulation. This assembly is taped in place on the coil face. The arrangement is shown in Fig. 7.19 before the lead dressing is covered with a final wrap. Impregnation makes this termination rugged enough to permit repeated soldering and unsoldering operations. The lugs must be free of impregnants for good connections. A variation of the terminal lug arrangement is the use of bobbins with integral lugs. The magnet wire is brought directly to the lugs and soldered. The thermoplastic materials commonly used to mold bobbins are subject to melting when the magnet

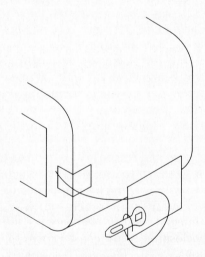

FIG. 7.19 Partially disassembled view of terminal lug mounting on coil periphery with insulating tape on margin under magnet wire.

wire is soldered to the lugs and when connections are made to the lugs on installation. Therefore, considerable care is required to limit the temperature and the duration of application of heat for solder connections.

Low-voltage terminations can be made by mounting terminal blocks on the transformer. The lamination screws and brackets can be adapted for this purpose. The use of screw-type terminations is most common in transformers that have optional connections.

Encapsulated Construction

The nature of encapsulants limits the choice of terminations. Flying leads are more vulnerable than in open construction and are seldom attempted. Terminal strips on the coil periphery are the most commonly used terminations. Thorough cleaning of the lugs is required before solder connections can be made.

Cast Construction

The mechanical control available in cast construction allows more termination options than in encapsulated construction. Flying leads are sometimes successful, especially if provision for stress relief is made, such as a well of resilient material surrounding the exit region of the flying lead. Good adhesion of the resin to the lead is needed for moisture protection. Many lead insulations have very poor adhesion properties. These can sometimes be enhanced by the use of primers. In addition to having the minimum required voltage and temperature ratings for the finished article, lead insulation must resist the solvent and chemical actions of the impregnants and the baking temperature of the process. If color coding is used, the colors of the insulation must be fast through the processing. To retain flexibility and restore appearance, it may be necessary to strip the processing materials from the leads after processing, a sometimes difficult task.

Inserts and screw terminals of many types are cast into place. Lug strips on the coil periphery may be used, but a higher order of precision in locating the lugs is needed than in open or encapsulated construction.Particularly advantageous in cast construction is the opportunity to mold in terminals integral with the coil structure, exploiting the existence of the insulating enclosure. In cast construction the fittings used to provide user connections are secured in place by the casting material. Connections to these fittings from the coil are made by extended or stranded leads. Since these leads are supported and protected after casting, the greatest abuse they will experience is during manufacture.

Terminations for Metal-Enclosed Devices

The terminations for metal-enclosed devices are brought through the grounded conducting surface of the enclosure with feed-through insulators. These insulators accommodate the voltage of the conductor to the enclosure and the current through the conductor. They also provide for sealing, resist the environmental conditions to which they are subjected, and provide a means for the user to make connections without damage to the insulator.

The breakdown voltage of feed-through insulators, upon which their operating voltage ratings are based, is determined by the weakest of three possible breakdown paths: along the outside surface of the insulator, along the inside surface of the insulator, or directly through the insulator between the conductor and the closest conductive surface at the enclosure potential. Most insulators are designed to break down along their outside surface at a voltage above the operating voltage, with the expectation that the device will still be usable at rated voltage. To achieve this, the insulator is dependent upon the filling medium or supplementary conductor insulation to compensate for the shorter inside creepage path. Some insulator constructions depend upon these same factors for protection directly through the insulator. The breakdown strength of the insulator becomes partly dependent upon the quality of the filling and insulation of the device. The installation of the insulator can influence the breakdown strength directly through the insulator by the electrode shape which the enclosure presents. The outside breakdown voltage will be affected by temperature, humidity, pressure, atmospheric contaminants, the condition of the surface of the insulator, and the electrode geometry at both ground and high-voltage surfaces.

Many ratings and breakdown data are based on 50/60 Hz rms sine wave voltages. When other than sine wave voltages are used, a conversion from rms to peak voltages must be made. Breakdown at dc voltage is usually higher than at peak ac voltages. Corona may develop on terminals under ac operation at voltages far below the breakdown voltage. Test conditions under which the rating applies are significant. Terminal ratings frequently apply to operations in dry, clean air at sea level pressure. Test specifications sometimes place arbitrary safety factors on terminals, forcing the selection of terminals to meet the test rather than the operating conditions. Electrical distribution impulse ratings are based on standards for overvoltage protection against transients from lightning. Ratings provided by insulator bushing suppliers are tentative. It may be necessary to test proposed insulators under the specific conditions required to confirm their selection.

Current-carrying capacity of conductors is determined by their cross-sectional area. The current-carrying capacity of terminations is often limited by the contact resistance of the fasteners, washers, and lugs used to make the connection. When overheating occurs, evidence can frequently be found on these surfaces. Generous contact areas and high-conductivity materials are needed for high-current-carrying terminations.

Feed-through insulators are often made of ceramic. Sealing may be accomplished by gasketing or by metalizing the surface of the ceramic. Gasketed terminals consist of a number of detail parts held together with fasteners. The adequacy of the seal depends upon the geometry of the gasketed surfaces, the gasket material, and the installation. The gasketing material, in addition to providing a satisfactory seal, must resist the required temperature extremes and the solvent action of filling materials. Assembled terminals to which user connections are made by screw threads must be capable of resisting the torque needed to secure those connections. The largest terminals as well as many small terminals are gasketed. A typical small assembled insulator is shown in Fig. 7.20. A typical feed-through distribution apparatus bushing is shown in Fig. 7.21.

When ceramic is used for sealing, metalizing is required on the ceramic insulators at the region where the ceramic passes through the enclosure and at the region where the conductor passes through the ceramic. Some type of fitting is soldered or brazed to the metalized surface at the conductor region. At the region where the ceramic passes

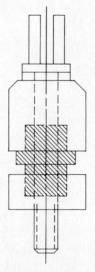

FIG. 7.20 Typical small gasketed feed-through insulator.

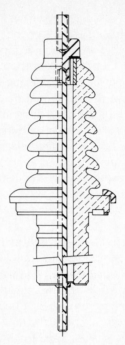

FIG. 7.21   Typical gasketed feed-through distribution apparatus bushing.

through the enclosure, either a flange is sealed to the metalized surface which is in turn soldered or brazed to the enclosure, or the metalized surface is soldered directly to the enclosure wall. When a flange is used, it is customarily attached by the terminal manufacturer. Fastening the flange to the enclosure wall is a less critical operation than soldering directly to the metalized surface, a connection which is easily destroyed by excessive heat. The flanged insulator is generally considered to be of higher quality and is more expensive.

A ceramic is an earthen material which may be molded or shaped and hardened by firing. First practiced as pottery, ceramic art has existed from antiquity. Today's engineering ceramics, though refined for high-performance properties, continue to use the processes of molding and firing. Electrical ceramics have good dielectric and high-temperature properties, but remain brittle materials. They have a high modulus of elasticity with no yield point, exhibiting linear elastic behavior to failure. They have high compressive strength but low tensile and impact strength. They have good resistance to moisture, although their surface is slightly porous. To provide clean and less pervious surfaces, the exterior portions of ceramic insulators are glazed.

Two ceramic-to-metal bonding processes are in regular use. One is the *sintered-metal process* in which a slurry, usually of silver, is placed on the surface of the ceramic which is then fired and electroplated. This process makes a mechanical bond. The second, called the *active alloy process,* uses titanium or zirconium to form a chemical bond with the ceramic at higher temperatures than used in the sintered-metal process.

Three ceramic materials are in common use in feed-through insulators. Porcelain is a mixture of clay (itself a mineral mixture of powdered aluminum silicate, sand, and other materials), flint (a form of silicon dioxide), and feldspar (a mineral consisting of silicates of aluminum and other metals). Its relative low cost makes porcelain the choice for many large bushings. Its thermal shock resistance is poor. Metal bonds on porcelain tend to be fragile, although proprietary methods claim to metalize porcelain by a low-temperature active metal process. When caps and flanges are attached by the terminal manufacturer, freeing the installer from the hazard of damaging the metalized surface, a much-improved sealed porcelain bushing results. Steatite (soapstone) is a magnesium silicate. It is easily formed and has superior high-frequency properties. Metal bonds are formed on steatite by the sintered-metal process. Soldering to the metalized surface must be done with a minimum use of heat. Alumina is far superior to either porcelain or steatite in mechanical strength as well as thermal and mechanical shock. High-temperature active alloy metalizing is done on alumina surfaces. In spite of its greater cost, alumina has been chosen to meet the difficult combination of mechanical and thermal stresses encountered in spark plugs. A typical alumina terminal is shown in Fig. 7.22.

Feed-through insulators may be made of epoxy resins with seals established either by gasketing or by embedding flanges and fittings in the epoxy casting. The epoxy surfaces cannot be metalized as can ceramic surfaces.

A good low-voltage feed-through insulator is made from glass. A glass bead through which passes a small-diameter conductor is mounted in a collar. The user solders the collar to the enclosure. By appropriate choices of materials and proper fusing and soaking temperatures during fabrication, the glass is maintained in compression over the operating temperature range of the finished article, providing an effective seal. This type of insulator is made with multiple conductors for use in miniaturized applications.

High-voltage terminations are sometimes made with coaxial cables. The cable has a center conductor surrounded by plastic insulation over which is placed a braided shield. The shield is usually protected with a thin plastic covering. The cable transit through the enclosure wall is made with a transition fitting which terminates the shield. The insulator and

FIG. 7.22 Typical metalized alumina feed-through insulator with mounting flange.

conductor extend into the filling material of the enclosure. On the exterior, the cable with its shield is extended to a convenient length and terminated as appropriate for the particular application. Cable terminations are not hermetically sealed, but they are frequently the most convenient and least expensive termination for high-voltage equipment.

7.5 MINIATURIZATION TECHNIQUES

Effective control over the size and weight of the magnetic device complement is needed to meet requirements for miniaturization in modern electronic equipment. Magnetics technology has not enjoyed the revolutionary breakthroughs typical of much of the rest of the electronics industry. Progress has been marked by the gradual introduction and acceptance of new materials combined with improvement in design and manufacturing techniques. Ingenuity is required to extract from the present state of the art the smallest size and weight consistent with price and reliability objectives. To reduce the size of magnetic devices, the following matters should be addressed: the specification requirements, materials selection, design techniques, and manufacturing processes.

Specifications control the size of magnetic devices through circuit function requirements, the power-handling capability, and mechanical needs. Among the most spectacular achievements in miniaturization has been the introduction of high-frequency inverters. The reduction in the size of transformers in inverters has been the result of circuit technique, not improvements in magnetics. Minimum operating frequency is a controlling factor in magnetic-device size. Other electrical requirements that often control size are regulation, exciting-current or open-circuit inductance, winding resistances, capacitance, and dielectric or high-voltage requirements. Specification control over such factors should not be exercised unless there is a real need. Power-handling capability, the product of rated output voltage and current, determines the core area and the size of the conductors. The area turn product is determined by the voltage and frequency by Faraday's law. There is an optimum balance between turns and core area for minimum size. This balance is affected by conductor size, which is governed by allowable temperature rise. Temperature rise is partly determined by copper loss, a function of true rms current. It is not necessary to place safety factors on current requirements for magnetic devices. They will operate safely at rated current. Their large thermal mass, in addition, usually permits temporary operation in excess of rated current. Advantage can be taken of duty cycles in reducing size when the duration of heavy loads is short compared to the thermal time constant of the device. In such cases the average power dissipated in the device determines the temperature rise. Accurate loading information is needed to obtain minimum size. The device temperature rise plus the maximum device ambient temperature is equal to the maximum insulation temperature. This number should equal but not exceed the safe operating temperature of the insulation system. The prediction of temperature rise is not precise. To determine whether an existing prototype is temperature rise–limited, a temperature rise measurement is often made. This is done by determining the cold and hot stabilized conductor temperatures. The test should be made under as nearly actual operating conditions as possible, including such features as heat sinks, airflow, and nearby heat sources. Actual loading or accurately simulated loading should be used. If the measured temperature rise under realistic conditions plus the maximum device ambient is less than the safe operating temperature of the insulation system, then there is an opportunity for size reduction. Alternatively, an insulation system with a higher maximum safe operating temperature may be substituted.

Magnetic devices are captive to available materials. A fundamental material choice is the core. Cores selected for minimum size may be costly because of both the material and the geometry. Correct choice of core material is prerequisite to size optimization. The core material with the highest saturation flux density and acceptable losses just below saturation will often yield the smallest size. The cobalt-iron alloy Supermendur[1] has this distinction. When its cost cannot be accepted, wound silicon steel cores provide a good alternative. The toroidal version of wound cores has the disadvantage of lost space in the center of the core. Cut cores permit the use of preformed coils with which the core window may be completely filled. Cut cores lack mechanical precision, a frequent disadvantage for reducing size, requiring careful design and manufacturing practice.

Available for operation over a wide range of temperatures, insulating materials are critical in temperature rise–limited devices. Manufacturing processes, insulation compatibility, and experience often limit the choices. The maximum safe operating temperature is usually selected to be consistent with a proven insulation system. The dielectric stress to which the insulation system is subjected is an important factor in size determination, especially when high voltage must be contained. Design stress levels contain experience factors relating to the degree of mechanical precision and quality of manufacturing processes maintained in the manufacturing facility which will produce the device. Size can sometimes be reduced by increasing dielectric design stress levels while also increasing mechanical precision and the quality of manufacturing processes. The impact of this choice on reliability must be considered. Successful miniaturization is often related to manufacturing quality.

Conventional coil-winding methods in which generous spaces are allocated to insulation are frequently inconsistent with size reduction objectives. Many low-voltage devices depend increasingly on magnet wire film alone for insulation. This places a greater burden on the quality of the magnet wire film. Modern plastic films often meet this need. Many small magnetic components are wound on molded plastic bobbins, which provide a measure of dimensional precision and insulation consistency. The use of bobbins can limit the choice of geometries and core materials. The insulation of compact windings can be enhanced by impregnation with polymerizing resins, providing good dielectric structure in a small size with improved reliability and environmental protection.

7.6 SPECIAL PROBLEMS IN HIGH-VOLTAGE APPLICATIONS

The nature of magnetic devices requires electrodes of opposing potential to be in close proximity. High voltage here makes severe demands upon

---

[1] Registered trade name of Arnold Engineering Co.

materials and processes of insulation systems to maintain reliable operation. The solutions to many high-voltage problems have been gained through sad experience. Past failures will have been analyzed and observed defects corrected in subsequent units. In this manner a reservoir of design and manufacturing practice is acquired which catalogs failures and successes. Tapping this reservoir for applications within its limits will usually give satisfactory results. For applications beyond the scope of this experience the outcome is less certain. Insulation systems depend upon processes which vary with the facility. The facility chosen should have experience which covers the application.

High-voltage insulation needs are frequently in conflict with other functional requirements, foremost among which is size. Many reliability problems start with attempts to package high-voltage devices in excessively small volumes. The only apparent solution to many dielectric problems is to provide adequate spacing between electrodes. The requirements of distributed parameters conflict with the demands of high voltage. Leakage inductance increases with the spacing between windings and with the ratio of coil height to length. These are the same factors which must be increased to provide adequate insulation between windings and control intrawinding stress. The product of leakage inductance and distributed capacitance, a figure of merit in high-frequency performance, is controlled by keeping the coil volume small, which is inconsistent with insulation needs. The conflicting needs of distributed parameters and high-voltage insulation converge on high-frequency and square wave devices, which often have preemptory requirements on distributed parameters. The resulting increase in dielectric stress levels may be accommodated through the use of superior construction and processing. Typical of this approach is the vacuum oil-filled hermetically sealed high-voltage transformer. Careful oil processing and filling will greatly increase the integrity of the insulation system. The exclusion of the atmosphere by hermetic sealing prevents subsequent deterioration.

High voltage in magnetic devices may be induced or externally applied. Often both exist in a single device. It is easier to insulate for externally applied dc voltages than for either induced or externally applied ac voltages. (See Sec. 7.3 on coil construction.) In addition to these easily identified voltages, there are transient voltages generated by inductive switching, sudden changes in load current, and the operation of control impedances. Resonances and transmission line effects also produce high transients. Receiving scant attention in low-voltage circuits, these transients become a major consideration in high-voltage devices because they reach destructive magnitudes. The ability of a device to withstand these transients is often the controlling factor in its reliability. The magnitudes of transient voltages cannot usually be predicted. The ability of high-voltage insulation to withstand overvoltages of great magnitude for short durations is uncertain. The effects of material and processing imperfections on insulation system integrity is indeterminate. These considerations lead to a conservatism in design that may appear excessive on the basis of the nominal functional voltages. Transient protection devices are often used to protect equipment against excessive transient voltages. Spark gaps are commonly used between points subject to high transients. Varistors are an effective way of preventing transients on low-voltage transformer windings from inducing destructive voltages on high-voltage windings. Certain circuits are recognized as initiators of dangerous transients, among which are arc discharge–type loads and silicon controlled rectifier (SCR) controllers. Experienced designers will take these circuits into account when they design insulation systems.

Corona, discussed in Sec. 8.5, is a frequent cause of failure in high-voltage devices. The failure mechanism with corona is unlike the immediate breakdown under transient voltages. Corona deterioration results from stresses of long duration, usually from ac voltages in regions with dielectric discontinuities between air and a solid or liquid dielectric. The presence of these discontinuities is often due to inadequate processing.

High-Voltage Terminations

High-voltage connections in air to magnetic devices have the problems of ionization of air adjacent to the conductors and breakdown along the creepage path of the air to solid dielectric boundary. Ionization at the conductors may be prevented by corona shields, which prevent sharp points on connecting hardware from being regions of high-stress concentration. Corona shields will also increase the breakdown voltage of terminals. Ceramic is the most common material used for high-voltage terminals. The creepage distance is increased by convolutions on the surface of the ceramic. Glazed ceramic has good moisture and arc-tracking resistance. Figure 7.23 illustrates a ceramic terminal bushing with corona shield.

Coaxial cables offer a convenient means of terminating high voltage. Transition fittings are required, and the cable ends usually enter a solid or liquid dielectric. The outer grounded conductor confines the dielectric stress to the cable insulation. Dielectric integrity is best preserved by terminating this conductor in the solid or liquid dielectric, as illustrated in Fig. 7.24. Coaxial cables are not hermetically sealed. The ac ratings of cables are modest. They are capable of operating at dc voltages far in excess of their ac ratings. Hermetically sealed cable connections may be made to oil-filled compartments by means of terminal wells which require mating cable fittings to plug into the well. Air is excluded from

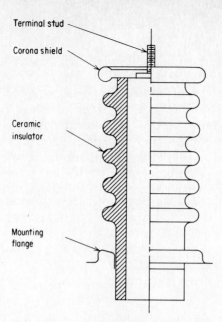

FIG. 7.23 Ceramic bushing with corona shield.

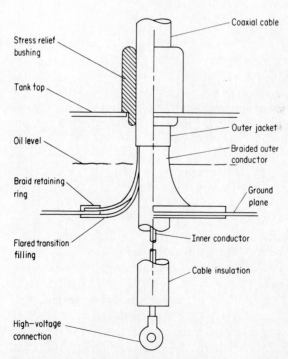

FIG. 7.24 High-voltage termination using coaxial cable.

the well by filling the voids with silicone grease. The outer conductor is terminated at the top of the well with a threaded cap. See Fig. 7.25.

Several types of high-voltage stranded lead wire allow connections to be brought out directly from devices potted in solid dielectric materials. The lead wire avoids the problem of terminating an outer shield. The lead termination inside the device must provide for adequate creepage if the lead passes through a metal case. An adequate voltage rating on the insulated lead does not prevent the development of excessive voltage stresses when the lead passes through air in close proximity to a grounded surface. Corona can develop in that region from the dielectric discontinuity, and breakdown can eventually occur. The epoxy cast coil (see Sec. 7.1) provides a nonconducting surface which may be shaped as needed to provide creepage surfaces. A cast coil with integral molded high-voltage terminal surface is illustrated in Fig. 7.26.

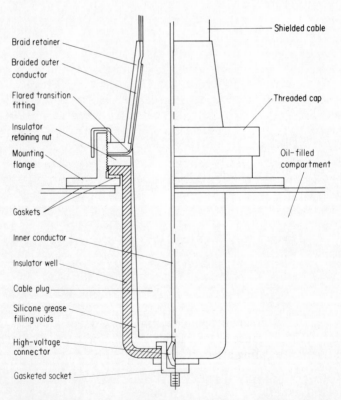

FIG. 7.25   Typical hermetically sealed high-voltage shielded cable connection.

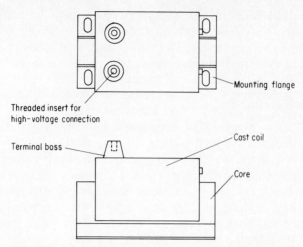

FIG. 7.26 Cast coil transformer with integral high-voltage terminal surface.

High-Voltage Impregnating and Potting

Most dielectric impregnating materials have adequate intrinsic dielectric strength. The problem with these materials is to exclude air. High-voltage insulation systems require vacuum impregnation. The most effective processes first draw a vacuum on the chamber containing the device to be impregnated, then admit the impregnant while the vacuum is being maintained. Finally pressure is applied to the surface of the impregnant to force the liquid into the interstices of the coil. The process is similar whether the impregnant remains a liquid or later solidifies. Degradation of insulation by air voids is especially severe if the pressure in the voids is reduced. This condition is sometimes created by incomplete vacuum filling. The combination of an imperfect vacuum and high-viscosity impregnant can seal off regions which are accessible through only a very small opening. The viscosity of the impregnant may prevent equalization of pressure between the entrapped air in the sealed-off region and that applied to the surface of the impregnant long enough to permit the impregnant to solidify. This leaves a permanent void of easily ionizable low-pressure air. Impregnants which remain liquid usually have superior void-filling ability because of their low viscosity and good wetting ability. They are also self-healing after dielectric puncture.

The tendency for solid potting materials to crack (see Sec. 7.1) is a more crucial and immediate problem in high-voltage devices than in low-voltage equipment. A crack in solid insulation which is a part of a high-voltage insulation path reduces a continuous barrier with good dielectric properties to a surface creepage condition with greatly reduced dielectric strength. Under high voltage, prompt failure along the creepage path provided by the crack can be expected. The use of solid high-voltage insulation systems is increasing with advances in materials and processes.

Coil-Winding Techniques

The typical high-voltage coil consists of a low-voltage winding placed next to the core with a high-voltage winding of lesser width insulated from the low-voltage winding and ground. The high-voltage winding must also be insulated for intrawinding stresses. Insulation between windings is usually sheet insulation. In liquid-filled units ducts may be placed between windings to allow for the flow of convection currents. In high-voltage coils, temperature is a consideration not only because of insulation life, but also because dielectric strength decreases with temperature. The supports used to hold the windings in position around the core present short creepage paths to the core and between windings. Dielectric strength along the creepage surface is improved when an ion traveling along the surface must move opposite to the field in order to continue along the surface to the opposing electrode. Such a surface is illustrated in Fig. 7.27. Ions traveling this path will be required to punc-

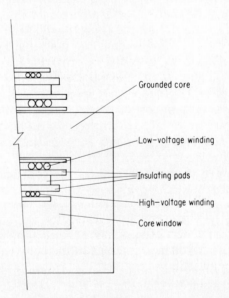

FIG. 7.27 Coil cross section showing creepage path opposing the voltage gradient.

ture the high-dielectric-strength impregnant to reach the opposing electrode, since they will not be accelerated against the voltage gradient.

Intrawinding stresses occur between neighboring turns of a single
winding. The lowest voltage that can occur here is the total induced voltage on the winding divided by the number of turns in that winding, that
is, the volts per turn. Neighboring turns which are not consecutive will
be stressed at a multiple of that voltage, the multiple being equal to the
turn interval between the neighboring turns. Close proximity between
consecutive turns will exist in all types of windings. Close proximity
between turns which are not consecutive also will occur in bobbin windings because of the random nature in which turns fall, in universal windings because a many-turn interval between neighboring turns is part of
the alternating winding pattern, and in layer-wound coils because the
winding direction reverses each layer, resulting in maximum stress
between each layer at alternating ends of the winding. The stresses
between neighboring turns in bobbin and universal windings as well as
consecutive turns of all types of windings must be resisted by the insulation on the wire, usually magnet wire with a thin plastic film. The layer-
wound coil has interlayer insulation whose thickness may be adjusted to
accommodate the stresses between end turns. The layer-wound coil provides a controlled geometry. In all these situations, the volts per turn help
determine the stress between neighboring turns. In layer-wound coils the
turns per layer also help to determine that stress. Large numbers of turns
per layer occur when the wire is of small diameter with small current-
carrying capacity. Thus high interlayer stresses occur more often in low-
current windings. The turns of very large coils with large current-carrying
capacity are individually supported to provide controlled spacing
between adjacent turns and allow the flow of convection currents in oil.
Volts per turn and turns per layer are considerations in high-voltage coil
design.

Devices with high intrawinding dielectric stresses which operate at
frequencies above 400 Hz are subject to dielectric heating. Dielectric
materials are rated by their dissipation factor, numerically equal to $R\omega C$,
where C and R are the capacitance and equivalent series resistance,
respectively, of a capacitor using the material as its dielectric. The dissipation factor, in most dielectrics a number less than 0.01, is also the ratio
of dissipated to reactive energy in the capacitor. In a magnetic device the
corresponding parameters of interest are the effective distributed capacitance across its high-voltage winding and the equivalent resistance representing the heat loss in the dielectric. Although the dielectric heat loss
may appear small when the dissipation factor and distributed capacitance
are used to estimate it, this heat is concentrated in regions of the coil
where the dielectric stress is highest. The dissipation factor increases with

temperature. The increase is rapid at temperatures near the maximum temperature rating of the insulation. This situation has the ingredients for thermal runaway. The use of materials with high-temperature ratings and low dissipation factors operating at reasonable stress levels will avoid this condition.

Equipment Adaptations for Reducing Dielectric Stresses

Favorable tradeoffs are often made by selecting circuits and mechanical arrangements which avoid some of the most difficult dielectric stresses in magnetic devices. High-voltage terminations in air can be eliminated by placing as much of the high-voltage circuitry as possible in a single compartment. The compartment is filled in a single operation with either solid or liquid dielectric. Especially useful in ac to dc conversions, this technique keeps the high-voltage ac circuits under oil or solid dielectric. Only the more manageable dc terminations are exposed to air. High induced voltages in rectifier transformers can be reduced by using multiplier circuits, with a frequent saving in size and weight. Dielectric strength can be improved between high stress points by interposing electrodes at intermediate guard potentials.

CHAPTER 8

Insulation Systems

The life span of a magnetic device depends upon its insulation system, an amalgam of different insulating materials, processes, and interactions. The characteristics of an insulation system are affected by electrode geometry, composite dielectrics, temperature gradients, and compatibility among different materials, making the prediction of system characteristics from the properties of individual materials uncertain. The selection of insulating materials requires a system perspective and experimental confirmation.

Insulation systems in magnetic devices are beset by ambiguity. Transient voltage stresses are imprecisely defined. Electrode geometry varies. The nature of composite dielectrics is uncertain. The operating temperature of the insulation is indeterminate. From a host of uncertain data on requirements, the insulation system designer selects materials using data which bear scant resemblance to actual operation conditions. As a result insulation system design contains generous safety factors to ensure reliability.

8.1 INSULATION AGEING

Insulation systems possess certain properties that are necessary for device performance. Initially the strengths of these properties are greater than the stresses to which they are subjected. Under stress these strengths deteriorate with time until they fall below the applied stress, resulting in failure. The stresses most often considered in insulation are voltage gradient and temperature. The rate of decline in the strengths of an insulation system is a function of temperature believed to follow the Arrhenius chemical reaction equation:

$$R = A \exp - \frac{B}{T} \tag{8.1}$$

In this equation A and B are constants, and R is the reaction rate at absolute temperature T. If R is the rate of the reaction causing deterioration of the critical property determining insulation system life and that rate is constant, then Eq. (8.1) leads to

$$L_2 = L_1 \, e \exp \left[-B\left(\frac{1}{T_1} - \frac{1}{T_2}\right) \right] \tag{8.2}$$

where L_1 and L_2 are the insulation lives at absolute temperatures T_1 and T_2. If L is known for several different temperatures T, the constant B may be approximately evaluated for that particular system. The life at other temperatures may then be estimated. This technique is used for evaluating individual insulating materials. A significant property of the material is monitored on samples maintained at different temperatures. End of life is considered to be the point where the monitored property has deteriorated by an arbitrary percentage, often one-half its original value. This permits reasonable estimates of insulation life at rated temperatures from accelerated life tests conducted at greater than rated temperatures. Such data are useful for comparison. Equation (8.2) may also be used to estimate the reduction in life of an insulation system of known characteristics which is operated at above rated temperatures.

A similar mathematical model, empirical in nature, has been proposed[1] for voltage gradient endurances:

$$L_G = L_0 \exp - KG \tag{8.3}$$

In Eq. (8.3) L_0 and L_G are the lives of the insulation under zero voltage gradient and at gradient G. K is a constant for a given insulation system. There appears to be a threshold voltage gradient below which aging due to voltage stress is minimal.

---

[1]L. Simoni, "A General Approach to the Endurance of Electrical Insulation under Temperature and Voltage," *IEEE Transactions on Electrical Insulation,* vol. EI-16, no. 4, August 1981.

8.2 FIELD ENHANCEMENT BY COMPOSITE DIELECTRICS

An approximate equivalent circuit of a single insulating material between two electrodes is a capacitor and resistance in parallel. If a second insulating material is included between the electrodes, the equivalent circuit becomes a series-parallel combination as shown in Fig. 8.1. The resistances in the equivalent circuit represent the insulation resistances of the two insulations. The capacitances are determined by the dielectric constants of the insulating materials and the geometry of the structure. The time constants of these circuits vary from a fraction of a second to many hours, but are typically a few seconds. When a dc voltage is suddenly applied between the electrodes, the voltage will initially divide inversely as the capacitances, then change exponentially to being in proportion to the resistances. If the electrodes are parallel planes, the steady-state dc voltage gradient across insulation no. 1 will be

$$G_1 = E_{dc} \left[\frac{1}{l_1 + (\rho_2/\rho_1)\, l_2} \right] \tag{8.4}$$

where l_1 = insulation thickness of insulation no. 1
 ρ_1 = volume resistivity of insulation no. 1
 l_2 = thickness of insulation no. 2
 ρ_2 = volume resistivity of insulation no. 2

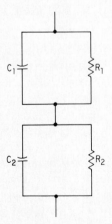

FIG. 8.1 Equivalent circuit of a composite dielectric.

All the quantities in Eq. (8.4) can be in any consistent set of units. Disparate values of insulation thicknesses and volume resistivities can increase the voltage gradient across either dielectric. In the typical composite dielectric, air is an unwanted constituent, which has a high-volume resistivity under low dielectric stress. If the thickness of the air layer is small and the insulation resistance of the solid dielectric relatively low, the voltage gradient in the air space may be high enough to initiate ionization in the air space. This will lower the resistivity of and decrease the gradient in the air space, causing more of the voltage to appear across the solid dielectric. This is a less severe condition than with ac voltage.

A composite dielectric under ac voltage stress will undergo field enhancement due to different dielectric constants of the two dielectrics. With plane-parallel electrodes the voltage gradient across dielectric no. 1 will be

$$G_1 = E_{ac} \left[\frac{1}{l_1 + (\varepsilon_1/\varepsilon_2)\, l_2} \right] \tag{8.5}$$

in which ε_1 and ε_2 are the dielectric constants of dielectrics 1 and 2, and l_1 and l_2 have the same meaning as in Eq. (8.4). E_{ac} is a time-variable voltage. Disparate values of insulation thicknesses and dielectric constants will increase the voltage gradient across the insulation with the lower dielectric constant.

8.3 FIELD ENHANCEMENT BY ELECTRODE GEOMETRY

The concept of parallel plane electrodes is a simplification useful for illustrative purposes. In practice, electrodes have complex shapes. An appreciation of the effect of electrode shape on electric fields can be obtained by considering a sphere. The voltage gradient at the surface of an isolated conductive sphere at an elevated potential is inversely proportional to the square of the radius of the sphere. A sharp point on an electrode can be likened to a sphere of small radius. Gradients of sharp surfaces can be many times greater than those which would exist for parallel plane electrodes. In high-voltage devices sharp electrodes must be avoided. Shields at electrode potential are useful for this purpose. A smooth conductive surface will prevent other surfaces within its shadow from affecting the voltage gradient.

8.4   SURFACE CREEPAGE

The interface between two dielectrics used to separate opposing electrodes is a region of dielectric weakness. The dielectric strength along the interface, often called the *creepage distance,* is less than the dielectric strength through either dielectric alone. Electric field enhancement occurs along this surface, causing the dielectric strength of the weaker dielectric to be exceeded more readily than would occur if the second dielectric were not present. Field enhancement is caused by differing resistivities and dielectric constants of the interfacing dielectrics, the accumulation of electric charges on the interface surface, and contamination by both conductive and nonconductive particles on the interface surface. A frequently occurring dielectric interface is the boundary between solid insulation and air. Here the presence of moisture has a major effect on the dielectric strength of the boundary.

8.5   CORONA

Corona is an insidious enemy of insulation. Insulation failure due to corona occurs prematurely and unexpectedly in functioning equipment. Life span expectations based on insulation temperature do not apply when insulation is subjected to corona deterioration. Failure may occur after operation for a few minutes to many months. Corona occurs most frequently under ac voltage stress in composite dielectrics, one of which is air. Field enhancement from electrodes of small radii frequently contribute to corona. The sensitive electrical detection schemes in use do not reveal the sites of corona, only its existence and approximate intensity. Precise knowledge of the site is needed to assess the gravity of the condition and to take corrective action. A considerable amount of detective work may be required to identify the site. Some insulating materials are more vulnerable to corona deterioration than others. Failure will occur with a vulnerable material under high stress. The elimination of corona involves one or more of the following:

1. Improvement in electrode geometry

2. Elimination of air or other ionizable dielectric

3. Increase in the spacing between opposing electrodes

4. Reduction in the applied voltage stresses

A method of predicting the corona-starting voltage in air-solid composite dielectrics between parallel-plane electrodes has been suggested[2] which is based on Eq. (8.5). The equation is rearranged to give the total applied voltage as a function of the breakdown voltage across the air space. The applied voltage calculated from these values is considered to be the corona-starting voltage because the corona-starting voltage and breakdown voltage in air in a uniform field are believed to be the same. The corona-starting voltage vs. air gap distance based on this premise is plotted in Fig. 8.2 for various values of the ratio of solid insulation thickness to dielectric constant of the solid insulation. Use of this curve must be tempered for practical conditions such as the presence of sharp electrodes and multiple dielectrics.

8.6 DIELECTRIC STRESSES IN MAGNETIC DEVICES

Dielectric stresses in coils are either intrawinding or interwinding, which includes winding to ground. Intrawinding stresses result from induced

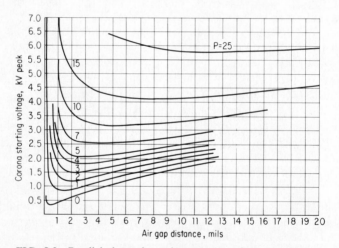

FIG. 8.2 Parallel-plane electrode corona-starting voltage at sea level vs. air gap distance in an air-solid composite dielectric with the ratio of solid dielectric thickness in mils to dielectric constant of solid as parameter: $p = l_s/\epsilon_s$. (M. C. Halleck, Calculation of Corona-Starting Voltage in Air-Solid Dielectric Systems, AIEE Transactions, April 1956. Copyright AIEE. Reproduced by permission of IEEE, successor to AIEE.)

[2]M. C. Halleck, "Calculation of Corona-Starting Voltage in Air-Solid Dielectric Systems," *AIEE Transactions,* April 1956.

voltages. Interwinding stresses result from both induced and externally applied voltages. When a coil is viewed as a pure inductance, any voltage developed across a winding will be equally distributed among all its turns instantaneously. Then the voltage between any two turns of the winding may be determined from the voltage induced across the entire winding. With this information an insulation structure can be provided to withstand the expected dielectric stress. For some applications this is an adequate procedure. In other cases the induced voltage is applied very rapidly so that the assumption of a pure inductance is not valid. The distributed capacitance between turns must be considered. The equivalent circuit of a winding with distributed capacitance across each turn approximates a transmission line. When a steep voltage front is applied to the winding, all the capacitances are initially uncharged. As the voltage charges the capacitance between the first two turns, the other capacitances remain uncharged, resulting in all the voltage appearing between the first two turns. As the wave front proceeds down the winding, each capacitance is charged in succession. The voltage becomes distributed across many turns, and the steep wave front becomes attenuated. The wave front may be a transient, not part of the steady-state transformation, but it generates dielectric stress concentrations which must be accommodated. Interwinding dielectric stress is due to the sum of the induced and externally applied voltages appearing across the interwinding insulation.

Coil geometry determines critical dielectric stress regions. Many small coils operating at low or moderate voltages are wound on bobbins. Bobbins are especially vulnerable to intrawinding stress because of the random placement of turns and the close proximity of interior leads to subsequent turns. These intrawinding stresses are resisted by the insulating film on the magnet wire, insulation placed over the leads (usually pressure-sensitive tape), and impregnating materials. The maximum stress to which the magnet wire film will be subjected is uncertain. When high induced voltages are to be encountered, bobbin construction is an infrequent choice. Interwinding insulation in bobbin construction is achieved by sheet insulation between adjacent windings or by the use of bobbins with two or more sections separated by a partition. The partition provides reliable interwinding insulation within the dielectric capabilities of the bobbin. Sheet insulation between adjacent windings in a single-section bobbin is subject to compromise by turns slipping around the insulation and making contact with the turns of the adjacent winding. The insulation of the outer winding to core can be compromised by overfilling the bobbin, leaving an inadequate clearance for the core. The core tends to abrade the insulation as it is installed around the bobbin, inviting premature failure. The space factors in magnet wire tables used for bobbin-wound coils neglect leads and insulating materials. In addition to occupying space, these objects disrupt the smooth flow of the wire on the bobbin, reducing the number of turns that can be placed in the space remaining.

In layer-wound coils the magnet wire insulating film must withstand only the voltage between adjacent turns. Resistance to wave front transients can be improved by space winding the end layers of the winding and increasing the layer insulation between end and adjacent layers. Insulation for a wide range of layer-to-layer voltages may be provided by varying the thickness and number of sheets of insulation between layers. The number of layers and turns per layer may also be varied to establish favorable voltage distributions. Layer-wound coils have excellent characteristics for high induced voltages. Interwinding insulation is also readily achieved by sheet insulation as needed. In large coils with high copper losses both layer and interwinding insulation may be supplemented with spaces which permit the flow of air or oil for cooling. Large coils may be wound on forms which support and separate individual turns. This construction provides support to resist mechanical stresses set up by high currents and also provides for insulation and cooling.

Universal coils, like layer-wound coils, enjoy a controlled placement of turns, but like bobbin coils they must depend upon the insulation of the wire itself. As in bobbin coils, many intervening turns will exist between turns in close proximity. Many universal coils operate without impregnation. High-frequency operation, which is the domain of universal coils, limits the impregnation options.

8.7 DIELECTRIC HEAT LOSS

Dielectric heating is usually neglected at power frequencies in magnetic devices. At frequencies above 1000 Hz, it often must be considered in coil design. Dielectric heat adds to the copper and core losses in regions where heat transfer is poor. Dielectric loss increases with temperature so that a runaway condition can readily occur. The dielectric loss of an insulating material is specified by the dissipation factor, which is the ratio of dissipated to reactive energy of a capacitor in which that insulation is the dielectric. If the equivalent circuit of a dielectric is as shown in Fig. 8.1, the dissipation factor of insulation no. 1 will be $1/R_1\omega C_1$. The dissipation factor of many insulating materials in common use at power frequencies increases rapidly above 1 kHz. In a magnetic device the capacitance related to the reactive energy in the dissipation factor definition is the intrawinding distributed capacitance. A reduction in intrawinding distributed capacitance will result in a reduction of dielectric loss. The dissipation factor of insulating materials tends to increase rapidly near their upper temperature limit. The use of high-temperature-insulation in high-frequency devices will often reduce the dielectric heat loss as well as permit operation at a higher temperature. Insulating materials with low dissipation factors at high frequencies may often be readily substituted for the familiar power frequency materials. Intrawinding sheet insulation and impregnants especially should be screened for high-frequency use. The dielectric loss can be estimated by calculating the intrawinding distributed capacitance and applying the dissipation factor definition. The resulting shunt resistance appears across the induced voltage of the winding for which the distributed capacitance was determined. From this voltage the dielectric loss may be calculated.

8.8 INSULATION CLASSES

The first classification of insulating materials was by origin: organic, inorganic, or a combination of both. Letter designations A, C, and B, respectively, were applied to these classifications. With a proliferation of insulating materials having intermediate thermal properties, these classifications were modified. Additional letters were added, temperature limits were assigned to each letter, and the description of classes was expanded to include materials with temperature limits appropriate to each class without regard to their origin. The U.S. Department of Defense Specification MIL-T-27 uses a different set of letters to classify the insulation temperature capabilities of magnetic components. While this letter system parallels industry usage, the designations indicate specific functional and legal requirements. Table 8.1 gives the insulation classifications and their upper temperature limits. Individual insulating materials,

TABLE 8.1 Letter Designations Used to Classify Insulations and Insulation Systems by Temperature

Classifications from IEEE Standards Publication 1		Classifications from Military Specification MIL-T-27	
Insulation Class	Limiting Temperature, °C (°F)	Temperature Class	Maximum Operating Temperature, °C (°F)
O	90 (194)	Q	85 (185)
A	105 (221)	R	105 (221)
B	130 (266)	S	130 (266)
F	155 (311)	V	155 (311)
H	180 (356)	T	170 (338)
C	220 (428)	U	>170

when used in combination, may result in an insulation system with either a higher or lower temperature classification than that of any of the individual materials. Insulation class may be specified by giving the limiting temperature in degrees Celsius rather than a letter. Insulation ratings are discussed extensively in IEEE Standards Publication 1.

8.9 MAGNETIC DEVICE INSULATION SYSTEMS

Terminal Bushing Materials

Most terminal bushings are ceramic. See Sec. 7.4. Ceramics have excellent high-temperature and dielectric properties. Problems with ceramics are likely to stem from their mechanical limitations. Fracture during handling is among the most common causes of failure. One popular style of bushing uses elastomer inserts which form a part of the dielectric structure of the bushing. In this type of bushing the insulation properties of the insert material become significant. The insert must withstand voltage stress, mechanical stress, and temperature extremes while exposed to the solvent action of fluid potting materials. Available insert materials control the usage of this type of bushing. The need to conserve space inside enclosures dictates that the portion of bushings inside the enclosure be as small as possible. To provide for adequate creepage distance, lead wire insulation and potting materials are sometimes called upon to supplement the dielectric strength of the bushing itself. This invokes the problems of composite dielectrics, air voids, and corona inception voltage.

8.10 POTTING MATERIALS

Potting material is used to fill an enclosure in which a device is mounted. The enclosure remains a permanent part of the assembly. Potting materials insulate, cool, seal, and mechanically support the device being potted. Potting materials are selected on the basis of their ability to serve these functions and the relative importance of these functions to the individual application. Potting materials may be gaseous, liquid, or solid.

Gaseous Potting Materials

Minimum weight will be achieved with gases provided their use does not necessitate heavier mechanical constructions. All gases have poor thermal conductivity compared with liquid and solid potting materials. The most commonly used gases for potting are air and sulfur hexafluoride. With a dielectric constant near 1, both of these gases have low dielectric loss at both low and high frequencies. SF_6 is a frequent choice for filling waveguides. Gas is used with liquid dielectrics where it fills an expansion volume. If the expansion volume is located in a region of dielectric stress, the dielectric properties of the gas in combination with solid and liquid dielectrics present may have to be considered.

The ease with which air ionizes makes it a poor dielectric for high voltage. Another serious limitation to air as a dielectric is the reduction in dielectric strength with altitude. See Fig. 8.3. The corona inception voltage of composite dielectrics which include air also decreases with altitude. Air at ambient pressure is not a suitable high-voltage dielectric in aircraft. The dielectric properties of this least costly of potting materials are improved if moisture is excluded.

Sulfur hexafluoride is a stable nontoxic gas with excellent dielectric properties. See Table 8.2 for breakdown-strength data in comparison with air and transformer oil. With increasing pressure the breakdown strength of SF_6 will reach and exceed that of oil. See Fig. 8.4. This gas does not ionize readily and does not contribute to corona inception as does air. SF_6 and air form a homogeneous mixture whose dielectric properties improve continuously with increasing concentrations of SF_6. The presence of a small percentage of air in SF_6 has only a minor effect on dielectric strength. This welcome property simplifies the filling procedure. Some of the advantages of SF_6 as a potting material are offset by the increased cost, weight, and complexity of pressurized containers used to achieve high dielectric strength.

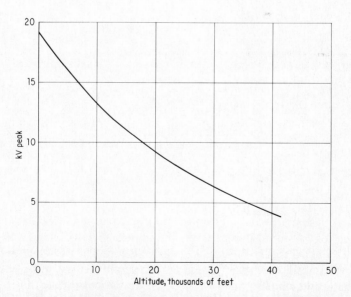

FIG. 8.3 Reduction in dielectric strength of air with altitude.

TABLE 8.2 Approximate Breakdown Strengths at Atmospheric Pressure of Air, Sulfur Hexafluoride, and Transformer Oil

	Spacing, cm	
	0.5	1.0
Air	19	32
SF$_6$	45	86
Transformer oil	110	218

Note: Values are peak kilovolt 60-Hz sine wave. Electrodes are 5-cm-diameter brass spheres.

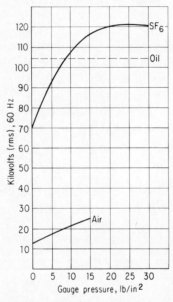

FIG. 8.4 Dielectric strength of SF$_6$ vs. pressure compared with transformer oil and air.

Liquid Potting Materials

Insulating systems employing liquid potting materials excel because of the ability of the liquid to replace air in the interstices of the device. Corona threshold voltage is thereby greatly increased. The degree of penetration depends upon the liquid, the magnetic assembly, and the filling process. The liquid should have low viscosity and good wetting ability. The assembly should contain insulating materials that are wetted by the liquid, and it should be free of air traps. Complete filling will be achieved only by a vacuum process. Liquid potting materials have good heat transfer characteristics by virtue of convection currents, and they have the valuable asset of being self-healing. When an arc occurs, the resulting decomposition products dissipate and fresh materials flow into the region where the arc occurred. The process is so effective that failure analysis is sometimes hampered by an inability to locate the site of the arc. Chosen for their dielectric excellence, liquid insulating materials degrade when they contain impurities. When they are of adequate purity, liquid dielectrics form with solid insulating materials composites which are superior to either material alone.

Compatibility of insulating materials is especially important in liquid systems. The solvent action of liquid dielectrics attacks adhesives, plastics, and elastomers. Gaskets used for sealing are especially vulnerable. Without special consideration, coils held together with pressure-sensitive tape and adhesives will disintegrate. Adhesives must be carefully chosen and correctly applied. Coils are sometimes secured by lashing to permit operation in liquid dielectrics. Some solvent-type varnishes after curing are impervious to liquid dielectrics. Varnish treatment prior to submersion will keep the coil in one piece, but the cured varnish will impede the penetration of the liquid dielectric, resulting in more voids than would exist if the liquid alone were used as an impregnant.

The most common liquid dielectric used with magnetic devices is transformer oil, a petroleum distillate with venerable credentials. The success of this mineral oil is due to a fortunate convergence of physical and electrical properties. A complex mixture of aliphatic and aromatic hydrocarbons, transformer oil does not ionize readily and has high dielectric strength. It has low viscosity and low surface tension, giving it high penetrating ability. Table 8.3 lists some of its essential properties.

The dielectric properties of transformer oil are quite vulnerable to degradation by impurities. Solid particles, conductive and nonconductive, can occur in suspension. Such materials are not usually present in oil as refined, but they may be readily introduced during the various manufacturing processes used with magnetic devices. Dissolved air and water vapor reduce the dielectric strength of oil. Transformer oil is stored and shipped in sealed containers. If left exposed to the atmosphere, its dielectric strength will decrease. Processing can restore it to its original condition or better. Industrial processing of transformer oil varies from careless to surgically precise. In the most rigorous procedure, filtering to remove suspended material, vacuum spraying to eliminate dissolved air and water, and vacuum filling of the device are executed in a single operation.

TABLE 8.3 Properties of Transformer Mineral Oil

	ASTM Test Method	Typical Values
Flash point	D 92	145°C
Specific gravity	D 1298	0.9
Viscosity		
cSt/SUS at 100°C (212°F)	D 445/D 88	3.0/36
40°C (104°F)	D 445/D 88	12.0/66
0°C (32°F)	D 445/D 88	76.0/350
Dielectric breakdown voltage (60 Hz, 1-in discs, 0.1-in spacing)	D 877	30 kV rms
Water, ppm	D 1533	35
Volumetric coefficient of expansion per °C	D 1903	0.00075
Dielectric constant at 25°C (77°F)	D 924	2.2
Power factor		
At 25°C (77°F)	D 924	0.05
At 100°C (212°F)	D 924	0.30
Specific heat, (g-cal)/g at 20°C (68°F)	D 2766	0.44
Thermal conductivity, cal/(cm·s·°C)	D 2717	0.0003

Transformer oil is a class A material by origin, tradition, and experience. In an oil-filled magnetic device with a class A insulation system, the copper temperature and hence the insulating materials in immediate contact with the copper will operate near 105°C (221°F). The average oil temperature will be in the vicinity of 85°C (185°F). Under these conditions transformer oil has a demonstrated great longevity. At temperatures of 105°C and above, transformer oil oxidizes slowly in the presence of air, forming precipitates referred to as *sludge*. Sludge formation may be retarded by the addition of inhibitors to the oil and by excluding air. At its flash point of 145°C (293°F), transformer oil presents an immediate explosion and fire hazard. The flash point is an absolute limit on its maximum operating temperature. Magnetic devices with large oil volumes constitute an inevitable fire hazard. They are usually exterior installations.

Various halogenated compounds have been used as liquid dielectrics. One group which has refrigerant properties has been used where both its cooling and insulating properties are exploited.[3] The most com-

[3]L. F. Kilham, Jr., and R. R. Ursch, "Transformer Miniaturization Using Fluorochemical Liquids and Conduction Techniques," *Proceedings of the IRE,* vol. 44, no. 4, April 1956.

mercially significant halogenated compounds used for liquid dielectrics have been the chlorinated biphenyls. Although they possess excellent dielectric properties, they are being phased out for environmental reasons.

Silicone fluids are used in high-temperature applications where the increased cost is justified. These fluids may operate up to 200°C (392°F). See Table 8.4 for characteristic properties.

Solid Potting Materials: Thermoplastic and Thermosetting

Wax (thermoplastic) has been for many years an inexpensive potting material for small transformers with low operating temperatures. Waxes are usually higher-order petroleum distillates. They are available over a range of melting temperatures, merging into the asphalts. Wax has good dielectric properties. When the component is preheated, good penetration of the melted wax can be achieved. Hot vacuum wax impregnation is sometimes used. The development of many insulating materials with higher maximum operating temperatures than the melting point of wax has resulted in the decline in the use of wax.

Asphaltic compounds are similar to waxes except that they have higher melting temperatures. These materials are often referred to as *pitch*. The addition of sand increases the thermal conductivity of asphalt but makes pouring more difficult. Asphaltic compounds are available over a range of melting temperatures. The choice is a compromise among the needs for high melting point, low viscosity for effective filling, and good thermal conductivity. The melting point of the thermoplastic potting materials places a ceiling on the operating temperature of copper.

Initially liquids, thermosetting materials are converted to solids by

TABLE 8.4 Typical Properties and Values of Silicone Transformer Liquid

Property	Value
Flash point	285°C
Specific gravity	0.96
Viscosity, cSt at 25°C (77°F)	50
Volumetric coefficient of thermal expansion, $cm^3/(cm^3 \cdot °C)$	0.00104
Water content, ppm	50
Dielectric strength, V/mil	350
Dielectric constant	2.71
Dissipation factor, 100 Hz	0.00002
Volume resistivity, $\Omega \cdot cm$	1×10^{15}

polymerization. After polymerization has occurred, they cannot be returned to the liquid state by heating. The process of filling voids and coil interstices is performed while the material is in the liquid state. Polymerization, which takes place after the filling process is complete, is accomplished by mixing in additives prior to filling and sometimes by the application of heat. As polymerization proceeds, the viscosity of the material increases until it assumes the solid state. The period between the mixing of additives and the time when the viscosity becomes so great that the filling process can no longer proceed is called the *pot life* of the material. Pot life varies from seconds to months, depending on the type of material. Thermosetting potting materials should have low viscosity for good penetration, good heat conduction after polymerizing, good dielectric properties, and good resistance to temperature extremes. They must be compatible with the rest of the insulation system. In the rich field of choices which polymer chemistry has provided magnetics technology, there are no materials which meet all these needs ideally.

Polyurethanes are a versatile class of compounds which are formulated to produce foams, elastomers, and plastics, all of which find use in potting applications. Rigid foams are useful in potting small high-frequency devices when little heat is to be dissipated. The foam has a limited effect on the electrical properties because of its low dielectric constant and because the foam does not subject the part to severe mechanical stress. At the same time it does a satisfactory job of supporting the component. Polyurethane foam is supplied as a two-component material. When the components are mixed together in the proper proportions, the reaction starts promptly, requiring that the liquid must be poured immediately. Liberated gas causes the rigid foam to assume a volume several times the volume of the liquid. Typical properties of a rigid polyurethane foam are shown in Table 8.5. In their elastomeric form, polyurethanes make a tough, resilient, and inexpensive substitute for silicone elastomers but without the latter's high-temperature properties. Usually two-component systems, polyurethane elastomers have a pot life of several hours, allowing opportunity to vacuum fill if needed. Both room temperature and elevated cure systems are available. Fillers are sometimes added to improve thermal conductivity. Typical properties of a polyurethane elastomer are shown in Table 8.5.

Epoxy resins have a combination of useful properties that have made them the most popular thermosetting potting and encapsulating materials. Properties vary with the type of epoxy, filling materials, curing agents, and modifiers. Most epoxies have good dielectric strength, high resistivity, and low dielectric loss at power frequencies. Caution is required at frequencies above 1 kHz. They have good mechanical strength and high maximum operating temperatures extending from class

TABLE 8.5 Typical Properties of Polyurethane Potting Materials

	Elastomer	Rigid Foam
Thermal conductivity, $(Btu \cdot ft)/(h \cdot ft^2 \cdot °F)$	0.14	0.05
Specific gravity	1.2	0.65
Hardness, Shore D	60	85
Tensile strength, lb/in^2	3000	2500
Volume resistivity, $\Omega \cdot cm$	5×10^{13}	2×10^{14}
Dielectric strength, V/mil	400	125
Dielectric constant	5.5	1.8
Dissipation factor	0.01	0.007
Maximum continuous operating temperature, °C (°F)	100 (212)	85 (185)

F to class H. Their resistance to moisture and many solvents is excellent. Epoxies have a high coefficient of thermal expansion compared to metals, and they become brittle at very low temperatures. These latter properties give them a tendency to crack at very low temperatures when they enclose metal objects. The tendency to crack can be reduced but not completely eliminated by carefully controlling the geometry of the enclosed object, by limiting curing temperatures, and by choosing fillers judiciously. Glass fiber is effective in imparting increased tensile strength. Silica and alumina are also commonly used as fillers. Alumina is notable for increasing thermal conductivity.

The variety of properties that epoxies possess does not fortuitously occur in any single material. Choices involve compromises. Epoxy resins may be made semiflexible to reduce the tendency to crack under thermal shock, but this is achieved at the price of mechanical strength and insulation resistance at high temperatures. The rigid resins have higher maximum operating temperatures, higher mechanical strength, and better moisture resistance, but poorer thermal shock resistance. The rigid resins have better adhesion to metals than the semiflexible resins.

The popularity of epoxy resins is due not only to their cured properties but also to their working properties. The liquid resins are convenient to use. The heat of polymerization is low enough to be controllable. The shrinkage rate during polymerization is small—about 3 percent by volume—compared with other thermosetting resins. As with the cured properties, a wide variety of working properties is available from which to choose. Favorable working properties are often secured at the sacrifice of cured properties. Low viscosity, an advantage in obtaining complete filling, can be achieved by the use of diluents, which injure the cured mechanical properties. Both room temperature and elevated temperature curing agents are available, but the convenience of room temperature cures is obtained at the expense of high maximum operating temperature and low exothermic heat.

Most epoxy systems are compounded as two components. The pot life varies from a few minutes to many days. Deaeration and filling with vacuum is a common practice. Chemical compatibility with other insulating materials is reasonably good. Polymerization of the resin is not easily inhibited. In this regard thorough prebaking of coils is required to cure thermosetting adhesives and to remove volatile solvents. Single-component epoxy systems are cured at elevated temperatures. The shelf life of single-component systems is limited, requiring closer monitoring than do two-component systems. Table 8.6 gives typical cured properties of an unfilled epoxy resin.

Silicones have the highest operating temperature ratings of the commonly used insulating plastics, and they are the most expensive. They have good dielectric properties over a wide frequency range, and both mechanical and dielectric properties are stable over wide temperature extremes. The commonly used potting materials are elastomers. The one-part silicone sealants which cure at room temperature to tough elastomers are not suitable for potting because they will cure only in thin sections. The potting elastomers are two-component materials. Various systems are available with different curing properties and different viscosities. They all have appreciable coefficients of thermal expansion

TABLE 8.6 Typical Properties of an Unfilled Rigid Epoxy Resin

	ASTM Test Method	Typical Value
Specific gravity	D 792	1.2
Water absorption (24 h, %)	D 570	0.1
Tensile strength, lb/in^2	D 638	5000
Hardness, Rockwell M	D 785	80
Thermal conductivity, $(Btu \cdot ft)/(h \cdot ft^2 \cdot °F)$	C 177	0.1
Linear coefficient of thermal expansion, °C	D 696	3.0×10^{-5}
Dielectric strength, V/mil	D 149	300
Dielectric constant	D 150	4.0
Dissipation factor	D 150	
60 Hz		0.007
1 mHz		0.03
Volume resistivity, $\Omega \cdot cm$	D 257	5×10^{15}

which must be accommodated in the case construction. Silicones have poor adhesion to many other materials. Primers are available for pre-coating surfaces to which adhesion is necessary. Polymerization of silicones is more readily inhibited than with some other resins. Investigation of chemical compatibility is necessary for untried applications. The excellent high-temperature and dielectric properties of silicones make them serious candidates for many applications, but they present problems that make investigation and testing necessary. Typical cured properties of a two-part silicone elastomer are shown in Table 8.7.

8.11 IMPREGNATING AND FILLING PROCESSES

Filling and impregnating may be accomplished by free pouring at atmospheric pressure, pouring under vacuum, or by the use of pressure. The choice of the process depends on the requirements for the completed article, the choice of impregnant and filling material, and the geometry of the object being treated. The ideal filling process would combine impregnation and potting in a single operation. The ideal filling material would be a low-viscosity liquid which would become a void-free solid with optimum electrical and mechanical properties.

TABLE 8.7 Typical Cured Properties of a Silicone Elastomer Potting Compound

	ASTM Test Method	Typical Values
Specific gravity	D 792	1.05
Water absorption (7 days, %)	D 570	0.1
Durometer, Shore A	D 2240	40
Tensile strength, lb/in^2	D 412	900
Thermal conductivity, $(Btu \cdot ft)/(h \cdot ft^2 \cdot °F)$		0.2
Volume expansion, $cm^3/(cm^3 \cdot °C)$		9.6×10^{-4}
Maximum operating temperature		180°C (356°F)
Dielectric constant	D 150	2.75
Dissipation factor	D 150	
60Hz		0.001
100 kHz		0.001
Volume resistivity, $\Omega \cdot cm$	D 257	2×10^{14}
Dielectric strength, v/mil	D 149	600

Varnish impregnation has long been used for many applications. Insulating varnish contains dielectric resins dissolved in volatile solvents. The solution is of low viscosity with good wetting action. Varnish is best applied by vacuum impregnation, which results in better penetration than the less costly dipping method. A typical insulating varnish contains about 40 percent volatile solvents by volume. The solvent content varies as the varnish is used. Solvent is lost by evaporation, especially during vacuum cycles. Solid material is removed with the varnished components. Addition of solvent will reduce both the viscosity and specific gravity. The addition of whole varnish will increase the specific gravity after it has been reduced by the addition of solvent. Both specific gravity and viscosity should be kept within the limits recommended by the varnish manufacturer.

The commonly used resins in varnish are phenolics, acrylics, polyurethanes, polyesters, and silicones. Varnishes have maximum operating temperatures from 105 to 180°C (221 to 356°F). The use of a high-temperature varnish with low-temperature sheet insulation raises the maximum operating temperature of the sheet insulation–varnish combination over that of the layer insulation alone. Some varnishes are air-drying. Others require an elevated temperature cure. The latter possess superior properties.

A properly maintained varnish charge has a very long pot life. This permits storing the varnish in tanks and moving the varnish through piping. A typical varnish-processing installation has a storage tank, impregnating tank, vacuum pump, solvent trap, associated plumbing, and possibly a means for applying gas pressure to the impregnating tank. A typical process involves prebaking the components to remove moisture and cure adhesives. While still hot, the components are placed in the impregnating tank in which a vacuum is established. The vacuum operates on the hot insulation to continue the drying process. A valve is then opened through which the varnish is drawn into the impregnating tank from the storage tank. After the components are completely submerged in varnish, the valve is closed. The vacuum is discontinued. Either air or gas under pressure is admitted to the space above the varnish. The resulting pressure forces the varnish into the interstices of the coil. The units are removed from the impregnating tank and oven-baked. Upon removal of the components from the impregnating tank, some varnish drains from the coils. Solvent evaporates from the varnish remaining in the coils. Draining and solvent loss cause voids in the coils. Voids allow moisture entry and dielectric stress enhancement which contribute to corona formation at high voltages. When used in the presence of moisture or high voltage, varnish-impregnated coils require additional processing and protection. The volatile solvents in varnish have constituents with relatively high vapor pressures. Since traces of the solvent will be present even in an empty impregnating tank, a high vacuum is not feasible in the varnish process nor would it serve a useful purpose since void formation is inevitable for other reasons than an insufficient vacuum. Varnish is effective in bonding the core and coil mechanically. Its convenience and low cost make it a frequent choice where its limitations can be tolerated.

Oil impregnation is performed in a manner similar to varnish impregnation. Since the coil remains in an oil-filled container after processing, the drain-off and solvent evaporation that detract from varnish impregnation do not occur. A higher vacuum is used with oil than with varnish because the applications are more critical and the lower vapor pressure of the oil constituents permits it. Treatment of oil to remove moisture is frequently done prior to impregnation. Transformer oil has superlative penetrating and wetting properties. It also has considerable solvent action, attacking many adhesives, plastics, and elastomers. Material compatibility with oil is a major consideration in the design and manufacture of oil-filled components.

The low vapor pressures of thermosetting materials permit their use as impregnants at relatively high vacuum. Their thermosetting property prevents void formation. The short pot life of most thermosetting materials precludes their use in storage tanks with plumbing, as varnish is used. Economics requires that large residues of unused liquid, left to polymerize, must be avoided when processing thermosetting materials. The best impregnation is obtained by first drawing a vacuum in the chamber containing the component, then admitting the impregnant. With short pot-life materials, this requires special equipment which is costly to operate and maintain. As a result many components enclosed in thermosetting materials are vacuumed after pouring. Reentrant surfaces and the hydrostatic pressure of the prepoured liquid tend to entrap air bubbles when this compromise is pursued. Vacuum casting with compromise thermosetting materials constitutes the closest approximation to the ideal solid insulation single-step filling process available.

8.12 SHEET INSULATION

Sheet insulation is used to insulate and provide mechanical support between layers of layer-wound coils and between windings. It is also inserted in core windows to insulate coils from the core and placed between coils and case walls. Factors influencing the choice of sheet insulation are temperature class, mechanical strength, flexibility, dielectric properties, porosity, insulation system compatibility, and cost.

Sheet insulation is available using materials with temperature classes from 105 to 180°C (221 to 356°F). The allowable operating temperature will be affected by the temperature class of the impregnant and the compatibility between the insulating material and the impregnant. For compatible materials the maximum allowable operating temperature can be expected to be intermediate between the temperature classes of the sheet insulation material and the impregnant.

Layer insulation must be sufficiently flexibie to conform to the coil contours under winding pressure and sufficiently strong to support the magnet wire wound on top of it. These requirements vary with the wire size and the size of the coil. To obtain a satisfactory balance between flexibility and strength, the thickness of layer insulation is varied with the wire size. Dielectric requirements may augment the layer insulation, and size limitations may diminish it. Sheet insulation is vulnerable to minute random defects. These defects are possible sites for initiating dielectric failure. The use of two thin sheets instead of a single thick sheet is superior dielectrically because the probability of two defects coinciding in adjacent layers is small. Two thin layers are more flexible but less strong than a single thick layer. The thicknesses in which various sheet insulations are available is a consideration in selecting a material.

Dielectric properties and porosity are related considerations. Materials used for sheet insulation vary from impervious to highly porous. Intrinsically impervious materials may be processed or filled with other materials to increase their porosity. An unimpregnated porous material will have lower dielectric strength than its nonporous equivalent. A satisfactory impregnation of porous material will greatly enhance its dielectric properties. A more homogeneous structure results which reduces the probability of a corona. Porous materials are selected when effective impregnation is expected. High porosity improves the impregnation.

Sheet insulation is subject to solvent and chemical action by impregnants. Processing introduces temporary but severe thermal and mechanical stresses. Sheet insulation must be capable of surviving this environment. The interface between impregnant and sheet insulation will be a surface of dielectric and mechanical weakness unless a good bond exists between the two materials.

The cost of the commonly used 105°C (221°F) class sheet insulations is low enough to have little effect on the overall cost of the component. The cost becomes significant as the temperature class increases. The use of aramid paper can have a substantial effect on final cost, and the cost of polyimide film is often prohibitive. Table 8.8. lists some of the properties of the most commonly used sheet insulations.

8.13 MAGNET WIRE COATINGS

Magnet wire is used in coils that generate magnetic fields. Space factor and dielectric requirements combine with high-temperature, mechanical,

TABLE 8.8 Properties of Some Commonly Used Sheet Insulations

Material	Temperature Class, °C (°F)	Available Thickness, in	Composition	Tensile Strength, lb/in²	Dielectric Strength, V/mil	Dielectric Constant	Relative Porosity	Relative Cost	Application Notes
Kraft paper	105 (221)	0.0003–0.031	Wood fiber	3,000	300	2.0	High	Low	Most popular layer-wound coil insulation. Flexible, compatible with impregnants.
Vulcanized fiber	105 (221)	0.004–0.187	Wood fiber	10,000	200	5.0	Low	Low	High strength, good abrasion resistance, low flexibility. Used as layer insulation with heavy wire.
Mylar (Du Pont trade name)	150 (302)	0.0005–0.014	Polyester film	20,000	6000	3.3	None	Moderate	Widely used low-cost high-temperature insulation. Lack of bond formation with impregnants can cause interface problems.
Aramid paper	220 (428)	0.002–0.030	Aromatic polyamide fiber	4,000	900	2.6	None	High	Tough, flexible. Outstanding high-temperature performance. Suitable for use as layer insulation for fine wire. Also available in a porous version.
Kapton (Du Pont trade name)	300 (572)	0.001–0.005	Polyimide film	10,000	3600	3.5	None	Very high	Very flexible and tough. Suitable for high-temperature layer insulation with finest wire sizes.

and compatibility needs to make magnet wire insulation a unique problem. The insulation used on hookup wire may easily increase the outside diameter by 50 percent. Magnet wire, in contrast, has insulation whose thickness is measured in thousandths of an inch.

Magnet wire varies from large rectangular cross sections capable of carrying hundreds of amperes to minute diameters with currents in microamperes. The largest sizes are often served with yarn or paper. There are many options available for served wire. The conductor may be round, square, or rectangular. A number of different serving materials may be placed over bare or film-coated wire, for which there are various bonding adhesives and varnishes. Kraft paper, glass yarn, and aramid paper are commonly used serving materials, providing a wide range of operating temperature ratings. Served wire improves reliability by providing generous physical spacing between turns. Smaller devices usually cannot afford the space occupied by served wire, in which case film-insulated wire is used.

Winding processes subject magnet wire insulation to stretching, flexing, and abrading. Modern insulating films have excellent resistance to these mechanical stresses. Standardized tests have been developed for evaluating the ability of films to resist mechanical stresses. The results of these tests are useful for making comparisons.

Impregnating and potting materials place chemcial and solvent resistance requirements on magnet wire films. Particularly significant are the short-time resistance to varnish solvents and the long-time resistance to transformer oil. Standardized tests have been developed for chemical and solvent resistance. Being short-term tests, the results cannot be directly related to the long-time effects of transformer oil. Industry does not readily accept new films for use in transformer oil.

Magnet wire insulating film is in intimate contact with the conductor, a source of heat. The film is the most vulnerable element to thermal overload in the insulation system. The insulation temperature rating of the film is critical. An assortment of films is available for temperature classes from 105 to 220°C (221 to 428°F).

Magnet wire is made with single-, double-, triple-, and quadruple-thickness films. Single-film wire, which has the best space factor and the least insulating value, is used mostly on the finer wire sizes. As the wire diameter increases, the percentage of the space occupied by the magnet wire which is devoted to insulating film decreases, reducing the advantage of single-film insulation. In the large sizes double film is most commonly used. Double films are also extensively used in the finer wire where there is a need for better dielectric properties. Triple and quadruple films are less used and are less readily available.

The film must be removed from magnet wire in order to make con-

TABLE 8.9 Properties of Some Commonly Used Magnet Wire Films

Property	Insulation			
	Polyvinyl Formal	Polyurethane with Polyamide Overcoat	Polyester	Polyimide
Temperature rating, °C (°F)	105 (221)	130 (266)	155 (311)	220 (428)
Abrasion resistance (single scrape average g to fail)	1350	1450	1200	1200
Elongation, %	35	34	35	36
Flexibility	Pass 1X	Pass 1X	Pass 1X	Pass 1X
Dielectric strength, V/mil	7000	7000	8300	10,000
Solvent Resistance (24 h at 25°C. Pass—No Effect. Fail—Severe Softening or Crazing.)				
VM&P naphtha	Pass	Pass	Pass	Pass
Toluol	Pass	Pass	Pass	Pass
Ethyl alcohol	Pass	Pass	Pass	Pass
Xylol	Pass	Pass	Pass	Pass
Carbon tetrachloride	Pass	Pass	Pass	Pass
Acetone (10 min)	Fail	Pass	Pass	Pass
Trichlorethylene	Pass	Pass	Pass	Pass
Solderability (minimum temperature to solder in 4 s)	Not solderable	750°C (1382°F)	Not solderable	Not solderable

nections. Some films are strippable by the heat of soft soldering. Connections may be made to solderable magnet wire by wrapping the wire around the termination, fluxing, and soft soldering through the melted film. Solderability is a great advantage for fine wire, which is weakened and easily broken by alternate stripping methods. The use of solderable wire does not necessarily prescribe a thermal disadvantage. Solderable films are available with insulation temperature classes of 130°C (266°F) and higher.

Magnet wire films may be stripped chemically. Commercial chemical strippers of varying effectiveness are available for most films. Chemical stripping is slow, and its efficiency decreases generally with increasing temperature ratings of the film. Caution is required when using some chemical strippers because they can irritate and burn the skin.

Mechanical stripping is effective when done correctly. Brush stripping machines are available which will strip a wide range of wire sizes, round, square, and rectangular. Skill is required in using brush strippers to avoid breaking fine wire. Collet strippers have rotating shafts upon which are mounted stripping collets of different sizes that receive the various wire sizes. Collet strippers will work only on round wire and cannot strip the finest sizes.

Magnet wire has been the object of extensive standardization. The industry standard publication is NEMA MW 1000. The U.S. government has two publications, Department of Defense Specification MIL-W-583 and federal Specification J-W-1177. The magnet wire industry is actively competitive and maintains good quality standards. Equivalent items are usually available from several sources. Table 8.9 gives some of the properties of several commonly used films.

CHAPTER 9

Thermal Considerations

The limitations on insulation and core operating temperatures ultimately determine the size of magnetic components. The operating temperature is the sum of the component ambient temperature and the component temperature rise. The slight dependence that component temperature rise has on ambient temperature is usually neglected. Reasonably accurate estimates of the component temperature rise and the component ambient temperature are needed in the design of the component. Heat transfer computations do not determine these values with precision. There are three modes of operation of particular interest: steady state in which the operating temperature is the final temperature under the heaviest continuous load and highest ambient temperature, short-term loading in which a heavy load is applied for an interval less than that required for the device to exceed the allowable insulation operating temperature, and adiabatic loading in which a very heavy load is applied for such a short interval that effectively no heat transfer occurs. In this section, rate of heat transfer will be in watts, energy in wattseconds (joules), temperature in degrees Celsius, absolute temperature in degrees Kelvin, length in inches, area in square inches, volume in cubic inches, and weight in pounds. Coefficients will be consistent with those units.

9.1 STEADY-STATE ANALYSIS

Heat is transferred from magnetic devices by conduction, convection, and radiation. Conduction is described by Fourier's law:

$$q = h_o A \frac{d\tau}{dx} = \frac{h_o A}{\Delta x} (\tau_1 - \tau_2) \tag{9.1}$$

where q = time rate of change of heat
A = cross-sectional area through which heat is being conducted
τ = temperature
x = length along conduction path
h_o = thermal conductivity of its material

The derivative $d\tau/dx$ is the gradient normal to the cross-sectional area.
Heat transfer by radiation is described by the Stefan-Boltzmann law:

$$q = \epsilon\sigma A T^4 \tag{9.2}$$

where ϵ = emissivity of radiating surface
σ = Stefan-Boltzmann constant
A = area of radiating surface
T = absolute temperature

The emissivity of painted surfaces of all colors is about 0.9. The emissivity of bright metal surfaces is in the range of 0.1. These latter surfaces should be avoided when good heat transfer is required. All bodies at temperatures above absolute zero radiate electromagnetic energy. Radiant energy is exchanged between a hot body enclosed within a cooler body in accordance with the following relationship:

$$q = \epsilon\sigma A_1(T_1^4 - T_2^4) \tag{9.3}$$

in which T_1 is the absolute temperature of the hot body and T_2 is the absolute temperature of the cooler enclosing body. The wavelength of temperature radiation is a function of temperature. For the surface temperatures of magnetic components, that wavelength is in the infrared range. The emissivity and absorptivity are functions of wavelengths and therefore temperature. For small temperature differences, the emissivity and absorptivity of a single body are approximately equal. Equipment exposed to the direct rays of the sun receives radiant energy in the visible light region and radiates energy in the infrared range. Thus the absorptivity and emissivity may be different. Flat-white paint has an absorptivity for solar radiation of about 0.2 and a low-temperature emissivity of about 0.9. Flat-black paint has an emissivity and absorptivity of about 0.95 for both conditions. The actual effect of color on the temperature rise of magnetic devices exposed to the direct rays of the sun depends on many factors, including the shape of the object, the angle of the sun's rays, the relative importance of radiation and convection in total heat dissipation, and the velocity of the wind.

Convection is a complex process involving conduction to the convecting fluid, changes in the density of the fluid with temperature, the viscosity of the fluid, and the motion of the fluid. Convection heat transfer is sometimes simplified to the form

$$q = h_c A(\tau_1 - \tau_2) \tag{9.4}$$

where q = rate of transfer of heat by convection
h_c = convection heat transfer coefficient
τ_1 = surface temperature of body from which heat is being convected
τ_2 = ambient fluid temperature
A = area of surface from which heat is being convected

The transfer of heat by radiation is sometimes simplified in a similar manner:

$$q = h_R A(\tau_1 - \tau_2) \tag{9.5}$$

in which h_R is the radiation heat transfer coefficient, which is defined as

$$h_R = \frac{G(T_1^4 - T_2^4)}{\tau_1 - \tau_2} \tag{9.6}$$

In Eq. (9.6) G is a proportionality constant determined by the emissivity of the body, the geometry of the radiating and absorbing system, and the temperature. Equations (9.1), (9.4), and (9.5) may be converted to an electrical analog in which temperature difference is represented by potential difference, the time rate of change of heat is represented by current, and the reciprocals of the products of area and coefficients of thermal conductivity are represented by resistance. An electrical analog of the heat flow in a transformer is shown in Fig. 9.1. Such analogs are useful in achieving a qualitative understanding of the heat transfer system. To obtain useful quantitative information, the coefficients, which are functions of temperature, must be evaluated. This is done by an iterative process.

Consider a body with two parallel faces within which heat is generated uniformly. The two faces extend indefinitely and are exposed to the same ambient temperature. Heat is transferred uniformly and equally from both faces. The maximum temperature rise will occur at the midpoint between the two faces and is given by

$$\theta_m = \frac{q_v}{2h_o} L^2 \tag{9.7}$$

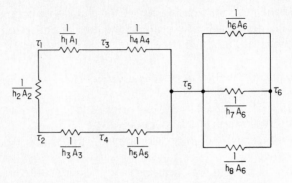

FIG. 9.1 Electrical circuit analog of heat transfer in a transformer.

τ_1 = average core temperature

τ_2 = average coil temperature

τ_3 = average core surface temperature

τ_4 = average exposed coil surface temperature

τ_5 = average surface temperature of enclosure

τ_6 = ambient temperature

$\dfrac{1}{h_1 A_1}$ = thermal resistance within core to surface of core

$\dfrac{1}{h_2 A_2}$ = thermal resistance between coil and core

$\dfrac{1}{h_3 A_3}$ = thermal resistance within coil to surface of coil

$\dfrac{1}{h_4 A_4}$ = thermal resistance between core surface and enclosure surface

$\dfrac{1}{h_5 A_5}$ = thermal resistance between coil surface and enclosure surface

$\dfrac{1}{h_6 A_6}$ = thermal resistance representing conduction between enclosure surface and ambient

$\dfrac{1}{h_7 A_6}$ = thermal resistance representing convection between enclosure surface and ambient

$\dfrac{1}{h_8 A_6}$ = thermal resistance representing radiation between enclosure surface and ambient

where θ_m = maximum temperature rise
q_v = time rate of heat generation per unit volume
h_o = coefficient of thermal conductivity
L = length as defined in Fig. 9.2

The temperature rise at any point x is given by

$$\theta = \theta_m(1 - x^2/L^2) \tag{9.8}$$

The average temperature rise through the body is given by

$$\theta_{av} = \tfrac{2}{3}\theta_m \tag{9.9}$$

If one surface of the body is warmer than the other, then the situation shown in Fig. 9.2 exists, in which the warmer surface is now at x_1. The point at which the maximum temperature rise exists is still defined by Eq. (9.7), but L in that equation no longer refers to the midpoint between the faces. The value of L is given by

$$R = \frac{L}{x_1 + L} \tag{9.10}$$

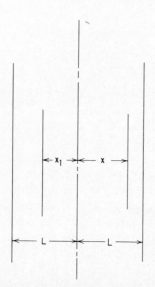

FIG. 9.2 Cross section of thermally conductive body with parallel faces between which heat is generated uniformly.

The value of R is determined by

$$R = \frac{h_o \theta_1}{q_v (L + x_1)^2} + \frac{1}{2} \tag{9.11}$$

in which θ_1 is the temperature difference between the warmer and cooler surfaces. The average temperature rise is given by

$$\theta_{av} = \theta_m (\tfrac{2}{3} + P - P^2) \tag{9.12}$$

in which P is

$$P = \frac{x_1}{L} \tag{9.13}$$

Relationships (9.7) through (9.13) are useful in approximating the temperature distribution in cores and coils. Equations (9.7) and (9.8) may be applied to the magnetic core. Heat is generated approximately uniformly through the core. Steel has a coefficient of thermal conductivity of approximately 1.6 W/(in · °C). The core loss is typically 0.28 W/in³. Consider a core with a dimension between exposed surfaces of 10 in. Use of these values in Eq. (9.7) gives

$$\theta_m = \frac{0.28}{2 \times 1.6} (5)^2 = 2.2°C$$

In this example the maximum internal temperature of the core is only slightly greater than the surface temperature, and little error results if the entire core is assumed to be at the same temperature.

Equations (9.9) and (9.12) are useful in estimating the relationship between maximum and average conductor temperatures, the latter being often obtained from the increase in resistance method of measuring temperature rise. The exposed portion of a coil mounted on a core is approximated by Fig. 9.2. The surface at x_1 corresponds to the contact surface between the coil and the core in which the core will be cooler than the coil because of the core's high thermal conductivity, but warmer than the core and coil ambient because of the core losses which must be dissipated to the ambient.

A coil with laminated structure of several different coefficients of thermal conductivity will have an average coefficient of thermal conductivity normal to the laminated structure given by

$$h_{av} = \frac{x_T}{\sum\limits_{n} x_n/h_n} \tag{9.14}$$

If the coil consists of alternate layers of metallic conductor and insulation, the disparate values for thermal conductivity of these materials cause Eq. (9.14) to reduce to approximately

$$h_{av} = \frac{x_T}{x_i} h_i \tag{9.15}$$

where x_T = total coil buildup
x_i = total insulation buildup
h_i = thermal coefficient of conductivity of insulation

Each conductor layer is at approximately uniform temperature. The temperature gradient exists mostly from layer to layer. The average value of the coefficient of thermal conductivity determined in Eq. (9.14) or (9.15) may be used in Eq. (9.7) to estimate the maximum conductor temperature.

The determination of temperature drop across the potting material in cased core and coils applies Fourier's law, Eq. (9.1). The average area, the average thickness of the potting compound, and the coefficient of thermal conductivity of the potting compound are needed. The average area is given by

$$A_{av} = \frac{1}{2}(A_c + A_{Fe} + A_{Cu}) \tag{9.16}$$

in which A_c, A_{Fe}, and A_{Cu} are the surface areas of the case, magnetic core, and exposed coil surfaces. The average thickness of the potting compound is estimated by determining the distance between two concentric spheres whose surface areas are equal to the surface area of the core and coil and the case:

$$x = \sqrt{\frac{A_c}{4\pi}} - \sqrt{\frac{A_{Fe} + A_{Cu}}{4\pi}} \tag{9.17}$$

Coefficients of conductivity of commonly used potting compounds are given in Table 9.1. The determination of temperature drop between core and coil and case is difficult when the impregnant is oil. The Armour Research Foundation[1] has suggested the following empirical formula:

$$\theta = 130 \frac{x}{A_{av}} (W_T)^{0.78} \tag{9.18}$$

where θ = temperature drop between core and coil surface and enclosure surface, °C
A_{av} = average area determinable from Eq. (9.16), in^2
x = average distance between the two surfaces, in
W_T = total heat loss of core and coil, W

The transfer of heat from the exposed surface of a device to the ambient by radiation and natural convection can be described by the following empirical formula:

$$W_A = 3.68 \times 10^{-11}\epsilon(T_1^4 - T_2^4) + 0.0014(T_1 - T_2)^{1.25} \tag{9.19}$$

where W_A = rate of heat transfer, W/in^2 surface area
T_1 = absolute temperature of radiating surface, K
T_2 = absolute temperature of ambient, K
ϵ = emissivity of radiating surface

The emissivity of commonly encountered surfaces is given in Table 9.2.

Some electronic equipment is totally enclosed. Heat is removed by the use of heat sinks, often finned aluminum extrusions upon which heat-dissipating components are directly mounted. The temperature of the heat sink is maintained at or below a specified maximum by an air stream, either natural or forced, flowing past the fins. With such an arrangement, most of the heat generated by the core and coil is transferred by conduction to the heat sink. Heat from the coil must be transferred to the heat sink through the core. The thermal resistance between the core and the heat sink is determined partly by the flatness of the surfaces. This resistance is reduced by the use of thermal joint compound. Fourier's law, Eq. (9.1), and Eq. (9.7), in which L now represents the total

---

[1]G. A. Forster, L. J. Stratton, and H. L. Garbarino, "Research and Development of New Design Method for Power Transformers," final report under U.S. government contract No. DA-36-039. SC-52656, Armour Research Foundation of the Illinois Institute of Technology, Chicago, March 1956.

TABLE 9.1 Conductivity and Specific Heat of Some Common Materials

Material	Conductivity, W/(in·°C)	Specific Heat, Ws/(lb·°C)
Aluminum	5.14	417
Copper	9.76	176
Silver	10.7	106
Tin	1.58	104
Steel	1.14–1.58	208
Kraft paper	0.003	700
Vulcanized fiber (fish paper)	0.007	
Epoxy resin		
Unfilled	0.0065	
Heavily filled	0.036	
Asphalt potting compound		
Unfilled	0.008	
Silica-filled	0.015	
Polyester film	0.004	530
Aramid paper	0.004	500
Polyimide film	0.004	496
Transformer oil	0.003	1140
Porcelain	0.024	494
Water	0.015	1898
Air	0.00075	460
Thermal joint compound	0.0189	
Ferrites		
MnZn	0.096	500
NiZn	0.096	340

coil buildup, apply. Table 9.1 gives the conductivities of some commonly encountered materials.

An approximate method of predicting temperature rise has been developed by the Armour Research Foundation.[2] The method uses empirical constants in the following formula:

$$\theta = K\left(\frac{W_{Cu}}{S_c}\right)^{0.8} \tag{9.20}$$

[2] *Ibid.*

TABLE 9.2 Radiation Properties of Common Surfaces

Emissivities to Infrared Radiation	
Polished aluminum	0.04
Polished brass	0.03
Polished copper	0.02
Sheet steel	0.55
Tinned sheet steel	0.04–0.06
Paint, all colors	0.92–0.96

Absorptivities to Solar Radiation	
Polished aluminum	0.26
Polished copper	0.26
Polished iron	0.45
Flat-black paint	0.98
Flat-white paint	0.2

where θ = average temperature rise of coil over air ambient, °C
$\quad\quad K$ = constant dependent on various conditions (values for which are given in Table 9.3)
$\quad W_{Cu}$ = copper loss, W
$\quad\quad S_c$ = exposed coil surface area, in^2

It is advisable to confirm predictions of temperature rise by this and other methods by actual tests.

9.2 TRANSIENT THERMAL RESPONSE

The increase in temperature of a magnetic device with time may be approximated with the following equation:

$$\theta = \theta_R\left(1 - \exp -\frac{t}{\psi}\right) \quad\quad (9.21)$$

in which θ is the temperature rise over ambient in degrees Celsius at time t, θ_R is the final value of the temperature rise in degrees Celsius, ψ is the thermal time constant, and t is time, both of the latter quantities being in the same units. If θ_R is known for rated conditions, then the steady-state temperature rise under an overload condition will be approximately

$$\theta_{ol} = \frac{q_{ol}}{q_R} \theta_R \tag{9.22}$$

where q_{ol} is the total loss in watts under that overload condition and q_R is the total loss in watts under rated conditions. By substituting Eq. (9.22) in Eq. (9.21), the temperature rise under overload conditions may be determined:

$$\theta = \frac{q_{ol}}{q_R} \theta_R \left(1 - \exp - \frac{t}{\psi} \right) \tag{9.23}$$

The time interval to reach rated temperature rise under the overload condition is given by

$$\Delta t = \psi \ln \frac{q_{ol}}{q_{ol} - q_R} \tag{9.24}$$

Temperature rise refers to the temperature of the coil and must be added to the initial stabilized coil temperature to obtain the transient coil-operating temperature.

The thermal time constant is

$$\psi = R_T C_T \tag{9.25}$$

in which R_T is the thermal resistance and C_T is the thermal capacitance, which is fully analogous with electrical circuits. The thermal resistance may be estimated by

$$R_T = \frac{\theta}{q_T} \tag{9.26}$$

in which θ is the average temperature rise in degrees Celsius and q_T is the total loss in watts. The thermal capacitance is the summation of the products of the weights and specific heats of all the materials in the device:

$$C_T = \sum_n H_n M_n \tag{9.27}$$

H_n is the specific heat of the nth material in wattseconds per pound $\cdot$ °C, and M_n is the weight of the nth material in pounds. These units leave the

TABLE 9.3 Constants for Calculation of Approximate Temperature Rise Using Eq. 9.20

Ambient Temperature, °C	Input Frequency, Hz	Construction								
		Open			Compound-Filled			Oil-Filled		
		Shell	Simple	Core	Shell	Simple	Core	Shell	Simple	Core
25	25	76	92	107	70	84	98	58	69	80
25	60	79	95	110	74	88	103	62	74	84
25	200	89	107	123	85	101	117	73	87	101
25	400	99	119	137	98	117	136	85	101	117
25	800	108	129	149	107	128	149	95	113	131
50	25	72	86	100	68	81	94	56	67	78
50	60	75	90	104	72	86	100	60	71	83
50	200	84	101	117	83	99	115	71	84	97
50	400	94	112	129	95	113	131	83	98	114
50	800	102	122	139	104	124	144	92	110	128
65	25	69	83	96	67	70	93	55	65	76
65	60	72	87	100	71	84	98	59	70	81

65	200	81	97	113	82	98	114	69	83	95
65	400	91	108	125	93	111	129	81	96	112
65	800	98	118	134	102	121	141	90	107	124
75	25	68	81	94	66	79	93	54	64	74
75	60	71	85	98	70	83	97	58	69	80
75	200	80	95	100	81	97	113	68	82	94
75	400	89	105	122	92	110	128	80	95	110
75	800	96	115	131	100	120	139	88	105	122
85	25	66	79	92	65	78	92	53	63	73
85	60	69	83	96	69	82	96	57	68	79
85	200	78	93	107	80	96	112	67	80	93
85	400	87	103	119	91	109	127	79	94	109
85	800	94	112	128	99	118	137	87	103	120
125	25	62	74	86	63	76	88	50	60	70
125	60	65	78	98	66	79	92	54	64	74
125	200	72	86	99	76	91	106	64	76	88
125	400	80	96	111	87	104	121	75	89	103
125	800	87	103	119	95	113	131	83	99	105

9.13

thermal time constant with the unit in seconds. The specific heats of some commonly used materials are given in Table 9.1.

9.3 ADIABATIC LOADING

When a large current flows in a winding for a short interval so that essentially no heat transfer occurs during the interval, all the energy dissipated by the winding resistances is absorbed by the wire whose temperature increases. The large current may safely flow until the insulation on the wire reaches the maximum allowable operating temperature. Under adiabatic conditions the following relationship exists:

$$I^2 R t = H_c \theta \tag{9.28}$$

where I = current flowing in wire
   R = resistance of wire
   t = time during which current flows
   H_c = thermal capacity of wire
   θ = temperature rise of wire

Since no heat transfer takes place, Eq. (9.28) may be generalized to show current density for any wire of unit length, provided R is expressed as resistivity in appropriate units. Equation (9.28), then, for copper wire in practical units becomes

$$\left(\frac{I}{CM}\right)^2 t = 4.163 \times 10^{-5}\theta \tag{9.29}$$

where I = current, A
   CM = wire cross section, circular mils
   t = time during which current flows, s
   θ = temperature rise of copper wire, °C

If the current is not constant, then its rms value during the time t should be used. The constant in Eq. (9.29) is the thermal capacitance of a copper wire of one circular mil cross section one foot long divided by the resistance per foot of that wire. The resistance used is for a temperature of 75°C (167°F), which is an average value for a winding initially at 20°C (68°F) that is heated to 130°C (266°F). Equation (9.29) is a useful relationship for determining safe operation under overload conditions in which circuit protection devices require a finite time to operate.

CHAPTER 10

Design Procedures

The design of transformers and inductors begins with the functional requirements contained in a user specification giving both mechanical and electrical requirements. The electrical requirements will usually be given in system notation. From this notation the properties of each equivalent circuit parameter that is significant to the application are determined. The complexity to which the equivalent circuit parameters must be pursued depends upon the application. In a simple power transformer operating over a narrow frequency range from a low-impedance source, the most significant parameter to consider is the open-circuit inductance. This inductance must support the highest voltage and lowest frequency to be experienced. The series resistance introduced by the windings is governed by temperature rise, size, economy, and regulation. Regulation may also require consideration of leakage inductance.

When source impedance and wideband frequency requirements are introduced, a much different situation is created. Typical is the need for reproducing a complex waveform. The circuits that generate complex waveforms often have finite-source impedances. It is then necessary to consider distributed capacitance operating in conjunction with leakage inductance, source impedance, and open-circuit inductance. Most transformers other than power transformers fall in this category. Refer to other chapters for specific applications. This chapter is devoted to converting circuit parameter and mechanical requirements into hardware designs using available materials and processes.

10.1 MECHANICAL DESIGN

With the detailed mechanical and electrical requirements settled, the first step is to determine the general mechanical construction. Severe environmental requirements are best met by hermetically sealed construction. Very high voltages usually require oil-filled construction. These constructions conflict with requirements for minimum size, weight, and cost. Moderate voltages with severe environmental requirements can be met with hermetically sealed construction and solid potting materials. Alternatives to hermetically sealed construction are the various open constructions. Open cast construction is useful for both high voltages and severe environmental conditions. Encapsulated construction is intermediate in value between cast and open varnished construction. Open varnished construction is the least expensive and least able to withstand severe environments and high-voltage stresses. See Sec. 7.1 for a further discussion of mechanical constructions.

10.2 ELECTRICAL DESIGN

From the electrical requirements the core type is chosen. See Secs. 6.4 and 7.2 for a discussion of core materials and geometry. A consideration in selecting core material is the range of sizes and shapes available. Limited availability of appropriate sizes frequently prevents the use of optimum core materials. Table 6.1 provides a summary of materials useful in design decisions. After the selection of construction and core material, the coil design can begin. From the selected construction and specific ambient termperature it is possible to estimate the core and coil ambient temperature. See Chap. 9 for a discussion of heat transfer and temperature rise. The insulation system must be selected. The maximum safe operating temperature of the insulation system less the component ambient is the allowable temperature rise, a principal determiner of size.

10.3 DETERMINATION OF CORE SIZE

The Armour Research Foundation has developed a method for making initial estimates of core size of power transformers from the allowable core and coil temperature rise, the frequency, operating flux density, and proportions of the core, using the following equation:

$$l = \left[\left(\frac{6.98 W_L}{f F_i B} \sqrt{\frac{bd}{e}} \right) \sqrt{\frac{\rho}{F_c(W_c/S_c)}} \right]^{2/7} \tag{10.1}$$

This expression is derived from Faraday's law and from geometric considerations in the core and coil. The terms in Eq. (10.1) are defined below. The characteristic linear dimension of the core is l, defined as

$$l = \sqrt[4]{A_c \times A_i} \tag{10.2}$$

in which A_c is the core window area in square inches, A_i is the cross-sectional area of the core in square inches, and l is in inches. The product $A_c \times A_i$ is known as the *core capacity*. W_L is the volt-ampere rating of the transformer in volt-ampere units; f is the lowest operating frequency in hertz; F_i is the core stacking factor. This factor is multiplied by the area occupied by the core cross section to obtain the actual mental area. This number varies between 0.8 and 0.95 and is usually supplied by the core manufacturer. B is the operating flux density in kilogauss. For minimum size this number should be as high as the magnetic material will allow, the limitation usually being the maximum acceptable exciting current. Factor b is defined by the following:

$$m_c = bl \tag{10.3}$$

in which m_c is the mean length of turn of the coil. Factor d is defined by

$$A_c = dl^2 \tag{10.4}$$

Factor e is defined by

$$S_c = el^2 \tag{10.5}$$

in which S_c is the exposed surface area of the coil in square inches. The square root term containing these factors is given a separate symbol:

$$K_o = \sqrt{\frac{bd}{e}} \tag{10.6}$$

The proportions of the core determine the value of K_o which is dimensionless. Values of K_o for several commonly used shapes are given in Table 10.1. The value of K_o for cores whose proportions vary widely from those in the table may be estimated from the proportions of cores used in other transformers of a design similar to the one under consideration. The term ρ is the resistivity of the conductor in microhm-inches. This value is about 0.9 for commercial copper wire at 100°C (212°F) and about 1.5 for aluminum wire. The factor F_c is the ratio of copper area to total core window area. This factor varies from as little as 0.05 for very high

TABLE 10.1 Constants for Estimating Core Size with Eq. (10.1)

Simple type, typical proportions

Scrapless EI shell type

Core type, average proportions

Constant	Simple Type, Typical Proportions	Scrapless EI Shell Type					Core Type, Average Proportions		
		S 1.0	1.5	2.0	2.5	3.0	1.0	1.5	2.25
$A_c \times A_i/L^4$	6.25	0.750	1.125	1.500	1.875	2.250	4.5	6.75	10.12
L/l	0.633	1.077	0.970	0.902	0.854	0.817	0.687	0.621	0.561
b	6.42	6.00	6.36	6.84	7.32	7.82	4.37	4.57	4.97
d	1.00	0.866	0.706	0.612	0.548	0.500	2.12	1.732	1.414
e	16.9	13.02	10.61	9.20	8.23	7.51	29.5	25.8	23.1
	0.616	0.620	0.640	0.675	0.606	0.730	0.560	0.554	0.552

10.4

voltage designs to 0.7 for high-current, low-voltage coils. For satisfactory results using this procedure, it is necessary to draw upon previous experience when estimating the value of F_c. W_c is the conductor loss in watts. The ratio W_c/S_c is determined from Eq. (9.20). The allowable core and coil temperature rise over its own ambient is needed to use that equation.

By the use of Eq. 9 in Table 1.1, a variation of Eq. (10.1) may be obtained which gives the value of l from the I^2L of an inductor:

$$l = \left[\left(\frac{15.5 I^2 L}{BF_i} \sqrt{\frac{bd}{e}} \right) \sqrt{\frac{\rho}{F_c(W_c/S_c)}} \right]^{2/7} \qquad (10.7)$$

In Eq. (10.7) I is in amperes and L is in henrys. All other terms have the same meaning and dimensions as in Eq. (10.1). The value of l for power transformers may be obtained from Eq. (10.1) and for inductors from Eq. (10.7). From l and Eq. (10.2) the required core capacity can be determined.

10.4   TERTIARY REQUIREMENTS

Equations (10.1) and (10.7) provide a means of estimating the size of temperature rise limited power transformers and inductors. They are a starting point for the design. They do not provide for a variety of additional requirements which commonly form a part of specifications. Regulation and short-circuit current specifications affect winding resistance and leakage inductance. Leakage inductance is sometimes an explicit specification requirement. Frequency response and exciting current specifications force an increase in open-circuit inductance. A high-efficiency specification will force an increase in both core area and conductor size. Direct current magnetization applied to the core requires changes in the core structure. A requirement for low capacitance places additional restrictions on minimum spacing between electrodes. Saturable devices have complex requirements for operating flux densities. They usually have tertiary windings which occupy a sizable percentage of the window area. The high core losses associated with saturation cause the thermal gradients in saturating devices to be different from linear devices, and so the experimental constants used for heat transfer calculations in linear devices do not apply. The consideration of these additional requirements invokes the need of the iterative process in core and coil design in which successive approximations are tested for all requirements until an acceptable design is obtained.

10.5 CALCULATING LEAKAGE INDUCTANCE

Leakage inductance is a function of the dimensions of the coil and the total number of turns in the winding. Figure 10.1a shows a cross section of a concentric two-winding coil with controlling dimensions labeled. The path of the leakage flux is shown by dotted lines. In Fig. 10.1a–c P is the mean perimeter of the insulation between windings in centimeters. N is the number of turns in the winding to which the leakage inductance is referred. Other dimensions are as indicated in the figures. The formulas for leakage inductance given in Fig. 10.1a–c are based on approximate calculations of the energy stored in the leakage magnetic field between and through the windings. Note that the permeability of the core does not appear in the formulas. The leakage inductance of any coil configuration is reduced by the term $[(n - 1)/n]^2$ when an autotransformer connection is used, where n is the step-up voltage ratio of the autotransformer.

10.6 THE DETERMINATION OF CAPACITANCE

The geometric capacitance between adjacent windings and windings to core in which the separation between electrodes is small compared with

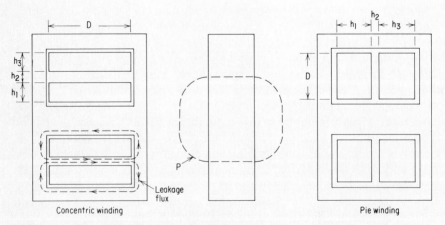

Concentric winding   Pie winding

FIG. 10.1a Controlling dimensions for single concentric and pie windings without interleaving. Approximate leakage inductance formula:

$$L_L = \frac{4\pi N^2 P}{D}\left(\frac{h_1 + h_3}{3} + h_2\right) \times 10^{-9}$$

Dimensions are in centimeters; L_L is in henrys.

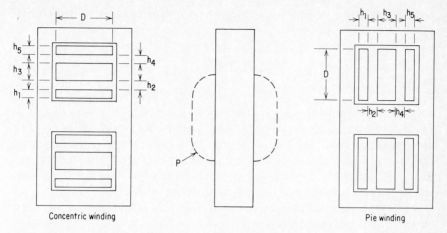

Concentric winding   Pie winding

FIG. 10.1b Controlling dimensions for single concentric and pie windings with interleaving. Approximate leakage inductance formula:

$$L_L = \frac{\pi N^2 P}{D} \left(\frac{h_1 + h_3 + h_5}{3} + h_2 + h_4 \right) \times 10^{-9}$$

Dimensions are in centimeters; L_L is in henrys.

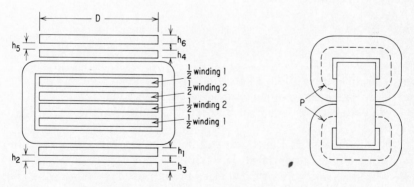

FIG. 10.1c Controlling dimensions for two concentric coils on a single core. Approximate leakage inductance formula:

$$L_L = \frac{\pi N^2 P}{D} \left(\frac{h_1 + h_3 + h_4 + h_6}{3} + h_2 + h_5 \right) \times 10^{-9}$$

Dimensions are in centimeters; L_L is in henrys. Leakage inductance is unaffected by any series or parallel connection of winding halves.

10.7

the dimensions of the surface area is given approximately by the parallel-plate capacitance formula:

$$C_G = \frac{0.225A\epsilon}{d} \quad \text{pF} \tag{10.8}$$

In this formula A is in square inches, ϵ is the dimensionless dielectric constant, and d is the distance between parallel electrodes in inches. For concentric surfaces, the area is approximately the mean perimeter multiplied by the winding length. if the parallel plates are equipotential surfaces, the geometric capacitance is the effective capacitance at the potential difference between the two surfaces. If the potential difference varies linearly along the winding length as it does in a solenoid winding, then the effective capacitance is given by

$$C_{\text{eff}} = C_G \frac{E_1^2 + E_1 E_2 + E_2^2}{3E^2} \tag{10.9}$$

where C_G = geometric capacitance between adjacent surfaces
E_1 and E_2 = potential differences at two ends of winding
E = potential across winding to which effective capacitance is referred

In a reciprocal solenoid winding consisting of two or more layers, the effective capacitance is given by

$$C_{\text{eff}} = \frac{4C_G}{3}\left(\frac{N_L - 1}{N_L^2}\right) \tag{10.10}$$

where N_L = number of layers in winding
C_G = average geometric capacitance between layers
C_{eff} = effective capacitance referred to same winding

In some low-capacitance transformers, the low-capacitance winding constitutes an equipotential surface with a large air space between it and other windings and core. A drawing of this construction is shown in Fig. 10.2. The parallel-plate capacitance formula is not applicable to this construction; instead the following empirical formula has been found useful:

$$C = \frac{1.35 m_c K}{\ln P_w/P_s} \quad \text{pF} \tag{10.11}$$

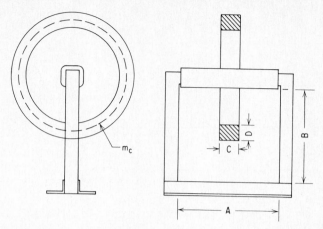

FIG. 10.2 Low-capacitance transformer construction with dimensions for calculating capacitance defined:

$$p_w = 2(A + B) \text{ and } p_s = 2(C + D).$$

In Eq. (10.11) m_c is the mean length of turn in inches of the low-capacitance winding, and K is a correction factor which compensates for the existence of insulation between the low-capacitance winding and ground which has a dielectric constant greater than 1. The correction factor typically has a value of about 1.2. P_w and P_s are the perimeters in any consistent units of the window and low-capacitance winding as defined in Fig. 10.2. This formula was proposed by the Armour Research Foundation in their final report under U.S. government contract No. DA-36-039 SC-52656.

10.7 DETAILED CORE AND COIL DESIGN

Core and coil design is facilitated by the use of design calculation sheets. A typical one is illustrated in Fig. 10.3. From the estimate of core capacity and the electrical requirements, a tentative selection of the core and core material is made. This choice from among many options available largely determines cost and performance. Unsatisfactory design approximations may force a revision of this choice. Information on dimensions and characteristics of cores and core material is best obtained directly from suppliers, and this information should be current. New materials and processes are regularly introduced and old ones dropped. Some items are available from only a single source. Similar items from several sources

WDG VOLTS	TURNS	WIRE SIZE DIA / T/IN	Ω/KFT / LB/KFT	AMPS OM/AMP	MARGIN VOLTS/MIL	T/L WDG LGTH	LAYER INS V/MIL	WRAP V/MIL	BUILD UP CL .031 / WF .040	Δ	MLT FT	R ___ °C	I^2R	CU WT	MLT P = .354
1	1005.	H30.	100.	.26	.125	95.	.002	4L.005	.131	.071	.452	47.2	3.19		
220		.0119	104				KP	KP	.020	.151	454.			.14	
		80.	.318	416.		11.	21V/M	50V/M	.020	.020					
2	26	HI4.	4110.	10.	.25	1.187	.010	4L.005	.136	.242	.564		3.70	.19	
		.0682	2.52			13.	KP	KP	.010	.146	14.7	.037			
5		13.9	12.6	411.	4V/M	2.	0.5V/M	50V/M	.020	.020					
3						.937			.408	.408			6.89		
													X1.41		
4													9.71	WATTS	
5															
6															

TOTAL .408

.500 = 81.6% BUILD UP

CORE E1-1 X 1 1/8   WINDOW 1/2 X 1 1/2

MTL M-19   GA 26

WF .1062 X 1.187 X 1.437 LG. $P_{FT} = \dfrac{STACK + TONGUE}{6} = \dfrac{2.12}{6}$

AREA 1 X 1 1/8   IN² 6.89 CM²

WT 1.74 LB $I_{EX} = \dfrac{V\,A_{EX}}{V_{PRI}}$

WATTS/LB 1.6 $= \dfrac{15.7}{220} = 0.07$ AMPS

CORE LOSS 2.8

VA/LB 9

VA EX 15.7

NOTES

MLT: $P + 2/3\,\Delta_1 + 1/3\,\Delta_2 =$

FIG. 10.3 Sample design sheet used with illustrative problem in Sec. 10.8.

may not be interchangeable and often vary in magnetic properties. Some items have limited availability.

The tentative selection of the core gives core cross section and window dimensions. The choice of core material places the ceiling on the operating flux density. A choice often made for flux density is a value just below the knee of the magnetization curve. With the chosen value of flux density, the core area, and the electrical specifications, the appropriate

Fig. 10.3 (*Continued*)

Flux Density:

$$B = \frac{E \times 10^5}{4.44NAf} = \frac{220 \times 10^5}{4.44 \times 1005 \times 6.89 \times 50} = 14.3 \text{ kg}$$

Regulation:

$$N_2 = N_1 \frac{(V_2 + I_2R_2)}{(V_1 - I_1R_1)} = 1005\left(\frac{5.0 + 10 \times 0.037}{220 - 0.26 \times 47.2}\right) = 26$$

Temperature Rise:

$$\Delta T = K\left(\frac{W_c}{S_c}\right)^{0.8} = 73\left(\frac{9.71}{11.3}\right)^{0.8} = 65°C$$

$$\begin{array}{r} 65 \\ +60 \\ \hline 125°C \end{array} = \text{average copper temperature}$$

Leakage Inductance:

$$L_L = \frac{4\pi N^2 P}{D}\left(\frac{h_1 + h_3}{3} + h_2\right) \times 10^{-9}$$

$$= \frac{4\pi(26)^2 \times 0.5 \times 12}{0.937}\left(\frac{0.151 + 0.146}{3} + 0.020\right) \times 2.54 \times 10^{-9}$$

$$= 16 \times 10^{-6}$$

$$\omega L_L = 377 \times 16 \times 10^{-6} = 6 \times 10^{-3} \qquad R_L = 0.5 \qquad \omega L_L \ll R_L \text{ neglect } L_L$$

Notes:

Increase in R with temperature

$$\left(\frac{234.5 + 125}{234.5 + 20}\right) = 1.41$$

$$Fc = \frac{cm \times \pi \times 10^{-6}}{4Aw} = \frac{(100 \times 1005 + 4110 \times 26)\pi \times 10^{-6}}{4 \times 0.75} = 0.217$$

formula in Table 1.1 is used to determine the required number of turns. For transformers the primary voltage magnitude and frequency or pulse width are used in one of the various forms of Faraday's law. For inductors the flux density–current relationship is used. The approximate number of turns required in all other windings is obtained from the voltage ratios.

A tentative selection of wire sizes is required. This is done by estimating the allowable current density in the conductors. The allowable current density is nearly the same in all the windings. A large increase in current density of any single winding over other windings in the same coil assembly leads to hot spots in the coil. The estimated allowable current density may be based on previous designs. It may be estimated from the following design formula developed by the Armour Research Foundation:

$$\frac{CM}{\text{amp}} = \sqrt{\frac{F_c l \rho}{W_c / S_c}} \sqrt{\frac{bd}{e} \frac{4000}{\pi}} \qquad (10.12)$$

This formula is related to Eq. (10.1), and the contained terms have the same meanings in both equations. The newly introduced symbol CM represents conductor cross-sectional area in circular mils. The rms currents in amperes flowing in each winding are determined from the specifications and the methods described in Sec. 1.7. These currents multiplied by the estimated allowable current density in circular mils per ampere give the required circular mils for each conductor. From the wire tables (Tables 10.2 to 10.4) sizes closest to the required circular mils are chosen for each winding. Designers accustomed to selecting hookup wire for equipment may consider the current density requirements of magnetic components too conservative. The lower current densities are necessitated by the unfavorable ratio of heat dissipated per unit area over heat generated per unit volume in compact coil assemblies compared with straight runs of wire.

The winding construction is usually suggested by the selection of the core. See Sec. 7.3 for a discussion of winding construction. Bobbin construction is used in small devices when available and the voltage stresses are low. The common alternative to bobbin construction is the layer-wound coil. If a bobbin is used, the available winding area is determined from the bobbin construction data. The number of turns for each winding and the wire size for each winding having been estimated, the wire table is now used to determine the turns per unit area for each wire size. The turns per unit area for each wire size divided into the number of turns required of that wire size gives the area occupied by each winding. The sum of the areas occupied by each winding is the total area occupied by all the windings. This figure is compared with the available area in the bobbin. Several factors prevent full utilization of the bobbin window area. Most wire tables are based on optimum conditions which seldom exist. In small bobbins leads and insulation over leads may constitute an indeterminate but sizable percentage of the available winding area. These leads and insulation interfere with the lay of the wire, reducing winding area utilization. Insulation between windings, usually pressure-sensitive tape, has the same effect. The skill of the winding operator also affects utilization efficiency. Area utilization will be about 70 percent depending on conditions and the conservatism of the wire table.

Greater precision in predicting winding area utilization is possible in layer-wound coils than in bobbin-wound coils. The turns of the windings are positioned in a controlled and repeatable manner. Most layer-wound coils are used with rectangular cores which allow a portion of their coils to be unrestricted by the core window. This region of layer-wound coils is used to locate leads, sheet insulation overlaps, and anchor tapes, leaving the portion of the coil in the core window unencumbered by these indeterminate and extraneous factors that increase coil size. A layer-wound coil is illustrated in Fig. 7.27. The foundation for a layer-wound coil is a rectangular winding tube. Laminated from sheet insulation for electrical and mechanical strength, the tube is supported during winding on a rectangular mandrel. The solenoid windings are placed on the winding tube in layers reciprocally wound with sheet insulation between each layer and between windings. The winding tube is cut to length slightly less than the core window length. The width of the subsequent layers of sheet insulation is made equal to the winding tube length which determines the coil length. The winding length of a layer of wire is equal to the coil length less a margin at each end of the coil. To prevent collapse of the end turns, the winding length of any winding is made equal to or less than the winding beneath. The minimum margins and minimum layer insulation are functions of the wire size and winding shop practice. The margin and sheet insulation data given in the wire tables are conservative. These minima must be increased for high voltage. Because of the reciprocating construction, the voltage between layers varies from zero to twice the voltage per layer. The voltage per layer is equal to the volts per turn multiplied by the number of turns in the layer. A commonly used rule of thumb for determining layer insulation thickness is 100 V/mil at the end of maximum stress. For margins 25 V/mil is regularly used. These stress levels are approximate. Allowable dielectric stresses are functions of many variables.

The wire tables give the number of turns per inch, the layer insulation, the minimum margins, and the diameter for each wire size. From

TABLE 10.2 Heavy-Film-Insulated Round Copper Magnet Wire*

AWG Size	Maximum Diameter over Insulation	Circular Mils (CM)	Pounds per 1000 ft	Ohms per 1000 ft at 20°C (68°F)	Turns per Inch	Minimum Layer Insulation Thickness, in	Turns per Square Inch, Random Wind	Turns per Square Inch, Layer Wind	Minimum Margin, in
4	0.2098	41,740	127	0.2485	4.53	0.020	20.4	17.6	0.312
5	0.1872	33,090	101	0.3134	5.07	0.020	25.7	21.9	0.312
6	0.1671	26,240	80	0.3952	5.68	0.020	35.2	27.2	0.312
7	0.1491	20,820	63.5	0.4981	6.37	0.010	40.5	35.8	0.312
8	0.1332	16,510	50.4	0.6281	7.13	0.010	50.7	44.6	0.250
9	0.1189	13,090	39.9	0.7925	7.99	0.010	63.7	58.7	0.250
10	0.1061	10,380	31.7	0.9988	8.95	0.010	79.9	69.0	0.250
11	0.0948	8,230	25.16	1.26	10.02	0.010	107.3	85.5	0.250
12	0.0847	6,530	20.03	1.59	11.2	0.010	134.2	105.9	0.250
13	0.0757	5,180	15.89	2.00	12.5	0.010	168	131.0	0.250
14	0.0682	4,110	12.60	2.52	13.9	0.010	207	159.3	0.250
15	0.0609	3,260	10.04	3.18	15.6	0.007	264	205	0.250
16	0.0545	2,580	7.95	4.02	17.4	0.007	324	253	0.250
17	0.0488	2,050	6.33	5.05	19.5	0.007	404	312	0.250
18	0.0437	1,620	5.03	6.39	21.7	0.005	504	399	0.250
19	0.0391	1,290	3.99	8.05	24.2	0.005	629	492	0.187
20	0.0351	1,020	3.18	10.1	27.0	0.005	781	603	0.187
21	0.0314	812	2.53	12.8	30.3	0.005	976	743	0.187
22	0.0281	640	2.00	16.2	33.9	0.005	1,219	972	0.187

Gauge									
23	0.0253	511	1.60	20.3	37.8	0.003	1,504	1,187	0.156
24	0.0227	404	1.26	25.7	42.0	0.003	1,868	1,457	0.156
25	0.0203	320	1.00	32.4	46.9	0.003	2,336	1,797	0.156
26	0.0182	253	0.794	41.0	52.2	0.003	2,907	2,202	0.156
27	0.0164	202	0.634	51.4	57.9	0.003	3,580	2,671	0.156
28	0.0147	159	0.502	65.3	64.6	0.002	4,456	3,462	0.125
29	0.0133	128	0.405	81.2	71.4	0.002	5,444	4,177	0.125
30	0.0119	100	0.318	104	79.8	0.002	6,800	5,138	0.125
31	0.0108	79.2	0.253	131	88.0	0.002	8,256	6,148	0.125
32	0.0098	64.0	0.205	162	97.5	0.001	10,027	8,031	0.125
33	0.0088	50.4	0.162	206	108	0.001	12,400	9,850	0.125
34	0.0078	39.7	0.127	261	122	0.001	15,800	12,400	0.125
35	0.0070	31.4	0101	331	136	0.001	19,600	15,100	0.125
36	0.0063	25.0	0.0805	415	152	0.001	24,200	18,400	0.125
37	0.0057	20.2	0.0655	512	168	0.001	29,600	22,200	0.125
38	0.0051	16.0	0.0518	648	186	0.0075	37,000	27,300	0.094
39	0.0045	12.2	0.0397	847	212	0.0075	47,500	35,900	0.094
40	0.0040	9.61	0.0312	1080	238	0.0075	60,100	44,700	0.094
41	0.0036	7.84	0.0254	1320	264	0.0075	74,300	54,200	0.094
42	0.0032	6.25	0.0203	1660	297	0.0075	94,000	67,200	0.094
43	0.0029	4.84	0.0158	2140	327	0.0075	114,000	80,300	0.094
44	0.0027	4.00	0.0131	2590	352	0.0075	132,100	91,200	0.094

*Multiply ohms per 1000 ft for copper wire by 1.61 to obtain ohms per 1000 ft of aluminum in same wire gauge. Multiply pounds per 1000 ft for copper wire by 0.304 to obtain pounds per 1000 ft of aluminum in same wire gauge.

TABLE 10.3 Single-Film-Insulated Round Copper Magnet Wire*

AWG Size	Maximum Diameter over Insulation	Circular Mils (CM)	Pounds per 1000 ft	Ohms per 1000 ft at 20°C (68°F)	Turns per Inch	Minimum Layer Insulation Thickness, in	Turns per Square Inch, Random Wind	Turns per Square Inch, Layer Wind	Minimum Margin, in
14	0.0666	4110	12.6	2.52	14.3	0.010	217	166.6	0.250
15	0.0594	3260	9.96	3.18	16.0	0.007	273	215.5	0.250
16	0.0531	2580	7.89	4.02	17.9	0.007	341	267	0.250
17	0.0475	2050	6.28	5.05	20.0	0.007	426	328	0.250
18	0.0424	1620	4.98	6.39	22.4	0.005	535	423	0.250
19	0.0379	1290	3.95	8.05	25.1	0.005	670	523	0.187
20	0.0351	1020	3.14	10.1	27.1	0.005	781	604	0.187
21	0.0303	812	2.50	12.8	31.3	0.005	1,049	795	0.187
22	0.0270	640	1.97	16.2	35.2	0.005	1,320	984	0.187
23	0.0243	511	1.58	20.3	39.1	0.003	1,630	1,263	0.156
24	0.0217	404	1.24	25.7	43.8	0.003	2,045	1,586	0.156
25	0.0194	320	0.989	32.4	48.9	0.003	2,558	1,858	0.156
26	0.0173	253	0.781	41.0	54.9	0.003	3,217	2,420	0.156
27	0.0156	202	0.623	51.4	60.9	0.003	3,924	2,929	0.156

Gauge								
28	0.0140	159	0.493	65.3	67.8	0.002	4,913	3,795	0.125
29	0.0126	128	0.397	81.2	75.4	0.002	6,065	4,620	0.125
30	0.0112	100	0.211	104	84.8	0.002	7,676	5,749	0.125
31	0.0100	79.2	0.246	131	95.0	0.002	9,630	7,083	0.125
32	0.0091	64.0	0.200	162	104	0.001	11,630	9,248	0.125
33	0.0081	50.4	0.157	206	117	0.001	14,680	11,530	0.125
34	0.0072	39.7	0.124	261	132	0.001	18,580	14,400	0.125
35	0.0064	31.4	0.0978	331	148	0.001	23,500	17,950	0.125
36	0.0058	25.0	0.0783	415	164	0.001	28,600	21,550	0.125
37	0.0052	20.2	0.0632	512	183	0.001	35,600	26,360	0.125
38	0.0047	16.0	0.0501	648	202	0.00075	43,600	33,180	0.125
39	0.0041	12.2	0.0382	847	232	0.00075	57,280	42,750	0.125
40	0.0037	9.61	0.0301	1080	256	0.00075	70,340	51,600	0.125
41	0.0033	7.84	0.0245	1320	288	0.00075	88,400	63,600	0.125
42	0.003	6.25	0.0196	1660	317	0.00075	107,000	75,560	0.125
43	0.0026	4.84	0.0152	2140	365	0.00075	142,460	97,590	0.125
44	0.0024	4.00	0.0125	2590	396	0.00075	167,200	112,400	0.125

* Multiply ohms per 1000 ft for copper wire by 1.61 to obtain ohms per 1000 ft of aluminum in same wire gauge. Multiply pounds per 1000 ft for copper wire by 0.304 to obtain pounds per 1000 ft of aluminum in same wire gauge.

TABLE 10.4 Heavy-Film-Insulated Square Copper Magnet Wire*

AWG Size	Maximum Diameter over Insulation	Circular Mils (CM)	Pounds per 1000 ft	Ohms per 1000 ft at 20°C (68°F)	Turns per Inch	Minimum Layer Insulation Thickness, in	Turns per Square Inch, Random Wind	Turns per Square Inch, Layer Wind	Minimum Margin, in
0	0.3329	132,700	403.1	0.07815	2.85	0.020	8.1	7.23	0.375
1	0.2972	104,800	318.4	0.09895	3.19	0.020	10.1	9.01	0.375
2	0.2652	182,740	251.4	0.1253	3.58	0.020	12.8	11.2	0.375
3	0.2367	65,250	198.4	0.1589	4.01	0.020	16.1	14.8	0.312
4	0.2113	51,400	156.4	0.2018	4.50	0.020	20.1	17.4	0.312
5	0.1887	40,380	122.9	0.2568	5.03	0.020	25.3	21.6	0.312
6	0.1686	32,300	98.39	0.3211	5.63	0.020	31.7	26.7	0.312
7	0.1507	25,390	77.42	0.4085	6.30	0.010	39.6	35.1	0.312
8	0.1348	19,940	60.74	0.5212	7.05	0.010	49.5	43.5	0.250
9	0.1205	15,920	48.66	0.6514	7.88	0.010	61.9	54.1	0.250
10	0.1079	12,480	38.19	0.8310	8.80	0.010	77.3	66.8	0.250
11	0.0967	10,040	30.74	1.033	9.82	0.010	96.2	82.3	0.250
12	0.0868	7,880	24.16	1.316	10.9	0.010	119.4	101.2	0.250
13	0.0780	6,320	19.41	1.641	12.1	0.010	147.9	123.8	0.250
14	0.0701	4,950	15.24	2.095	13.5	0.010	183.1	151.4	0.250

*Multiply ohms per 1000 ft for copper wire by 1.61 to obtain ohms per 1000 ft of aluminum in same wire gauge. Multiply pounds per 1000 ft for copper wire by 0.304 to obtain pounds per 1000 ft of aluminum in same wire gauge.

these data the number of turns per layer and the number of layers for each winding are determined. The insulation between windings and over the last winding is determined by dielectric and mechanical requirements. The sum of all these numbers is the coil buildup. The allowable coil buildup is about 85 percent of the core window height. This percentage is partly dependent upon the absolute value of window height and on shop practice. Small coils, coils with many turns of fine wire, and coils with many leads have relatively low window area utilization efficiency.

The first try at the coil design may be a poor fit in the core window. If the coil buildup is too great, the core size can be increased or the conductor current densities can be increased by decreasing the sizes of the conductors. Increasing the core size may mean increasing the core area, increasing the window area, or increasing both. Increasing the core area allows a reduction in the number of turns in all the windings. If the coil buildup is too small for the core window, the core can be reduced or the conductor current densities can be reduced by increasing the size of the conductors in all the windings. The direction to take for this next approximation will depend upon an estimate of whether the temperature rise is high or low. The temperature rise will vary directly with the number of turns and inversely with the current density. This process will converge to the point where the design appears sufficiently reasonable to justify a temperature rise calculation. The winding resistances are determined by calculating the mean length of turn (MLT) of each winding, multiplying the MLT by the number of turns to find the wire length, finally multiplying the length by the resistance per unit length. The resistance is multiplied by the square of the current for each winding to obtain the conductor loss for each winding. The sum of these losses is multiplied by a factor of approximately 1.3 to compensate for the increase in resistance with temperature. A design sheet simplifies the calculations. The design calculation sheet of Fig. 10.3 contains a series of rows and columns. Each set of rows contains the data for one winding. The column headings describe the data. Two columns are used to calculate the buildup. In the left column are entered all the items in sequence that contribute to the buildup: first the winding tube clearance and winding tube thickness, next the wire diameter multiplied by the number of layers followed by the layer insulation thickness multiplied by 1 less than the number of layers, finally the top insulation. This process is repeated for each winding. The sum of these numbers is entered at the bottom of the column. This addition is repeated in the right-hand column in a different form that simplifies the calculation of mean length of turn. The sum of the clearance and winding tube thickness is entered in the top row of the first winding. The sum of the wire and layer insulation buildup for that winding is entered in the second row. The top insulation thickness is entered in the third row. The sum of these three numbers is entered in the first row of the following winding. This process is continued through all the windings, the final sum being the same as obtained in the left-hand column. The perimeter of the core alone in feet is calculated by dividing the sum of the two core cross-section dimensions by 6. The mean length of turn of any winding in feet is obtained from the sum of the core perimeter in feet plus two-thirds of the total previous buildup plus one-third of the buildup of the present winding. This method of calculating the MLT assumes a rectangular form with no bulge. The departure of the turn from rectangular form and the occurrence of bulging are compensating effects making this method fairly accurate. Further columns on the design sheet provide space for extending the MLT to feet, resistance, and power loss. The MLT is also useful in calculating the distributed capacitance and leakage inductance.

With the size and shape of the core and coil tentatively established, the exposed coil area may be calculated. This is done most accurately from the dimensions of the coil. Equation (10.5) and Table 10.1 are useful for approximations. The total conductor loss at operating temperature divided by the exposed coil area gives the wattage loss per square inch. This ratio and the temperature rise constant from Table 9.3 may be used in Eq. (9.20) to obtain the temperature rise of the conductor over the core and coil ambient. The conductor and core sizes are adjusted to obtain the estimated allowable core and coil temperature rise, and the process is repeated.

The core loss is determined from data supplied by the core manufacturer. This is usually given in watts per pound for various flux densities and frequencies in the form of curves. Similar curves give the exciting volt-amperes per pound from which the exciting current may be determined. The core loss per pound is multiplied by the weight of the core to determine the total core loss. Ideally the core loss and conductor loss are equal. This may be used as an approximate test of the design. Conflicting requirements frequently prevent this relationship from being achieved.

With the core and copper loss known, a more accurate estimate of the temperature rise is possible. The approximate size of the case, when required, can be determined, and the temperature differential between component ambient and coil ambient can be estimated by the methods discussed in Sec. 9.1. If no case is required, the coil ambient is the same as the component ambient. If the temperature drop through the potting material and case wall to component ambient plus the coil temperature rise differs from the allowable total temperature rise, a new approximate core and coil design is required.

After the first or second approximation, the design should be tested for the additional requirements. The elements of the equivalent circuit that relate to those requirements should be determined, the adjustments made as needed. If there is an inductance requirement, Eq. (1.20) may be used to predict the inductance of the approximate design. If the predicted inductance is low, Eq. (1.20) shows that an increase may be obtained by increasing turns, permeability, core area, or a combination of those factors. The inductance requirement may force a change in core or core material. Changes made to increase inductance frequently cause a decrease in flux density, and so the core loss can be expected to be relatively low. Capacitance between windings and windings to ground [Eq. (10.8)] can be decreased by increasing the spacing between electrodes and by changing to a dielectric material with a lower dielectric constant. The effective capacitance across a winding [Eq. (10.10)] may be reduced by increasing the number of layers and reducing the winding length of each layer. Leakage inductance (Fig. 10.1) may be reduced by reducing the spacing between windings, increasing the winding length, and reducing the number of layers. Reducing the number of turns will also reduce leakage inductance. For a given core capacity, both the leakage inductance and the open-circuit inductance are decreased by decreasing the ratio of window to core area. Attempts to improve high-frequency performance will generally hurt low-frequency performance and vice versa. Winding resistance is also reduced by reducing the ratio of window to core area. The weight of the conductor may be determined by multiplying the weight per unit length, obtainable from the wire tables, by the length of conductor for each winding. The weight of the core and coil is very nearly the sum of the weights of the conductors and the weight of the core.

Excessive size or weight may be reduced by choosing an alternative core material that permits operation at higher flux densities or which has higher permeability. Current densities may be increased and the core window reduced if the higher temperature rise is accommodated by using high-temperature insulating materials.

Once a satisfactory design has been obtained, an adjustment must be made in the turns for regulation. A sample regulation calculation is given in Sec. 3.3. The turns ratio must be adjusted to compensate for the drop through the series circuit elements. If the device is an inductor, the permeability calaculations shown in the illustrative problem of Sec. 1.6 must be made. The winding resistance of inductors is often critical. Lowering resistance and increasing inductance are conflicting requirements. Resistance can be controlled in production more closely than inductance. Failure to provide a reasonable margin of safety for inductance can create difficult production problems with what should be a simple component.

10.8 STEP-BY-STEP DESIGN PROCEDURE

Given below is the step-by-step summary of a transformer design procedure. Detailed information may be found in the sections listed next to
various items. Following this or any other transformer design procedure
is greatly facilitated by using a design calculation sheet. See Fig. 10.3.

1. Assemble specification data (Sec. 18.2).
   a. Primary voltage, frequency, waveform, duty cycle, repetition rate,
      pulse width
   b. Secondary voltage
   c. Secondary rms current (Sec. 1.7)
   d. Externally applied noninduced voltages
   e. Equivalent circuit parameters (Chaps. 2 to 5)
      Winding resistances
      Leakage inductance
      Distributed capacitance
      Shunt inductance
      Shunt resistance
   f. Evironmental conditions
      Temperature extremes
      Humidity
      Shock and vibration
   g. Maximum dimensions and weight
2. Select construction to be used (Sec. 7.1).
3. Select magnetic material for core (Chap. 6).
4. Select insulation system (Chap. 8).
5. Estimate the allowable core and coil temperature rise.
6. Calculate the allowable winding loss per unit area of exposed coil surface (Chap. 9).
7. Estimate core capacity required (Sec. 10.2).
8. Determine availability of core in material and size required (Secs. 6.4
   and 7.2).
9. Make tentative selection of core.
10. Estimate maximum operating flux density (Sec. 6.4).
11. Calculate required number of primary turns (Sec. 1.5).
12. Calculate the required number of secondary turns from the number
    of primary turns and the secondary-to-primary voltage ratio (Sec.
    2.1).
13. Estimate the required current density for both primary and second-
    ary windings (Sec. 10.7).
14. Calculate the primary current from the secondary current and the
    secondary-to-primary voltage ratio (Sec. 2.1).

15. Select the primary and secondary wire sizes from the known currents and required current densities.
16. Select the coil construction to be used (Sec. 7.3).
17. Determine the insulation required between windings, between windings and ground including margins, and between layers of windings (Sec. 7.3 and Chap. 8).
18. Determine the winding length of the core selected in step 9 and the required margins in step 17.
19. Determine the buildup of the windings, the interwinding insulation, and the insulation to ground.
20. Compare calculated buildup to available window height of core selected in step 9. In most cases the calculated buildup should be about 85 percent of the available window height. If the buildup is grossly off, go back to step 9 with a new core selection and repeat all subsequent steps.
21. Calculate required equivalent circuit parameters.
    Winding resistance (Sec. 10.7)
    Leakage inductance (Sec. 10.5)
    Distributed capacitance (Sec. 10.6)
    Shunt inductance (Sec. 1.6)
    Shunt resistance (Sec. 6.2)   → *calculate core loss, copper loss & exciting current*
22. Calculate temperature rise (Sec. 9.1).
23. Compare size and weight of core and coil to maximum size and weight requirements.
24. Go back to step 9 or step 4 if necessary to meet all requirements.
25. Adjust turns so that correct full-load voltage is obtained (Sec. 3.2).
26. Revise and check all calculations. Compare all calculated values to requirements.

Illustrative problem

Design the core and coil of a transformer to provide 5 V rms at 10 A rms with the 5-V winding at 1000 V above ground. The maximum ambient temperature will be 60°C (140°F). The primary will be 220 V 50 Hz. Design for ease of manufacture and low cost.

Solution

Choose open-type construction with a paper section coil impregnated with a high-temperature varnish, since there are no severe environmental requirements and this type of construction can be safely used at 1000 V. For the core choose the scrapless series of laminations because they provide low cost and ready availability. Use M-19 grade silicon steel for low cost. The cost of stacking is also less than for better grades of silicon steel because this grade is of heavier-gauge material (26 gauge). If material cost were not a first consideration, stocking convenience or availability might suggest other grades of silicon steel.

Choose an insulation system with an operating temperature of 130°C (266°F). This temperature instead of the frequently used 105°C (221°F) rating still permits the use of easily heat strippable magnet wire, a great convenience in lead dressing. Kraft paper impregnated with high-temperature varnish is also acceptable for use at 130°C. There is little cost penalty from using these materials over materials rated for 105°C operating temperature, and there is cost and size saving when the higher operating temperature is utilized. Operating temperatures above 130°C preclude easily heat strippable magnet wire and require more costly sheet insulation.

The above choices are the basic design decisions. They are of concern to user as well as designer because they are the major determiners of cost and quality.

The lamination size may be estimated by using Eq. (10.1). To enter that equation, the following data are needed:

$W_L = 5 \times 10 = 50$ VA $\qquad$ given

$f = 50$ Hz $\qquad$ given

$F_i = 0.95$ $\qquad$ given lamination supplier

$B = 15$ kG $\qquad$ from magnetization curve

$\rho = 0.9$ $\qquad$ resistivity of copper, a material property

$F_c = 0.4$ $\qquad$ ratio of copper to window area, estimated

$\dfrac{W_c}{S_c} = \left(\dfrac{\Delta T}{K}\right)^{1.25} = \left(\dfrac{70}{75}\right)^{1.25} = 0.92$ $\qquad$ from given data, Table 9.3, and Eq. (9.20)

$K_o = \sqrt{\dfrac{bd}{e}} = 0.65$ $\qquad$ from Table 10.1

Substituting these data in Eq. (10.1) yields

$$l = \left(\frac{6.98 \times 50}{50 \times 0.95 \times 15} \, 0.65 \, \sqrt{\frac{0.9}{0.4 \times 0.92}}\right)^{2/7} = 0.82$$

The current density may be estimated by means of Eq. (10.12):

$$\frac{CM}{amp} = \sqrt{\frac{0.4 \times 0.82 \times 0.9}{0.92} \frac{0.65 \times 4000}{\pi}}$$

$$= 468$$

From Table 10.1 it is seen that the ratio of center-leg width to characteristic linear dimension l is approximately 1 for scrapless laminations with nearly square cross sections. This would suggest the EI-⅞ scrapless lamination. It is now necessary to make a trial design using the estimated lamination size and the wire sizes chosen from the current ratings and the current density calculated. This was done, and the EI-⅞ lamination was found to be too small. Successive approximations of cores and wire sizes were made until a 1⅛ stack of the EI-1 laminations was chosen. The calculations on the final design are shown in Fig. 10.3. The acceptability of the design is based on the temperature rise, the percentage buildup of the coil, the exciting current, and the relationship of core to copper losses. The design is acceptable except that the core loss is small compared with the copper loss. This is common in small transformers and justifies the use of lower grades of silicon steel. Equalizing core and copper losses would result in an excessive exciting current. The original estimate of core size was off mainly because of a high estimate of F_c. That factor for the final design is 0.217 compared with the estimate 0.4.

10.9   WIDEBAND AND SWITCHING TRANSFORMER DESIGN

In Chaps. 4 and 5 the effects of equivalent-circuit parameters on wideband and pulse circuits are discussed. In this section construction features and circuit approaches that can enhance performance are discussed.

Satisfactory wideband performance is a compromise. Any superior feature is paid for by a loss of another feature. Success is more likely if all characteristics are designed for minimum acceptable performance with as little safety margin as economics will allow.

In Chap. 4 it was shown that the low and high cutoff frequencies are

$$\omega_H = \frac{1}{\sqrt{L_L C_D}}$$

$$\omega_L = \frac{R_L R_G}{L_e(R_L + R_G)}$$

Figure 4.4 shows that best high-frequency performance is obtained when

$$R_G = R_L = \sqrt{\frac{L_L}{C_D}}$$

These relationships show that open-circuit inductance should be high for good low-frequency response, and leakage inductance should be low for good high-frequency response. This requirement conflict can be only partially resolved. The requirement for high open-circuit inductance can be moderated by keeping the impedance of the load and generator small. This permits lowering both L_L and L_e which helps the high-frequency response. Direct current magnetization of the core should be avoided. If that is not possible, then the core should be optimally gapped for maximum inductance at the rated dc current. The open-circuit inductance is increased by increasing the number of turns, the core area, and the effective permeability of the core. Permeability is a matter of selecting the most suitable core material. Materials with the highest permeability do not have the best high-frequency properties. The toroidal core provides the highest permeability. Leakage inductance can also be kept low in toroidal windings. Many wideband transformers are designed on toroids. The design of toroidal transformers is circumscribed by winding-machine limitations. The turns are placed on the core in a somewhat random fashion. Turn-to-turn insulation is limited to magnet wire film, and the voltage between adjacent turns is uncertain, making the value of the distributed capacitance uncertain. Winding-to-winding insulation usually consists of overlapped layers of tape. The capacitance between windings is also uncertain. The winding operation requires a clearance hole in the middle of the toroid to allow room for the toroidal shuttle in which the magnet wire is preloaded. The size of this hole depends upon the cross section of the shuttle, the diameter of the shuttle, and the dimensions of the toroidal core upon which the winding is placed. The design of a toroidal winding cannot be considered complete until the winding has been completed successfully. Toroidal windings are not suitable for high voltage.

Increasing the number of turns to increase the open-circuit inductance also reduces the flux density. This has the advantage of increasing the core selection options to include materials which saturate at lower flux densities. These materials include the high-nickel alloys and the ferrites. Ferrites have the best high-frequency properties among currently available materials. Ferrites are available in toroidal and rectangular shapes.

The choice of the area helps to determine both the open-circuit and the leakage inductance. Increasing the area allows a reduction in the number of turns for the same open-circuit inductance and would decrease the leakage inductance if other factors remained constant. Increasing the area increases the perimeter of the interwinding space which increases the leakage inductance. This perimeter is least for a given enclosed area when that area is a circle. If the area is four-sided, then the perimeter is least when the enclosed area is a square.

The product of leakage inductance and distributed capacitance determines the high-frequency cutoff. This product is proportional to coil volume. For this reason many wideband power-handling devices operate at high temperatures. This involves the use of insulating materials whose required properties must not only include high operating temperatures but low dielectric losses at high frequencies and low dielectric constants. Some insulating materials are available whose allowable operating temperature exceeds the temperature at which the magnetic properties of ferrites seriously degrade. Caution is required in high-temperature, high-frequency devices using ferrites because of the combination of core losses and the high temperature of the coil. Many common insulating materials are unsuitable for high-frequency use. This is particularly true of the impregnants because of both high dielectric losses and high dielectric constants. The application of impregnants to a coil can radically degrade its high-frequency performance. The selection of insulating materials with low dielectric constants is one of the few options open for lowering the leakage inductance-distributed capacitance product without hurting other characteristics.

The ratio of L_L to C_D may be varied for a constant $L_L C_D$ product by varying the interwinding insulation and the ratio of coil height to coil length. This is done to obtain a match between transformer characteristic impedance and the impedance of load and generator. The transformer turns ratio is used to match generator and load impedance and to obtain desired voltage ratios. High stepup turns ratios are damaging to high-frequency performance because stray capacitance in the high-voltage section reflects to the lower-voltage section as the square of the turns ratio which lowers the high-frequency cutoff.

Some high-frequency transformers are foil-wound with sheet insulation between each turn. The width of the foil and the thickness of the insulation determine the characteristic impedance. This construction is well adapted to bobbin windings and ferrite pot cores. Excellent performance is achievable with this technique, but control of characteristics in production is difficult.

Leakage inductance may be reduced by interleaving the primary and secondary windings, but this increases distributed capacitance. Core-type construction in which two coils are placed on a single core yields lower leakage inductance than either shell-type or simple single coil–type construction. Push-pull-type switching circuits usually require that the two halves of the primary transformer winding have very low leakage inductance between them. Because the two halves have the same number of turns and the same wire size, low leakage can be achieved by winding with two conductors simultaneously. The start of one conductor is connected to the finish of the other to form the center tap. This arrangement places the full primary voltage between adjacent turns at one end of the winding. Some sacrifice in reliability is thus made for improved performance. Winding two conductors simultaneously is sometimes referred to as *bifilar* or *two-in-hand*. Leakage inductance is decreased if an autotransformer connection is used. The closer the voltage ratio of the autotransformer approaches unity, the greater is the reduction in leakage inductance. Reducing the ratio of coil height to coil length reduces the leakage inductance at the expense of the distributed capacitance.

Pulse and switching transformers need wideband response, but their characteristics are usually expressed in terms of the effect on waveshapes. The rise time is analyzed by the same equivalent circuit as that used to analyze high-frequency performance of wideband transformers. The effect on the top of rectangular waves is predicted with the same equivalent circuit that is used to analyze the low-frequency performance of wideband transformers. Thus the rise time of a square wave is associated with the high-frequency performance, and reproduction of the top of the pulse is associated with the low-frequency performance. Neglected in this analysis is the effect on switching of the energy stored due to load current flowing in the leakage inductance. This energy is dissipated during the switching interval, usually in the switching device. Leakage inductance consequently represents a hazard to the switching components. This problem is attacked in several ways. The leakage inductance is made as small as possible. The switching frequency may be made equal to the resonant frequency of the transformer circuit so that the load current is passing through zero at the moment of switching. Various snubbing, damping, and protective circuits may be used to prevent damage to the switches. The problem can be reduced by transformer design but not eliminated.

10.10 Inductor design added

CHAPTER 11

Multiple-Phase Transformers

Three-phase circuits have significant advantages in the generation, distribution, and use of electrical energy. Three-phase rotating machinery is superior in various ways to the single-phase equivalents. Three-phase circuits make better use of the current-carrying capacity of distribution lines than single-phase circuits. Most ac electrical energy is generated as three-phase. This is true of equipment, aircraft, and shipboard machinery as well as fixed ground installations. Polyphase circuits other than three-phase are usually the province of the user. The versatile transformer, which in its three-phase configuration has intrinsic advantages, sees extensive use in polyphase conversions as well as in regular three-phase transformation.

11.1 DESCRIPTION OF THREE-PHASE CIRCUITS

The nature of three-phase circuits can be shown with rotating vectors representing sine wave functions of time. Figure 11.1 shows a three-phase circuit and vector diagrams of the associated voltages, currents, and fluxes. Figure 11.1a shows a three-phase transformer whose primary is arranged in what is known as a delta connection and whose secondary is arranged in what is known as a wye connection. The neutral of the transformer is tied to the neutral of a wye-connected load. Figure 11.1b shows the line to neutral voltages in the wye connection. The three voltages are symmetrically disposed around 360°, making the phase displacement

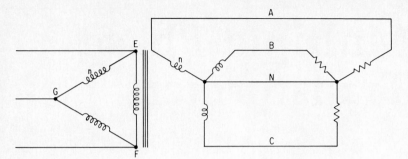

FIG. 11.1a Three-phase delta-to-wye-connected transformer with four-wire wye-connected load.

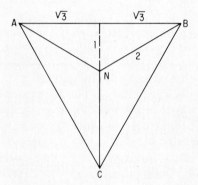

FIG. 11.1b Line-to-neutral and line-to-line voltages in wye-connected circuit of Fig. 11.1a.

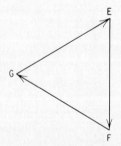

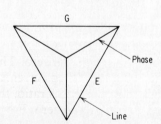

FIG. 11.1c Primary line voltages in delta-connected primary of transformer in circuit of Fig. 11.1a.

FIG. 11.1d Primary line and phase currents in delta-connected primary of transformer in circuit of Fig. 11.1a.

11.2

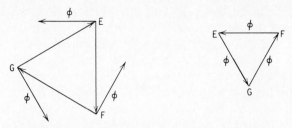

FIG. 11.1e Primary line voltage vector triangle and associated flux vector triangle for circuit of Fig. 11.1a.

between adjacent voltages 120°. The line-to-line voltages are equal to the vector sums of the line-to-neutral voltages. This addition is illustrated in the same figure. In accordance with Kirchhoff's law, the vector sum of the line voltages is zero. Note that the line-to-neutral voltages are displaced 30° from the line-to-line voltages. The primary and secondary voltages of corresponding legs of the transformer are in phase with each other independent of the type of connection. In the delta connection the line and leg voltages are the same, and so the vector triangle of the primary voltages in Fig. 11.1c resembles the line voltage vector triangle of the secondary displaced by 30°.

From the geometry of Fig. 11.1b the amplitude of the line-to-line voltage is $\sqrt{3}$ times the line-to-neutral voltage. The transformer in the circuit of Fig. 11.1a has a turns ratio in each leg of 1:1. The line-to-line voltage in the primary is equal to the line-to-neutral voltage in the secondary. Thus the transformer with a 1:1 ratio in each leg is providing a line-to-line step-up voltage ratio of $\sqrt{3}$:1. The currents in a delta-wye transformation are transposed from the voltages. In the wye connection the line current, determined by the line-to-neutral voltage and the line-to-neutral impedance of the load, is the same as the line-to-neutral current. Because of the 1:1 turns ratio of the transformer, which is assumed to be lossless, the currents in the primary and secondary legs are equal. The primary line current flowing into any delta junction is equal to the vector sum of the two-phase currents flowing away from that junction. The vector triangle is shown in Fig. 11.1d. If the load on the secondary is balanced so that the neutral current is zero, then the sum of the secondary line currents is zero. This is also true of the primary by Kirchhoff's law. Since the leg currents in the primary and secondary are equal or all differ by the turns ratio, then the vector sum of the phase currents in the primary is also zero. From the geometry of the vector triangle the line currents in the primary are equal to $\sqrt{3}$ times the phase currents in the primary. If E_{LN} is the line-to-neutral voltage in the secondary and I_L is the line current, then the total volt-amperes of the load is $3E_{LN}I_L$. Since the secondary line current is equal to the primary line current over $\sqrt{3}$, then the volt-amperes expressed in the primary line current and voltage are $\sqrt{3}E_pI_p$. The volt-amperes in the primary and secondary are equal.

The fluxes generated by the volt-time integral of the primary are phase-shifted by 90° from the voltages. These fluxes may also be represented by rotating vectors and may be added vectorially the same as the voltages and currents. The flux vectors are shown in Fig. 11.1e. The vector addition of fluxes is used in the design of three-phase cores.

11.2. PHASE TRANSFORMATIONS

Given two voltages in a polyphase circuit, it is possible to construct a third voltage of arbitrary phase by appropriate choice of magnitude and polarity of the other two voltages. This construction is shown in Fig. 11.2. There are two common transformations using this construction. The Scott transformer is used to transform three-phase to two-phase and vice versa. The Scott transformation is illustrated in the circuit and vector diagram of Fig. 11.3. The delta-delta-wye three-phase transformer is used to construct six-phase voltages from three-phase. This transformer has both a delta and a wye secondary winding with line-to-line voltages of the same magnitude but shifted by 30°. The schematic and vector diagram of the transformer are shown in Fig. 11.4. The 30° phase displacement between the delta and wye windings is described under three-phase circuits. This transformation is used in rectifier circuits to reduce the amplitude of the ripple voltage and increase its frequency for ease in filtering. See Chap. 12.

The phase relationships obtained in the delta-delta-wye transformation are also obtainable in the extended delta transformation. In this

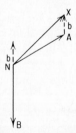

FIG. 11.2 Vector construction of a third voltage, *NX*, of arbitrary phase and magnitude from two-phase voltages, *NA* and *NB*, of a three-phase circuit.

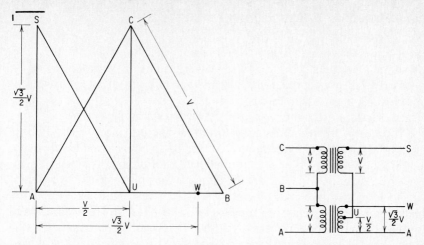

FIG. 11.3 Scott transformer connection and vector diagram for transforming three-phase to two-phase.

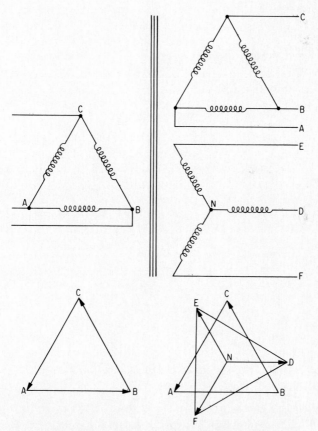

FIG. 11.4 Delta-delta-wye transformer schematic and vector diagram showing three-phase to six-phase transformation.

11.5

transformation two secondary windings are placed on each leg as in the delta-delta-wye transformation. The two windings in the extended delta transformation are identical except that the polarity of one is reversed with respect to the other. Each winding has a tap. The tap and one end of each winding are connected with the corresponding windings of the other phases to form a delta configuration. The portion of the winding beyond the tap adds additional wye voltage to that developed by the delta connection. By properly selecting the position of the tap, it is possible to develop a secondary line voltage which leads the phase voltage by a phase angle of 15°. The line voltage developed by the phase winding of opposite polarity will lag the phase voltage by 15°. The phase difference between the two secondary line voltages will be 30°, as with the delta-delta-wye connection. The advantage of the extended delta is the use of two identical secondary coils with the polarity being obtained by reversing the

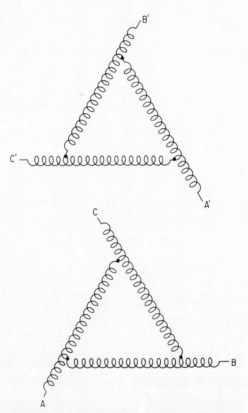

FIG. 11.5 Schematic of two extended delta windings for producing line voltages with a 30° phase displacement.

position of one coil on the leg with respect to the other. In the extended delta the ratio of the tap turns to the entire winding is an irrational number, as is the ratio of the turns in the wye winding to the turns in the delta winding in the delta-delta-wye connection. In windings with few turns a measurable imbalance may result. The effect of the irrational ratios will be negligible in windings with large numbers of turns. In most cases the primary input imbalance will be greater than that introduced by the transformer regardless of the circuit selected. A schematic and a vector diagram of the extended delta are shown in Figs. 11.5 and 11.6.

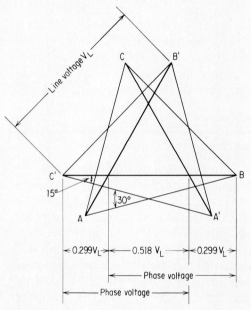

FIG. 11.6   Vector diagram of two extended delta windings.

It is sometimes necessary to generate polyphase circuits from single-phase voltages. This is regularly done in single-phase motors which contain a starting winding in the stator to which is applied a phase-shifted-voltage. A circuit for generating three-phase from single-phase is shown in Fig. 11.7. A series capacitor is used to shift the phase of the voltage as in a capacitor motor. A transformer is then used to apply the voltage at proper polarity and amplitude to the load. When the load is fixed, it is possible to obtain with this circuit a very accurate three-phase output. The difficulties arise from the change in phase and magnitude of the voltage with changing load. A common application for this circuit is the oper-

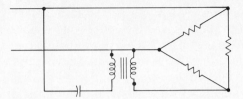

FIG. 11.7 Circuit used to generate a three-phase voltage from a single-phase source.

ation of a three-phase motor from a single-phase voltage. The success of this circuit in this application is limited by the changing load characteristic of the motor on start-up. If the capacitor and turns ratio of the transformer are selected for the run condition, the motor will see badly unbalanced voltages when starting and will have low starting torque. If the capacitor and turns ratio are selected for the start condition, the voltages will be unbalanced when the motor is running at rated speed.

11.3 THREE-PHASE TRANSFORMERS

A three-phase transformer functions like three single-phase transformers, but a single three-phase transformer occupies less volume and weighs less than three single-phase transformers designed for the same purpose. The unit cost of most three-phase transformers is less than the unit costs of three equivalent single-phase transformers. However, the economics is more complex than the unit cost alone. The quantity being procured, the existence of commercial standards, the maintenance of spare parts, and other factors often affect an economical choice. A three-phase core-type transformer is illustrated in Fig. 11.8 together with three equivalent single-phase transformers. A weight saving of from 10 to 30 percent can be expected, dependent upon the geometry of the particular transformers. As suggested by the figure, neither the cross section of the core nor the coil is changed by the choice of single- or three-phase. The flux density in the core is determined by the voltage and frequency applied to the primary winding in either case. The fluxes in the yokes connecting the three legs of the three-phase transformer add vectorally to give the total flux. While the flux in the yokes is ideally no greater in magnitude than the flux in the legs, departures from this ideal are one of the complications in three-phase construction.

Each leg and coil of a three-phase transformer is associated with one phase. The voltages developed on the secondary winding(s) on each leg are in phase with the primary leg voltages independent of the type of connections made external to the windings. The voltage current relationships

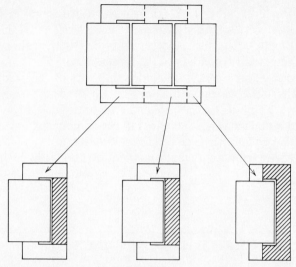

FIG. 11.8 Geometry of single-phase vs. three-phase core-type transformers. Core volume saved in three-phase transformer is indicated by shaded areas required in single-phase equivalents.

in the individual coils are the same as in a single-phase transformer. The polarity of the secondary voltages in relation to the primary is a function of the way the two windings tend on the coil, this being determined by the direction in which the winding machine rotates and the position of the coil on the core. In most transformer shops all the winding machines rotate in the same direction. This permits a commonly used convention of referring to the two leads of a winding as a start and a finish for the purpose of establishing polarities. With multiple secondary windings, wye-delta connections, and options for positioning coils, there is considerable opportunity for error. An important rule in three-phase devices is symmetry. The phases must be connected symmetrically, and the coils must be symmetrical in construction and placement on the core. The starts of each winding on a single leg will all be of like polarity if winding and placement conventions are observed. Then for a delta connection, a start is connected to a finish until the delta is closed, preserving symmetry. In a wye connection all the starts or all the finishes are connected together, again preserving symmetry.

The choice of whether to connect the starts or finishes together to form the neutral of a wye connection can be significant. If the neutral will operate at ground, then a consideration is whether it is easier to insulate the start or finish from ground. This is very important if the secondary induced voltage is high. If the finishes are connected together in a core-type transformer (see Fig. 11.8), then minimum insulation is required between adjacent coils. The starts of the secondaries must then be insulated for full line-to-ground voltage, assuming that primary or shield between primary and secondary are at ground. If the starts are grounded, minimum insulation is required under the start of the secondary, but adjacent coils must be insulated for full line-to-line voltage which is higher than the line-to-neutral voltage and sometimes higher than the line-to-ground voltage. These considerations do not arise in a delta winding. Both start and finish of a delta-connected winding reach the same peak voltage. Appropriate insulation is required both under the start and over the finish.

11.4   TAPS ON THREE-PHASE WINDINGS

Taps on three-phase windings require special consideration. If the winding being tapped is wye connected, then the tap with a lower voltage rating must have the appropriate neutral to tap position. A change in voltage is accomplished by moving the three connections from the end of the windings to the three taps. A delta winding may be tapped in the same manner, except that the delta must be opened and six leads must be adjusted instead of three. This is sometimes a disadvantage. It is more convenient to move three leads to three different terminals than it is to open and remake three connections in addition. By proper choice of tap positions it is possible to leave the delta closed and simply move the leads to the taps. The maximum reduction in voltage that can be achieved in this manner is 50 percent. That this connection is possible is demonstrated in the vector diagram of Fig. 11.9. Triangle ABC represents the voltages on the windings of a delta-connected transformer. Points A', B', and C' represent the position of taps symmetrically placed on their respective windings. Then the triangle $A'B'C'$ formed by the voltages on these taps is equilateral, as is the triangle in which it is inscribed. This allows the application of a balanced three-phase voltage to terminals A', B', and C'. The given information in this configuration is usually the magnitude of the voltages AB, BC, and CA for the full winding and the voltages $A'B'$, $B'C'$, and $C'A'$ for the tap. The position of each tap on its own winding can be determined from an application of the law of cosines. The following formula is obtained:

$$\frac{AA'}{AC} = \frac{1 - \sqrt{1 - \dfrac{4[1 - (A'C'/AC)^2]}{3}}}{2} \tag{11.1}$$

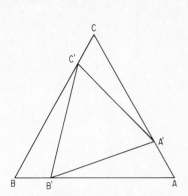

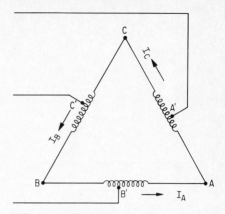

FIG. 11.9 Vector representation of taps on a closed delta winding, the inscribed triangle representing a reduced and balanced three-phase voltage.

FIG. 11.10 Closed delta winding with taps showing phase currents.

In this expression AA' is less than or in the limit equal to AC. The schematic for this connection is shown in Fig. 11.10.

The currents in a closed delta winding with taps may be determined with the aid of the vector diagram in Fig. 11.11. The three vectors $\mathbf{I}_A$, $\mathbf{I}_B$, and $\mathbf{I}_C$ are the currents defined in Fig. 11.10. Symmetry requires that they be of equal magnitude and phase displaced by 120°. Current I_A flows in that part of the winding represented by vector $\mathbf{B'A}$ in Fig. 11.11. Current I_B flows in that part of the winding represented by vector $\mathbf{BB'}$. The net effect of these two currents on the total winding of n turns can be determined by vector addition of these two currents weighted by the turns through which they flow. Refer to the vector diagram in Fig. 11.11. If a resistive load is assumed, vector $\mathbf{B'Y}$ is in phase with applied voltage $B'A'$. The magnitude of vector $\mathbf{B'Y}$ is proportional to $nI_A \, AB'/AB$ where n is the number of turns in the total winding and the ratio AB'/AB represents the position of the tap B'. Vector $\mathbf{YX}$ is in phase with applied voltage $C'B'$. The magnitude of vector $\mathbf{YX}$ is proportional to $nI_B \, B'B/AB$. For convenience the scale is chosen to make $B'Y$ equal to $B'A$. Then the length of YX is equal to the length of BB'. Triangle XYB' is then congruent to the triangle $A'AB'$, and $B'X$ is equal to $B'A'$. Thus the magnitude of the net current in the winding is

$$|I_{\text{net}}| = \left| I_A \frac{A'B'}{AB} \right|$$

The input volt-amperes per phase by inspection are

$$V_{A'B'}I_A$$

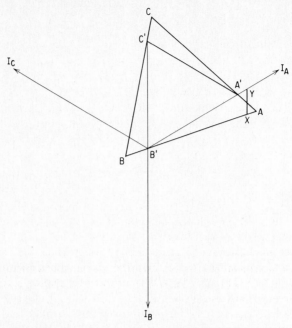

FIG. 11.11 Vector diagram showing voltage and current relationships in a tapped delta winding.

The volt-amperes net in the winding from the full winding voltage and net current are

$$I_{net}V_{AB} = V_{AB}\frac{A'B'}{AB}I_A = V_{A'B'}I_A$$

This preserves the volt-ampere equality. The net current is used to determine power delivered to the load. The phase current I_A is used to determine conductor size.

11.5 FLUX DISTRIBUTION IN THREE-PHASE TRANSFORMERS

If the voltages applied to the windings of a three-phase transformer are balanced, then the fluxes induced in the legs of the cores within those windings will also be balanced. The values are determined by Faraday's law. Where the fluxes exit from the legs and enter the yoke of the core, they are separated into vector components so that the sum of these components produces the balanced three-phase fluxes shown in Fig. 11.12. If the reluctance between the two flux paths in the legs of the core is low, then the flux distribution across the area of the core is uniform and the flux density can be accurately predicted. This is not necessarily true in the yokes of the core. As shown in Fig. 11.12, the magnitude of the flux in either of the two components in a balanced transformer is 57.7 percent of the net flux in the leg. If the core is of uniform cross-sectional area, there will be regions of the yoke where the flux density will be higher than in the legs of the core. This can be demonstrated with wound cores where the physical separation of the component flux paths makes them available for examination. When search coils are placed around the individual paths, as shown in Fig. 11.13, the voltages induced in the windings are 57.7 percent of the voltages developed by search coils of equal turns placed around the legs when the fluxes are balanced. Usually the fluxes are not balanced, and the voltages induced in the coils around the individual component flux paths vary from one another by as much as a factor of 2, reflecting a proportionate variation in the flux density. This imbalance is the result of unequal reluctances in the three flux paths. The flux path linking the two outboard coils is longer than the other two and hence can be expected to have a greater reluctance. Perhaps more significant are the unequal air gaps in the three paths. It is unrealistic to expect these air gaps to be exactly equal. Since there is an effort to make these gaps small, minor variations constitute a large percentage of the difference. In the case of stamped laminations, the flux distribution in the yoke is less obvious. The problem is avoided in some three-phase laminations by making the yoke of greater width than the legs. (See Sec. 7.2.)

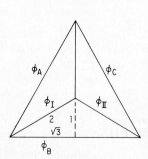

FIG. 11.12 Vector diagram of fluxes in a balanced three-phase core.

$$|\phi_I| = \frac{2}{2\sqrt{3}} |\phi_B| = 0.577 |\phi_B|$$

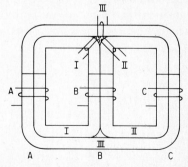

FIG. 11.13 Three-phase wound core showing flux paths and search coils for measuring flux.

11.6 HARMONIC CURRENTS IN THREE-PHASE TRANSFORMERS

The rotating vector representation of voltages and currents in three-phase circuits neglects the effect of harmonics which are present whenever magnetic devices are used. If the magnetic device is symmetrically magnetized as is most often the case, only odd harmonics are present. The harmonics in the exciting currents of a three-phase device can be represented in a Fourier series as follows:

$$
\begin{aligned}
i_\mathrm{I} = {} & A_1 \sin \theta + B_1 \cos \theta + A_3 \sin 3\theta \\
& + B_3 \cos 3\theta + A_5 \sin 5\theta + B_5 \cos 5\theta + \cdots
\end{aligned}
\tag{11.2}
$$

$$
\begin{aligned}
i_\mathrm{II} = {} & A_1 \sin \left(\theta + \frac{2\pi}{3} \right) + B_1 \cos \left(\theta + \frac{2\pi}{3} \right) + A_3 \sin 3 \left(\theta + \frac{2\pi}{3} \right) \\
& + B_3 \cos 3 \left(\theta + \frac{2\pi}{3} \right) + A_5 \sin 5 \left(\theta + \frac{2\pi}{3} \right) + B_5 \cos 5 \left(\theta + \frac{2\pi}{3} \right) + \cdots
\end{aligned}
\tag{11.3}
$$

$$
\begin{aligned}
i_\mathrm{III} = {} & A_1 \sin \left(\theta + \frac{4\pi}{3} \right) + B_1 \sin \left(\theta + \frac{4\pi}{3} \right) + A_3 \sin 3 \left(\theta + \frac{4\pi}{3} \right) \\
& + B_3 \cos 3 \left(\theta + \frac{4\pi}{3} \right) + A_5 \sin 5 \left(\theta + \frac{4\pi}{3} \right) + B_5 \cos 5 \left(\theta + \frac{4\pi}{3} \right) + \cdots
\end{aligned}
\tag{11.4}
$$

Peculiar effects result from the addition of these currents. Consider, for example, a four-wire wye-connected primary. The current that flows in the neutral is equal to the sum of the phase currents in accordance with Kirchhoff's law. The sum of Eqs. (11.2) to 11.4)

$$
\begin{aligned}
i_\mathrm{I} + i_\mathrm{II} + i_\mathrm{III} = {} & A_1 \left[\sin \theta + \sin \left(\theta + \frac{2\pi}{3} \right) + \sin \left(\theta + \frac{4\pi}{3} \right) \right] \\
& + B_1 \left[\cos \theta + \cos \left(\theta + \frac{2\pi}{3} \right) + \cos \left(\theta + \frac{4\pi}{3} \right) \right] \\
& + A_3 \left[\sin 3\theta + \sin 3 \left(\theta + \frac{2\pi}{3} \right) + \sin 3 \left(\theta + \frac{4\pi}{3} \right) \right] \\
& + B_3 \left[\cos 3\theta + \cos 3 \left(\theta + \frac{2\pi}{3} \right) + \cos 3 \left(\theta + \frac{4\pi}{3} \right) \right]
\end{aligned}
$$

$$+ A_5 \left[\sin 5\theta + \sin 5 \left(\theta + \frac{2\pi}{3} \right) + \sin 5 \left(\theta + \frac{4\pi}{3} \right) \right]$$

$$+ B_5 \left[\cos 5\theta + \cos 5 \left(\theta + \frac{2\pi}{3} \right) + \cos 5 \left(\theta + \frac{4\pi}{3} \right) \right]$$

$$(11.5)$$

In Eq. (11.5) the sum of the fundamental terms is equal to zero since these terms represent three equal voltages at 120° phase angles. The sum of the fifth harmonic terms is also zero for the same reason, since

$$\sin 5 \left(\theta + \frac{2\pi}{3} \right) = \sin \left(5\theta + 2\pi + \frac{4\pi}{3} \right) = \sin \left(5\theta + \frac{4\pi}{3} \right)$$

$$\sin 5 \left(\theta + \frac{4\pi}{3} \right) = \sin \left(5\theta + 6\pi + \frac{2\pi}{3} \right) = \sin \left(5\theta + \frac{2\pi}{3} \right)$$

$$\sin 5\theta \qquad\qquad\qquad\qquad = \sin 5\theta$$

The sum of the third harmonic terms is not zero. The value of that sum is as follows:

$$\sin 3\theta + \sin 3 \left(\theta + \frac{2\pi}{3} \right) + \sin 3 \left(\theta + \frac{4\pi}{3} \right)$$

$$= \sin 3\theta + \sin (3\theta + 2\pi) + \sin (3\theta + 4\pi)$$

$$= 3 \sin 3\theta$$

The third-harmonic term of each phase is in phase with the third-harmonic terms of the other two phases.

The third harmonic is typical of odd harmonics that are multiples of 3. The fifth harmonic is typical of odd harmonics that are not multiples of 3. The third-harmonic currents that can flow in the neutral line are often objectionable. If the primary is connected in delta, the odd harmonics which are multiples of 3, being in phase, will flow as a circulating current around the delta connection and not in the lines supplying power. If the neutral point in a wye-connected primary is left unconnected, Eq. (11.5) must equal zero including the odd harmonics that are multiples of 3:

$$3A_3 \sin 3\theta + 3B_3 \cos 3\theta = 0$$

Thus there is no odd harmonic content in the line current. With a sinusoidal applied voltage and a sinusoidal current, the nonlinear behavior of the inductance requires that series circuit elements such as the resistance of the windings absorb a part of the applied voltage, allowing the voltage across the inductance of the winding to change with magnetization. The voltage on the neutral will rotate around the neutral point in response to the third harmonics, ensuring output voltages that contain third-harmonic components. Harmonics will also appear in the output voltage of a transformer with a delta-connected primary due to the flow of harmonic current through the series circuit elements. However, the preemptive condition for harmonic content that exists in the three-wire wye-connected primary is absent. The harmonics in the output voltage of a delta-connected primary transformer can be reduced by reducing the series circuit elements. The delta-connected primary is generally better. Wye-connected secondaries offer the option of two different single-phase voltages, can operate with neutral either connected or not, and have other advantages in rectifier transformers. See Chap. 12.

CHAPTER 12

RECTIFIER TRANSFORMERS

Rectifiers are among the most prevalent applications for transformers. One of the classic challenges in transformer design is the precise determination of sine wave voltage and rms current from the dc voltage and current obtained from innumerable rectifier circuits. Rectifier transformers are found in traction devices, electroplating, battery chargers, electronic and communications systems, computer power supplies, and x-ray equipment.

Rectifier transformers increase or decrease ac voltages for rectification, filtering, and use in equipment. They operate from a narrow band of sine wave frequencies or from a fixed waveform. Fidelity is not an intrinsic requirement. Economy in establishing the required dc voltage at the required current is the objective. The transformer secondary voltage and current ratings must be established from the dc requirements. The interrupted and nonsinusoidal currents associated with rectifier circuits make this an involved task, but optimum performance can be achieved only when the task is performed correctly. The available voltage and frequency determine the primary ratings of rectifier transformers. The voltage and frequency standards for the principal countries of the world are given in Table 12.1. Not listed in the table is the standard aircraft power of 120/208 V three-phase 400 Hz. Equipment intended for use with more than one voltage or frequency requires special consideration. Transformers designed for 50 Hz will operate satisfactorily at 60 Hz, but the converse is not true. The same is true of filter circuits. The ripple voltage on the dc output will be greater at 50 Hz than at 60 Hz. Therefore, rec-

TABLE 12.1 Voltages and Frequencies of Electric Power in Major Countries

Country	Frequency, Hz	Voltage
Afghanistan	50 and 60	220/380
Algeria	50	127/220/380
Angola	50	220/380
Antigua	50	230/400
Argentina	50	220/380/440
Australia	50	240/415/440
Austria	50	220/380
Azores	50	220/380
Bahamas	60	120/208/240
Bahrain	50 and 60	110/220/230/400
Bangladesh	50	230/400
Barbados	50	110/200/120/208
Belgium	50	127/220/380
Belize	60	110/220/440
Bermuda	60	115/230/120/208
Bolivia	50 and 60	110/220/380
Botswana	50	220/380
Brazil	50 and 60	127/220/380/440
Bulgaria	50	220/380
Burma	50	120/208/220/380/440
Cambodia	50	120/208/220/380
Cameroon	50	127/220/380/400
Canada	60	120/240
Canary Islands	50	127/220/380
Central African Republic	50	220/380
Chad	50	220/380
Chile	50	220/380
China, People's Republic of	50	220/380
Colombia	60	110/220
Congo, Republic of	50	220/380
Costa Rica	60	120/240
Cyprus	50	240/415
Czechoslovakia	50	220/380
Denmark	50	220/380
Dominican Republic	60	110/220
Ecuador	60	120/208/127/220
Egypt	50	110/220/380
El Salvador	60	115/230

12.2

TABLE 12.1 (*continued*)

Country	Frequency, Hz	Voltage
Ethiopia	50	220/380
Fiji	50	240/415
Finland	50	220/380
France	50	127/220/380
Germany, Federal Republic of	50	220/380
Ghana	50	220/400
Gibraltar	50	240/415
Greece	50	220/380
Greenland	50	220/380
Grenada	50	230/400
Guadeloupe	50	220/380
Guatemala	60	110/220/120/240
Guinea	50	220/380
Guyana (under revision)	50 and 60	110/220
Haiti	60	110/220/380
Honduras	60	110/220
Hong Kong	50	200/346
Hungary	50	220/380
Iceland	50	220/380
India	50 (some direct current)	230/400
Indonesia	50	127/220
Iran	50	220/380
Iraq	50	220/380
Ireland (Eire)	50	220/380
Israel	50	230/400
Italy	50	127/220/380
Ivory Coast	50	220/380
Jamaica	50	110/220
Japan	50 and 60	100/200
Jerusalem	50	220/380
Jordan	50	220/380
Kenya	50	240/415
Korea	60	100/200
Kuwait	50	240/415
Laos	50	220/380
Lebanon	50	110/190/220/380
Liberia	60	120/208/240
Libya	50	127/220/230/400

TABLE 12.1 (*continued*)

Country	Frequency, Hz	Voltage
Luxembourg	50	120/208/220/380
Majorca Island	50	127/220/380
Malagasy Republic	50	127/220/380
Malaysia	50	240/415
Malta	50	240/415
Mauritania	50	220
Mexico	60	127/220
Monaco	50	127/220/380
Morocco	50	115/200/127/220/380
Mozambique	50	220/380
Nepal	50	230/440
Netherlands	50	220/380
New Caledonia	50	220/380
New Zealand	50	230/400
Nicaragua	60	120/240
Niger	50	220/380
Nigeria	50	230/415
Norway	50	230
Okinawa		
Military facilities	60	120/240
Nonmilitary	60	100/200
Oman	50	220/440
Pakistan	50	230/380/400
Panama	60	110/220/120/240
Paraguay	50	220
Peru	60	110/220
Philippines	60	110/220
Poland	50	220/380
Portugal	50	220/380
Puerto Rico	60	120/240
Qatar	50	240/415
Rhodesia (see Zimbabwe)		
Romania	50	220/380
Rwanda	50	220/380
Saudi Arabia	50 and 60	127/220/380/400
Senegal	50	110/220
Sierra Leone	50	220/440
Singapore	50	230/400

TABLE 12.1 (*continued*)

Country	Frequency, Hz	Voltage
Somalia	50	110/220/380/440
South Africa, Republic of	50 (some direct current)	220/380/400
Spain	50	127/220/380
Sri Lanka	50	230/400
Sudan	50	240/415
Surinam	60	110/220/115/230
Swaziland	50	230/400
Sweden	50	127/220/380
Switzerland	50	220/380
Syria	50	115/200/220/380
Tahiti	60	127/220
Taiwan	60	110/220
Tanzania	50	230/400
Thailand	50	220/380
Togo, Republic of	50	127/220/380
Trinidad	60	115/230/400
Tunisia	50	127/220/380
Turkey	50	220/380
Uganda	50	240/415
Union of Soviet Socialist Republics	50	127/220
United Kingdom	50	240/415
United States	60	115/220/120/208
Upper Volta, Republic of	50	220/380
Uruguay	50	220
Venezuela	60	120/240
Viet Nam	50	127/220/380
Virgin Islands (American)	60	120/240
Western Samoa	50	230/400
Yemen (Aden)	50	230/400
Yemen (Arab Republic)	50	220
Yugoslavia	50	220/380
Zaire, Republic of	50	220/380
Zambia	50	220/380
Zimbabwe	50	220/380

Source: U.S. Department of Commerce, *Electric Current Abroad,* 1975. The table is a guide only. Equipment suppliers should obtain from the end user the electric power characteristics in the area where the equipment will be used.

tifier circuits for use in different areas of the world requiring operation at both 50 and 60 Hz must be designed for 50 Hz. Transformers that are designed for operation at 400 Hz are subject to immediate destruction if operation at 50 or 60 Hz is attempted. Transformers designed for 50 or 60 Hz will operate safely at 400 Hz, as will the filter circuits, but the leakage reactance of the transformer will cause the output to drop at the higher frequency. Investigation of individual cases is required to determine whether satisfactory performance can be achieved. The percent ripple will be much lower at 400 Hz than at 50 or 60 Hz. After frequency compatibility is achieved, different primary voltages may be accommodated with taps and series-parallel arrangements of the primary winding or with separate auxiliary transformers. The size and weight of these auxiliaries can be reduced if autotransformer connections are used. Rectifier transformers are also used in high-frequency inverters in which switching circuits generate square wave voltages that are transformed and rectified. This application is discussed further in Chap. 13.

12.1 RECTIFIER CIRCUIT DESCRIPTION

Rectifier circuits convert ac voltage to dc voltage, a process of fundamental importance because of the need for dc voltage in equipment and the universal availability of ac electric power. Rectification is accomplished in modern rectifier circuits with solid-state diodes, usually silicon. Silicon diodes are available for a wide range of currents. The inverse voltage rating of single diodes is limited to a few hundred volts. The use of avalanche diodes permits operation of many diodes in series for high-voltage applications. The dc voltage produced by rectifier circuits from sine wave ac voltages contains a substantial ripple voltage which is harmonically related to the line frequency. Filter circuits are used to reduce this ripple voltage to an acceptably low value.

Rectifier circuits are classified as half wave, full wave, bridge, and multiplier; by the number of phases; and by the type of filter used. The most common rectifier circuits are shown in Table 12.2 and Fig. 12.1. Most of the information in this chapter applies to sine wave voltage input. Modification of some of the data is required for application to square wave voltages.

Filter circuits are classified as either choke input or capacitor input. Choke input filters are analyzed on the assumption that the choke has an infinite inductance. In such a filter the output voltage has zero ripple. The only current that can pass through the inductance is the dc component. The output voltage from the filter will be equal to the average value of the voltage input to the filter less the dc drop through the filter.

TABLE 12.2 Data on Various Choke Input Filter Rectifier Circuits

---

Single-Phase Half-Wave Circuit

Schematic

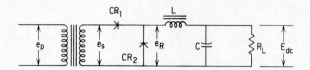

Note: Free-wheeling rectifier CR_2 is used to maintain current during interval when rectifier CR_1 is biased negatively.

Voltage and current relationships

Primary voltage

$$e_p = E_{pk} \sin 2\pi ft$$

Secondary voltage

$$e_s = E_{pk} \sin 2\pi ft$$

RMS voltages

$$(E_p)_{rms} = (E_s)_{rms} = \frac{E_{pk}}{\sqrt{2}} = 2\left(\frac{\pi}{2}\frac{E_{dc}}{\sqrt{2}}\right) = 2.22E_{dc}$$

Unfiltered rectified voltage

$$E_{pk} = \pi E_{dc} = 3.14E_{dc}$$

Transformer secondary current

$$(I_s)_{rms} = \frac{I_{dc}}{\sqrt{2}} = 0.707I_{dc}$$

Free-wheeling rectifier current

$$(I_{CR2})_{rms} = \frac{I_{dc}}{\sqrt{2}} = 0.707I_{dc}$$

TABLE 12.2 (*continued*)

Single-Phase Half-Wave Circuit

Transformer primary current

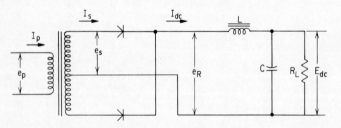

$$(I_p)_{rms} = \frac{I_{dc}}{2}$$

Transformer volt-ampere rating

Primary

$$(VA)_p = (E_p)_{rms}(I_p)_{rms} = \frac{\pi}{\sqrt{2}} E_{dc} \frac{I_{dc}}{2} = 1.11 E_{dc} I_{dc}$$

Secondary

$$(VA)_s = (E_s)_{rms}(I_s)_{rms} = \frac{\pi}{\sqrt{2}} E_{dc} \frac{I_{dc}}{\sqrt{2}} = 1.57 E_{dc} I_{dc}$$

Power factor

$$\frac{P_{dc}}{(VA)_p} = \frac{E_{dc} I_{dc}}{1.11 E_{dc} I_{dc}} = 0.90$$

Fundamental ripple frequency

$$f$$

Note: This circuit results in a dc magnetomotive force being applied to the transformer core equivalent to half the dc load current multiplied by the number of turns in the secondary.

Single-Phase Full-Wave Center-Tap Circuit

Schematic

Voltage and current relationships

Primary voltage

$$e_p = E_{pk} \sin 2\pi ft$$

TABLE 12.2 *(continued)*

Single-Phase Full-Wave Center-Tap Circuit

Secondary voltage

$$e_s = E_{pk} \sin 2\pi ft$$

RMS voltages

$$(E_p)_{rms} = (E_s)_{rms} = \frac{E_{pk}}{\sqrt{2}} = \frac{\pi}{2\sqrt{2}} E_{dc} = 1.11 E_{dc}$$

Unfiltered rectified voltage e_R

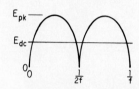

$$E_{pk} = \frac{\pi}{2} E_{dc} = 1.57 E_{dc}$$

Diode and transformer secondary current

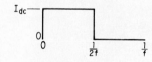

$$(I_s)_{rms} = \frac{I_{dc}}{\sqrt{2}} = 0.707 I_{dc}$$

Transformer primary current

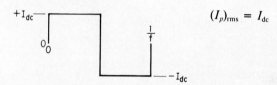

$$(I_p)_{rms} = I_{dc}$$

Transformer volt-ampere rating

Primary

$$(VA)_p = (E_p)_{rms} = (I_p)_{rms} = 1.11 E_{dc} I_{dc}$$

Secondary

$$(VA)_s = 2(E_s)_{rms}(I_s)_{rms} = 2 \frac{\pi}{2\sqrt{2}} E_{dc} \frac{I_{dc}}{\sqrt{2}} = 1.57 E_{dc} I_{dc}$$

Power factor

$$PF = \frac{P_{dc}}{(VA)_p} = \frac{E_{dc} I_{dc}}{1.11 E_{dc} I_{dc}} = 0.90$$

Fundamental ripple frequency

$$2f$$

TABLE 12.2 (*continued*)

Single-Phase Bridge Circuit

Schematic

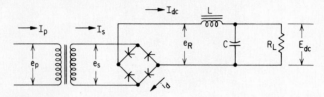

Voltage and current relationships

Primary voltage

$$e_p = E_{pk} \sin 2\pi ft$$

Secondary voltage

$$e_s = E_{pk} \sin 2\pi ft$$

RMS voltages

$$(E_p)_{rms} = (E_s)_{rms} = \frac{E_{pk}}{\sqrt{2}} = \frac{\pi}{2\sqrt{2}} E_{dc} = 1.11 E_{dc}$$

Unfiltered rectified voltage e_R

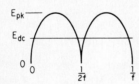

$$E_{pk} = \frac{\pi}{2} E_{dc} = 1.57 E_{dc}$$

Diode current i_d

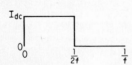

$$(I_d)_{rms} = \frac{I_{dc}}{\sqrt{2}} = 0.707 I_{dc}$$

Transformer currents

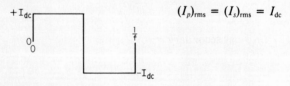

$$(I_p)_{rms} = (I_s)_{rms} = I_{dc}$$

Transformer volt-ampere rating

$$(VA)_p = (VA)_s = (E_p)_{rms}(I_p)_{rms} = 1.11 E_{dc}I_{dc}$$

12.10

TABLE 12.2 *(continued)*

Single-Phase Bridge Circuit

Power factor

$$\text{PF} = \frac{P_{dc}}{(VA)_p} = \frac{E_{dc}I_{dc}}{1.11E_{dc}I_{dc}} = 0.90$$

Fundamental ripple frequency

$$2f$$

Three-Phase Half-Wave Circuit Wye-Wye Transformation

Schematic

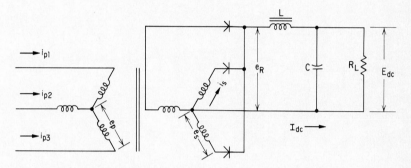

Voltage and current relationships

Primary voltage

$$e_p = E_{pk} \sin 2\pi ft$$

Secondary voltage

$$e_s = E_{pk} \sin 2\pi ft$$

Unfiltered rectified voltage e_R

$$E_{dc} = \frac{3\sqrt{3}}{2\pi} E_{pk} = \frac{3\sqrt{3}}{\pi\sqrt{2}} (E_s)_{rms}$$

$$E_{dc} = 1.17(E_s)_{rms} = 1.17(E_p)_{rms}$$

Diode and transformer secondary current i_s

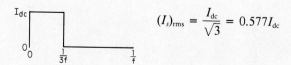

$$(I_s)_{rms} = \frac{I_{dc}}{\sqrt{3}} = 0.577I_{dc}$$

TABLE 12.2 (*continued*)

Three-Phase Half-Wave Circuit Wye-Wye Transformation

Transformer primary currents line and phase

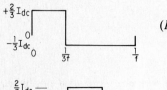

$$(I_p)_{\text{rms}} = \sqrt{\left(\frac{2}{3}\right)^2 \frac{1}{3} + \left(\frac{1}{3}\right)^2 \frac{2}{3}}\, I_{\text{dc}}$$

$$= 0.471 I_{\text{dc}}$$

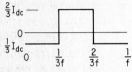

Transformer volt-ampere ratings

Primary

$$(VA)_p = 3(E_p)_{\text{rms}}(I_p)_{\text{rms}} = 3\,\frac{E_{\text{dc}}}{1.17}\,(0.471)I_{\text{dc}} = 1.21 E_{\text{dc}}I_{\text{dc}}$$

Secondary

$$(VA)_s = 3(E_s)_{\text{rms}}(I_s)_{\text{rms}} = 3\,\frac{E_{\text{dc}}}{1.17}\,\frac{I_{\text{dc}}}{\sqrt{3}} = 1.48 E_{\text{dc}}I_{\text{dc}}$$

Power factor

$$\frac{E_{\text{dc}}I_{\text{dc}}}{(VA)_p} = \frac{1}{1.21} = 0.83$$

Fundamental ripple frequency

$$3f$$

Note: This circuit results in a dc unbalance current of ⅓I_{dc}. See three-phase half-wave circuit delta-wye connection.

TABLE 12.2 (*continued*)

Three-Phase Half-Wave Circuit Delta-Wye Transformation

Schematic

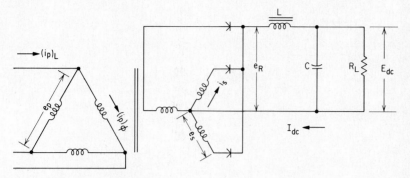

Voltage and current relationships

Primary voltage

$$e_p = \sqrt{3}\, E_{pk} \sin 2\pi ft$$

Secondary voltage

$$e_s = E_{pk} \sin 2\pi ft$$

RMS voltages

$$(E_p)_{rms} = \sqrt{3}\,(E_s)_{rms}$$

$$(E_s)_{rms} = \frac{E_{pk}}{\sqrt{2}}$$

Unfiltered rectified voltage e_R

$$E_{dc} = \frac{3\sqrt{3}}{2\pi} E_{pk} = \frac{3\sqrt{3}}{\sqrt{2}\,\pi} (E_s)_{rms}$$

$$E_{dc} = 1.17(E_s)_{rms}$$

Diode and transformer secondary current i_s

$$(I_s)_{rms} = \frac{I_{dc}}{\sqrt{3}} = 0.577 I_{dc}$$

Transformer primary phase current $(i_p)_\phi$

$$(I_p)_{\phi,rms} = \frac{1}{\sqrt{3}} \sqrt{\left(\frac{2}{3}\right)^2 \frac{1}{3} + \left(\frac{1}{3}\right)^2 \frac{2}{3}}\, I_{dc}$$

$$(I_p)_{\phi,rms} = 0.272 I_{dc}$$

TABLE 12.2 (*continued*)

Three-Phase Half-Wave Circuit Delta-Wye Transformation

Transformer primary line current $(i_p)_L$

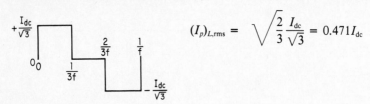

$$(I_p)_{L,\text{rms}} = \sqrt{\frac{2}{3}\frac{I_{dc}}{\sqrt{3}}} = 0.471 I_{dc}$$

Transformer volt-ampere ratings

Primary

$$(VA)_p = 3(E_p)_{\text{rms}}(I_p)_{\phi,\text{rms}} = 3\frac{\sqrt{3}E_{dc}}{1.17}(0.272 I_{dc})$$

$$(VA)_p = 1.21 E_{dc}I_{dc}$$

Secondary

$$(VA)_s = 3(E_s)_{\text{rms}}(I_s)_{\text{rms}} = 3\frac{E_{dc}}{1.17}(0.577 I_{dc})$$

$$(VA)_s = 1.48 E_{dc}I_{dc}$$

Power factor

$$\frac{E_{dc}I_{dc}}{(VA)_p} = \frac{1}{1.21} = 0.83$$

Fundamental ripple frequency

$$3f$$

Note: This circuit results in a dc unbalance current of $\frac{1}{3}I_{dc}$. Cores will saturate if three individual transformers are used. DC fluxes will cancel as shown in diagram below if one three-phase transformer with three equal reluctance flux paths is used. AC ampere-turn equality is preserved in this circuit.

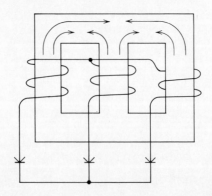

TABLE 12.2 (*continued*)

Three-Phase Full-Wave Circuit Wye-Wye Transformation

Schematic

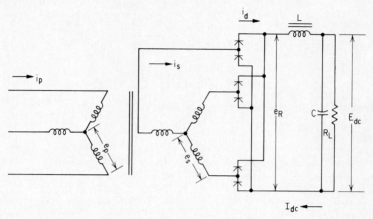

Voltage and current relationships

  Primary voltage

$$e_p = E_{pk} \sin 2\pi ft$$

  Secondary voltage

$$e_s = E_{pk} \sin 2\pi ft$$

  Unfiltered rectified voltage e_R

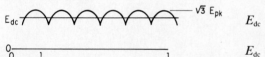

$$E_{dc} = \frac{3\sqrt{3}}{\pi} E_{pk} = \frac{3\sqrt{6}}{\pi} (E_s)_{rms}$$

$$E_{dc} = 2.34(E_s)_{rms} = 2.34(E_p)_{rms}$$

  Diode current i_d

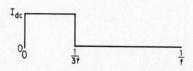

$$(I_d)_{rms} = \frac{I_{dc}}{\sqrt{3}} = 0.577 I_{dc}$$

Transformer primary and secondary circuits i_p and i_s

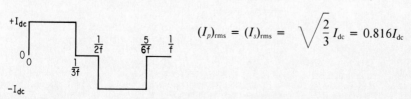

$$(I_p)_{rms} = (I_s)_{rms} = \sqrt{\frac{2}{3}}\, I_{dc} = 0.816 I_{dc}$$

12.15

TABLE 12.2 (*continued*)

Three-Phase Full-Wave Circuit Wye-Wye Transformation

Transformer volt-ampere ratings

$$(VA)_p = (VA)_s = 3(E_p)_{rms}(I_p)_{rms} = \frac{3 \times 0.816}{2.34} E_{dc}I_{dc} = 1.05 E_{dc}I_{dc}$$

Power factor

$$\frac{E_{dc}I_{dc}}{(VA)_p} = \frac{1}{1.05} = 0.95$$

Fundamental ripple frequency

$$6f$$

Three-Phase Full-Wave Circuit Delta-Wye Transformation

Schematic

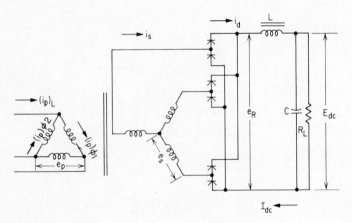

Voltage and current relationships

Primary voltage

$$e_p = \sqrt{3}\, E_{pk} \sin 2\pi ft$$

Secondary voltage

$$e_s = E_{pk} \sin 2\pi ft$$

Unfiltered rectified voltage e_R

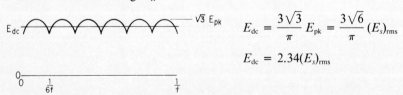

$$E_{dc} = \frac{3\sqrt{3}}{\pi} E_{pk} = \frac{3\sqrt{6}}{\pi} (E_s)_{rms}$$

$$E_{dc} = 2.34(E_s)_{rms}$$

12.16

TABLE 12.2 (*continued*)

Three-Phase Full-Wave Circuit Delta-Wye Transformation

Diode current i_d

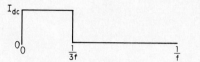

$$(I_d)_{rms} = \frac{I_{dc}}{\sqrt{3}} = 0.577I_{dc}$$

Transformer secondary current i_s

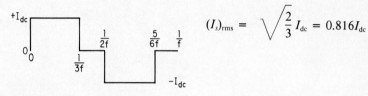

$$(I_s)_{rms} = \sqrt{\frac{2}{3}} I_{dc} = 0.816I_{dc}$$

Transformer primary phase currents

Phase 1 current $(i_p)_{\phi 1}$

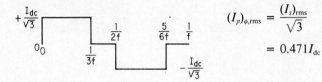

$$(I_p)_{\phi,rms} = \frac{(I_s)_{rms}}{\sqrt{3}}$$

$$= 0.471I_{dc}$$

Phase 2 current $(i_p)_{\phi 2}$

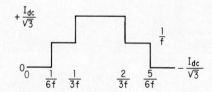

Transformer primary line current

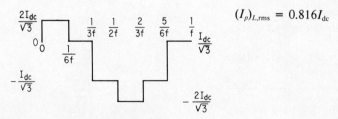

$$(I_p)_{L,rms} = 0.816I_{dc}$$

Transformer volt-ampere ratings

$$(VA)_p = (VA)_s = 3(E_s)_{rms}(I_s)_{rms} = 1.05E_{dc}I_{dc}$$

TABLE 12.2 (*continued*)

Three-Phase Full-Wave Circuit Delta-Wye Transformation

Power factor

$$\frac{E_{dc}I_{dc}}{(VA)_p} = \frac{1}{1.05} = 0.95$$

Fundamental ripple frequency

$$6f$$

Three-Phase Full-Wave Circuit Wye-Delta Transformation

Schematic

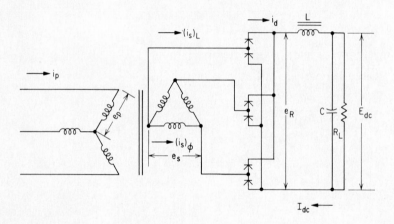

Voltage and current relationships

Primary voltage

$$e_p = E_{pk} \sin 2\pi ft$$

Secondary voltage

$$e_s = \sqrt{3}\, E_{pk} \sin 2\pi ft$$

Unfiltered rectified voltage e_R

$$E_{dc} = \frac{3\sqrt{3}}{\pi} E_{pk} = \frac{3\sqrt{6}}{\pi} (E_p)_{rms}$$

$$E_{dc} = 2.34(E_p)_{rms}$$

$$E_{dc} = \frac{2.34}{\sqrt{3}} (E_s)_{rms} = 1.35(E_s)_{rms}$$

TABLE 12.2 (*continued*)

Three-Phase Full-Wave Circuit Wye-Delta Transformation

Diode current i_d

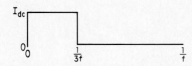

$$(I_d)_{rms} = \frac{I_{dc}}{\sqrt{3}} = 0.577I_{dc}$$

Transformer secondary line current $(i_s)_L$

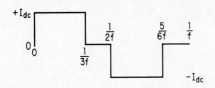

$$(I_s)_{L,rms} = \sqrt{\frac{2}{3}}\, I_{dc} = 0.816I_{dc}$$

Transformer secondary phase current $(i_s)_\phi$

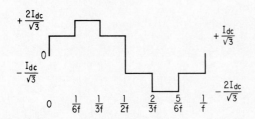

$$(I_s)_{\phi,rms} = \sqrt{\frac{1}{3}\left(\frac{2}{3}\right)^2 + \frac{2}{3}\left(\frac{1}{3}\right)^2}\, I_{dc}$$

$$(I_s)_{\phi,rms} = 0.471I_{dc}$$

Transformer primary current

$$(I_p)_{rms} = \sqrt{3}\,(I_s)_{\phi,rms}$$

$$(I_p)_{rms} = 0.816I_{dc}$$

Transformer volt-ampere ratings

Primary

$$(VA)_p = 3(E_p)_{rms}(I_p)_{rms} = 3\,\frac{E_{dc}}{2.34}\,(0.816I_{dc}) = 1.05E_{dc}I_{dc}$$

Secondary

$$(VA)_s = 3(E_s)_{rms}(I_s)_{\phi,rms} = 3\,\frac{E_{dc}}{1.35}\,(0.471I_{dc}) = 1.05E_{dc}I_{dc}$$

TABLE 12.2 (*continued*)

Three-Phase Full-Wave Circuit Wye-Delta Transformation

Power factor

$$\frac{E_{dc}I_{dc}}{(VA)_p} = \frac{1}{1.05} = 0.95$$

Fundamental ripple frequency

$$6f$$

Three-Phase Full-Wave Circuit Delta-Delta Transformation

Schematic

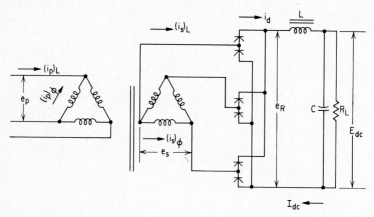

Voltage and current relationships

Primary voltage

$$e_p = E_{pk} \sin 2\pi ft$$

Secondary voltage

$$e_s = E_{pk} \sin 2\pi ft$$

Unfiltered rectified voltage e_R

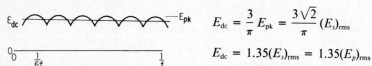

$$E_{dc} = \frac{3}{\pi} E_{pk} = \frac{3\sqrt{2}}{\pi} (E_s)_{rms}$$

$$E_{dc} = 1.35(E_s)_{rms} = 1.35(E_p)_{rms}$$

Diode current i_d

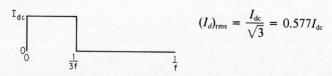

$$(I_d)_{rms} = \frac{I_{dc}}{\sqrt{3}} = 0.577 I_{dc}$$

TABLE 12.2 (*continued*)

Three-Phase Full-Wave Circuit Delta-Delta Transformation

Transformer primary and secondary line currents $(i_p)_L = (i_s)_L$

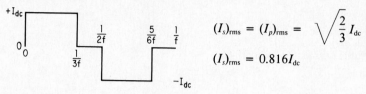

$$(I_s)_{rms} = (I_p)_{rms} = \sqrt{\frac{2}{3}}\,I_{dc}$$

$$(I_s)_{rms} = 0.816 I_{dc}$$

Transformer primary and secondary phase currents $(i_p)_\phi = (i_s)_\phi$

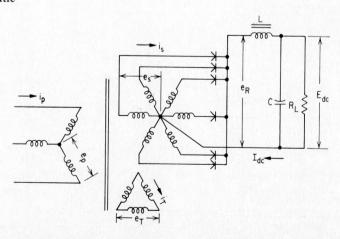

$$(I_s)_{\phi,rms} = (I_p)_{\phi,rms}$$

$$= \sqrt{\left(\frac{2}{3}\right)^2 \frac{1}{3} + \left(\frac{1}{3}\right)^2 \frac{2}{3}}\,I_{dc}$$

$$(I_s)_{\phi,rms} = 0.471 I_{dc}$$

Transformer primary and secondary volt-ampere ratings

$$(VA)_p = (VA)_s = 3(E_s)_{rms}(I_s)_{\phi,rms} = 3\left(\frac{0.471}{1.35}\right) E_{dc}I_{dc} = 1.05 E_{dc}I_{dc}$$

Power factor

$$\frac{E_{dc}I_{dc}}{(VA)_p} = \frac{1}{1.05} = 0.95$$

Fundamental ripple frequency

$$6f$$

Three-Phase Full-Wave Center-Tap Circuit Wye-Wye Transformation

Schematic

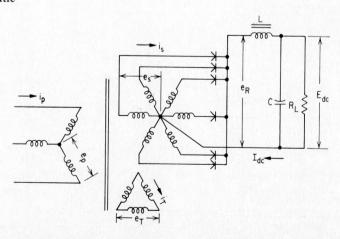

12.21

TABLE 12.2 (*continued*)

Three-Phase Full-Wave Center-Tap Circuit Wye-Wye Transformation

Voltage and current relationships

Primary voltage

$$e_p = E_{pk} \sin 2\pi ft$$

Secondary and tertiary voltages

$$e_s = e_T = E_{pk} \sin 2\pi ft$$

Unfiltered rectified voltage

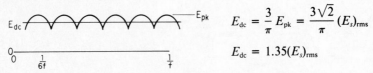

$$E_{dc} = \frac{3}{\pi} E_{pk} = \frac{3\sqrt{2}}{\pi} (E_s)_{rms}$$

$$E_{dc} = 1.35(E_s)_{rms}$$

Diode and transformer secondary current

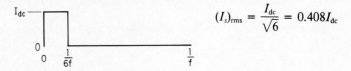

$$(I_s)_{rms} = \frac{I_{dc}}{\sqrt{6}} = 0.408 I_{dc}$$

Primary current

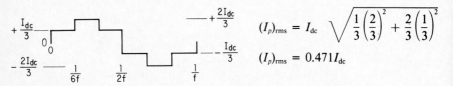

$$(I_p)_{rms} = I_{dc} \sqrt{\frac{1}{3}\left(\frac{2}{3}\right)^2 + \frac{2}{3}\left(\frac{1}{3}\right)^2}$$

$$(I_p)_{rms} = 0.471 I_{dc}$$

Tertiary winding current

$$(I_t)_{rms} = \frac{I_{dc}}{3}$$

Transformer volt-ampere ratings

Secondary

$$(VA)_s = 6(E_s)_{rms}(I_s)_{rms} = 6\frac{E_{dc}}{1.35}(0.408)I_{dc} = 1.81 E_{dc}I_{dc}$$

Primary

$$(VA)_p = 3(E_p)_{rms}(I_p)_{rms} = 3\frac{E_{dc}}{1.35}(0.471I_{dc}) = 1.05 E_{dc}I_{dc}$$

TABLE 12.2 (*continued*)

Three-Phase Full-Wave Center-Tap Circuit Wye-Wye Transformation

---

Tertiary

$$(VA)_t = 3(E_t)_{rms}(I_t)_{rms} = 3\frac{E_{dc}}{1.35} = \frac{I_{dc}}{3} = 0.74 E_{dc}I_{dc}$$

Power factor

$$\frac{E_{dc}I_{dc}}{(VA)_p} = \frac{1}{1.05} = 0.95$$

Fundamental ripple frequency

$$6f$$

Tertiary delta-connected winding required in this circuit to maintain ac ampere-turn balance.

Three-Phase Full-Wave Center-Tap Circuit Delta-Wye Transformation

---

Schematic

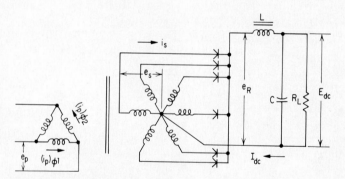

Voltage and current relationships

Primary voltage

$$e_p = \sqrt{3}\, E_{pk} \sin 2\pi ft$$

Secondary voltage

$$e_s = E_{pk} \sin 2\pi ft$$

Unfiltered rectified voltage

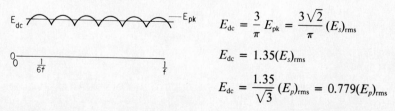

$$E_{dc} = \frac{3}{\pi} E_{pk} = \frac{3\sqrt{2}}{\pi}(E_s)_{rms}$$

$$E_{dc} = 1.35(E_s)_{rms}$$

$$E_{dc} = \frac{1.35}{\sqrt{3}}(E_p)_{rms} = 0.779(E_p)_{rms}$$

12.23

TABLE 12.2 (*continued*)

Three-Phase Full-Wave Center-Tap Circuit Delta-Wye Transformation

---

Diode and transformer secondary current

$$(I_s)_{rms} = \frac{I_{dc}}{\sqrt{6}} = 0.408 I_{dc}$$

Primary phase current

Phase 1

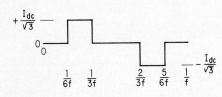

$$(I_\phi)_{rms} = \frac{I_{dc}}{\sqrt{3}} \sqrt{\frac{1}{3}} = \frac{I_{dc}}{3}$$

Phase 2

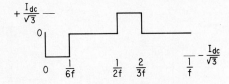

Primary line current

$$(I_p)_{rms} = \frac{I_{dc}}{\sqrt{3}} \sqrt{\frac{2}{3}} = 0.471 I_{dc}$$

Transformer volt-ampere ratings

Primary

$$(VA)_p = 3(E_p)_{rms}(I_\phi)_{rms} = 3 \frac{E_{dc}}{0.779} \frac{I_{dc}}{3} = 1.28 E_{dc} I_{dc}$$

Secondary

$$(VA)_s = 6(E_s)_{rms}(I_s)_{rms} = 6 \frac{E_{dc}}{1.35} (0.408 I_{dc}) = 1.81 E_{dc} I_{dc}$$

Power factor

$$\frac{E_{dc} I_{dc}}{\sqrt{3} \, (E_p)_{rms}(I_p)_{rms}} = \frac{E_{dc} I_{dc}}{\sqrt{3}} \frac{0.779}{E_{dc}} \frac{1}{0.471 I_{dc}} = 0.95$$

12.24

TABLE 12.2 (*continued*)

Three-Phase Full-Wave Center-Tap Circuit Delta-Wye Transformation

---

Fundamental ripple frequency

$$6f$$

Three-Phase Full-Wave Circuit Delta-Delta-Wye Transformation

Schematic

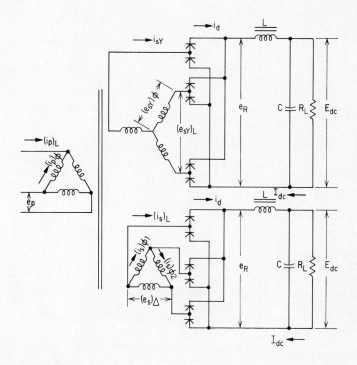

Voltage-current relationships

  Primary voltage

$$e_p = \sqrt{3}\, E_{pk} \sin 2\pi ft$$

Secondary voltages

  Wye connection

$$(e_{sY})_\phi = E_{pk} \sin 2\pi ft \qquad (e_{sY})_L = \sqrt{3}\, E_{pk} \sin\left(2\pi ft + \frac{\pi}{6}\right)$$

Delta connection

$$(e_s)_\Delta = \sqrt{3}\, E_{pk} \sin 2\pi ft$$

12.25

TABLE 12.2 *(continued)*

Three-Phase Full-Wave Circuit Delta-Delta-Wye Transformation

Unfiltered rectified voltages

  Wye output

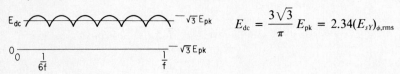

$$E_{dc} = \frac{3\sqrt{3}}{\pi} E_{pk} = 2.34(E_{sY})_{\phi,rms}$$

  Delta output

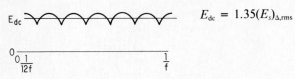

$$E_{dc} = 1.35(E_s)_{\Delta,rms}$$

Diode current i_d

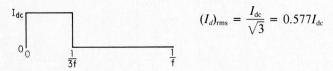

$$(I_d)_{rms} = \frac{I_{dc}}{\sqrt{3}} = 0.577 I_{dc}$$

Transformer secondary currents

  Wye current i_{sY}

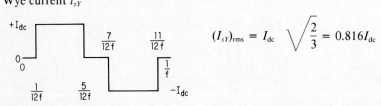

$$(I_{sY})_{rms} = I_{dc}\sqrt{\frac{2}{3}} = 0.816 I_{dc}$$

  Delta current phase 1 $(i_s)_{\phi 1}$

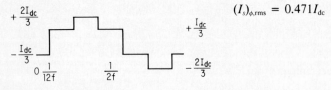

$$(I_s)_{\phi,rms} = 0.471 I_{dc}$$

  Delta current phase 2 $(i_s)_{\phi 2}$

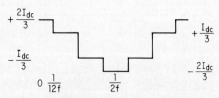

12.26

TABLE 12.2 (*continued*)

Three-Phase Full-Wave Circuit Delta-Delta-Wye Transformation

Delta line current $(i_s)_L$

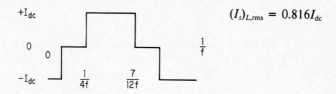

$(I_s)_{L,\text{rms}} = 0.816 I_{\text{dc}}$

Transformer primary currents

Phase current $(i_p)_\phi$

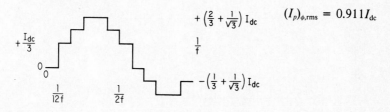

$+\left(\frac{2}{3} + \frac{1}{\sqrt{3}}\right) I_{\text{dc}}$

$(I_p)_{\phi,\text{rms}} = 0.911 I_{\text{dc}}$

$\frac{1}{f}$

$-\left(\frac{1}{3} + \frac{1}{\sqrt{3}}\right) I_{\text{dc}}$

Line current $(i_p)_L$

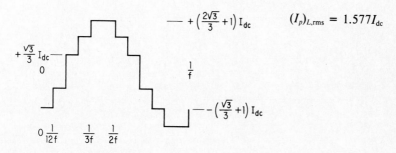

$+\left(\frac{2\sqrt{3}}{3} + 1\right) I_{\text{dc}}$

$(I_p)_{L,\text{rms}} = 1.577 I_{\text{dc}}$

$\frac{1}{f}$

$-\left(\frac{\sqrt{3}}{3} + 1\right) I_{\text{dc}}$

Transformer volt-ampere ratings

Secondary wye winding

$$(VA)_{sY} = 3(E_s)_{\text{rms}}(I_{sY})_{\text{rms}} = 3\,\frac{E_{\text{dc}}}{2.34}\,(0.816 I_{\text{dc}}) = 1.05 E_{\text{dc}} I_{\text{dc}}$$

Secondary delta winding

$$(VA)_{s\Delta} = 3\sqrt{3}\,\frac{E_{\text{dc}}}{2.34}\,(0.471 I_{\text{dc}}) = 1.05 E_{\text{dc}} I_{\text{dc}}$$

Primary winding

$$(VA)_p = 3\sqrt{3}\,\frac{E_{\text{dc}}}{2.34}\,(0.911 I_{\text{dc}}) = 2.02 E_{\text{dc}} I_{\text{dc}}$$

TABLE 12.2 (*continued*)

Three-Phase Full-Wave Circuit Delta-Delta-Wye Transformation

Power factor

$$\frac{2E_{dc}I_{dc}}{(VA)_p} = \frac{2}{2.02} = 0.99$$

E_{dc} and I_{dc} are voltage and current outputs for either delta or wye secondary. The two secondaries may be connected in parallel to provide $2I_{dc}$ at E_{dc} with the use of a balancing transformer, or they may be connected in series to provide I_{dc} at $2E_{dc}$ by insulating the secondary windings for the developed voltages. The fundamental ripple frequency will be $12f$ for either connection.

Source: The information in this table is taken in part from J. Schaefer, *Rectifier Circuits: Theory and Design,* Wiley, New York, 1965, used by permission.

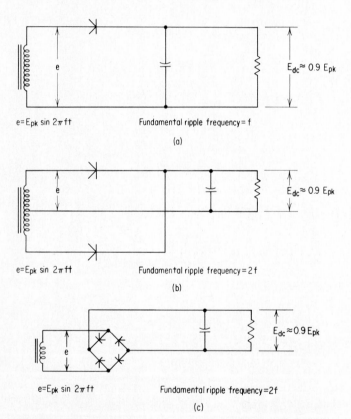

FIG. 12.1 Frequently used capacitor input rectifier circuits. (*a*) single-phase half-wave; (*b*) single-phase, full-wave, center-tap; (*c*) single-phase full-wave bridge; (*d*) three-phase-wye full-wave bridge; (*e*) single-phase full-wave doubler; (*f*) single-phase cascade doubler; (*g*) single-phase cascade quadrupler. Additional sections may be added to the single-phase cascade quadrupler (*g*) to increase the voltage multiplication but with reduced multiplying efficiency.

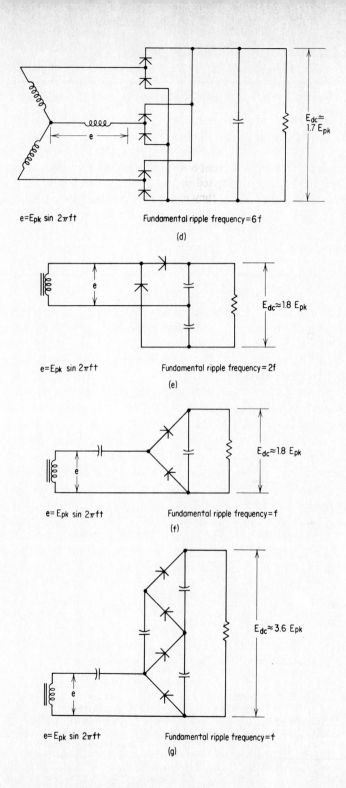

$e = E_{pk} \sin 2\pi ft$ Fundamental ripple frequency $= 6f$

(d)

$E_{dc} \approx 1.7 E_{pk}$

$e = E_{pk} \sin 2\pi ft$ Fundamental ripple frequency $= 2f$

(e)

$E_{dc} \approx 1.8 E_{pk}$

$e = E_{pk} \sin 2\pi ft$ Fundamental ripple frequency $= f$

(f)

$E_{dc} \approx 1.8 E_{pk}$

$e = E_{pk} \sin 2\pi ft$ Fundamental ripple frequency $= f$

(g)

$E_{dc} \approx 3.6 E_{pk}$

12.29

12.2 CHOKE INPUT FILTER CIRCUITS

A choke input filter is a ripple-reducing filter used with rectifier circuits in which a filter inductor, or choke, in series, forms the input element. Some of the properties of choke input circuits are given in Table 12.2. Choke input filters provide better regulation than capacitor input filters, and they allow better utilization of the volt-ampere capacity of the rectifier transformer. Filter chokes are bulky. They often make the equipment size, weight, and cost greater than for an equivalent capacitor input filter circuit.

12.3 CRITICAL INDUCTANCE

When the inductance of the choke in a choke input filter is relatively low, so that the assumption of infinite inductance cannot be made, the current through the choke will consist of two components, a dc component equal to the output voltage divided by the load resistance and an ac component controlled by the reactance of the choke, which in most cases is sufficiently high to neglect the effect of the filter capacitor and load. These currents are shown in Fig. 12.2. When the dc load current is greater than the peak value of the ac current, the average value of the sum of these two currents is unaffected by the value of the ac component. This is shown in Fig. 12.2a. If the inductance is continuously decreased, the ac component of current will increase until its peak value is equal to and then greater than the average value of the filter input voltage divided by the load resistance. These conditions are illustrated in Fig. 12.2b and c. The diodes in the rectifier circuit prevent the flow of negative current. In Fig. 12.2c the average value of the current in the choke, which includes the contribution of the ac component of current, must flow through the load. This causes the voltage across the load to rise above the average

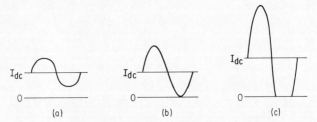

FIG. 12.2 Sum of ac and dc components of current in the inductance of a choke input filter. (*a*) Inductance above critical value; (*b*) inductance equal to critical value; (*c*) inductance less than critical value.

value of the filter input voltage. The circuit no longer operates as a choke input filter, but displays the characteristics of a capacitor input filter. The threshold for this change from inductive to capacitive input is the point where the peak ac current equals the average filter input voltage divided by the load resistance. This value of current can be determined from the formulas in Table 1.1. By equating formula 9 and either formula 6, 7, or 8, an expression for the inductance at the threshold is obtained. This inductance is called the *critical inductance.* The derivation for the full-wave single-phase circuit is as follows:

$$B_{max} = \frac{LI_{max} \times 10^8}{NA}$$

$$= \frac{V_{dc} \times 10^8}{19NAf}$$

$$V_{dc} = R_L I_{dc}$$

$$= R_L I_{max}$$

$$L = \frac{R_L}{19f} \tag{12.1}$$

Table 12.3 gives the formulas for the critical inductance of rectifier circuits that are most commonly used with choke input filters.

The critical inductance behavior of choke input filters places a lower limit on load current which, if passed, will cause the output voltage to rise excessively. To prevent this, a bleeder resistor may be placed across the filter output to ensure that the dc current through the choke is always above the critical limit.

TABLE 12.3 Formulas for Critical Inductance in Choke Input Filter Circuits

Rectifier Circuit	General Formula*	Formula for 60-Hz Line Frequency
Single-phase full-wave center-tap or bridge	$L = \dfrac{R_L}{19f}$	$L = \dfrac{R_L}{1140}$
Three-phase half-wave	$L = \dfrac{R_L}{75.9f}$	$L = \dfrac{R_L}{4554}$
Three-phase full-wave	$L = \dfrac{R_L}{664f}$	$L = \dfrac{R_L}{3.98 \times 10^4}$

*In the above formulas L is the inductance in henrys, R_L is the load resistance in ohms, and f is the line frequency in hertz.

12.4 RIPPLE REDUCTION IN CHOKE INPUT FILTERS

The attenuation of a choke input filter is approximately equal to the ratio of the reactance of the filter capacitor to the algebraic sum of the reactances of the filter choke and the filter capacitor:

$$r = \frac{1/j\omega c}{j\omega L + 1/j\omega c}$$

$$= \frac{1}{1 - \omega^2 LC} \tag{12.2}$$

In Eq. (12.2), $\omega = 2\pi f$ where f is in hertz, L is in henrys, and C is in farads. The frequency f is the frequency being attenuated. Since the filter attenuation increases with frequency, it is generally not necessary to consider frequencies above the fundamental ripple frequency when choosing the values of L and C. Care must be taken to make $\omega^2 LC$ larger than 2 regardless of the ripple reduction required to avoid the resonant condition where the ripple voltage is actually increased instead of attenuated.

The attenuation of the filter is dependent upon the LC product. Any value of inductance above the critical value will work. The ratio L/C is chosen for economy. The size of the filter choke for a given inductance varies with the square of the dc current. A high-current, low-voltage supply will have a lower L/C ratio than a higher voltage supply with the same power and ripple requirement. Sometimes the total size of the filter can be reduced by using two stages of filtering. The total ripple reduction is equal to the product of the ripple reduction factors of each individual stage. In a two-stage filter only the input stage requires an inductance above the critical value. The fundamental ripple frequencies given in Table 12.1 and Fig. 12.1 are based upon ideal conditions. These conditions include perfect symmetry in the transformer and rectifiers and perfectly balanced three-phase voltages. Lack of symmetry in the rectifier will cause an additional ripple voltage of frequency equal to the line frequency. Asymmetry in the rectifier transformer causes an additional component in the ripple voltage of line frequency or twice line frequency, depending on the circuit. These asymmetries are controllable within limits. The line voltage unbalance in three-phase circuits is generally not controllable or predictable with accuracy. The greater the sophistication of the rectifier circuit is in reducing ripple voltage amplitude and increasing ripple frequency, the more sensitive it is to asymmetry from any source. The design of the filter must generally consider the estimated value of these aberrant components of ripple voltage. This consideration may govern the filter characteristics.

12.5 SWINGING CHOKES

In choke input filter supplies having varying load currents, the current rating of the choke must be the highest value of load current, while the inductance of the choke must be greater than the critical value determined from the lowest value of load current. The size penalty resulting from rating a choke in this manner is sometimes avoided by using a swinging choke. This is a choke with dual ratings, a high inductance at the lowest load current and a low inductance at the highest current, both inductance values being above the critical value for their respective currents. The change in inductance is achieved by using a small air gap appropriate for the lower current and allowing the core to partially saturate at the higher current, causing a reduction in inductance. The core may be assembled with two different air gaps, creating in effect two magnetic paths, each path fully responsive to one current limit and partially responsive to the other current limit. The swinging choke permits a size reduction over an equivalent linear choke, but there is no improvement in electrical performance, the ripple attenuation being less at the higher current because of the reduced inductance.

12.6 OPERATING CONDITIONS AFFECTING FILTER CHOKES

The inductance of the filter choke will vary with the ac voltage and frequency developed across it and the direct current flowing through it. (See Sec. 1.6.) For low values of ac voltage, the inductance will be lower, corresponding to the initial permeability of the core material. As the ac voltage increases, the inductance will reach a maximum and then decline with the onset of saturation. The inductance will also gradually decrease with increasing direct current until the onset of saturation, beyond which the inductance drops rapidly.

A filter choke placed in the high side of a rectifier circuit must be insulated from ground for the dc voltage. It must also be insulated to withstand the induced ripple voltage. Sometimes the lead nearer the core is connected to the core, and the entire core and coil are insulated for the dc voltage. A choke connected in this manner may be placed in the ground return leg, eliminating the need for dc voltage insulation. Transient conditions subject filter chokes to greater voltage stresses than they experience in steady-state operation. If the power supply is turned on suddenly at full-line voltage, known as a hard start, the filter capacitor is initially uncharged, resulting in the full-peak-rectified voltage being applied across the choke. The same thing will happen more violently if the output of the supply is momentarily short-circuited. To prevent damage by these transients in high-voltage circuits, spark gaps are sometimes placed across the terminals of the filter choke.

12.7 REGULATION IN CHOKE INPUT FILTER CIRCUITS

The regulation of a choke input filter circuit is the ratio of the difference between the dc output voltage at minimum current and the dc output voltage at maximum current over the dc output voltage at maximum current, usually expressed in percent. Peaking effects caused by load currents below the critical limit are excluded by a bleeder resistor. The regulation is then determined by the sum of all the dc resistances of the circuit, the voltage drop across the diodes, and the leakage inductance of the transformer. The dc circuit resistance includes the resistance of the choke, the transformer winding resistance, and diode resistance (if not included in the diode forward voltage characteristic). The transformer winding resistance is the resistance of only one path in those circuits where the direct current switches among several paths and includes the primary resistance referred to the secondary winding. The dc voltage drop across the resistances is determined by the actual current when it is flowing, not an average or rms value. With all resistances which contribute to the regulation being in series, the contribution of the choke resistance is weighted the same as other resistances. Minimum size will usually result when the

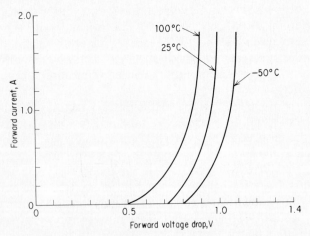

FIG. 12.3 Typical forward voltage characteristic of a silicon diode.

choke is designed first to be temperature rise–limited. The output voltage of the rectifier transformer can then be adjusted to compensate for the drop through the choke. The reverse procedure where the resistance of the choke is arbitrarily specified can sometimes lead to inductors of unreasonable size.

The voltage drop across the diodes during the time that current is flowing through them reduces the output voltage. The forward voltage drop is nonlinear. A typical characteristic is shown in Fig. 12.3. From this figure the drop across each diode element is seen to be approximately 1.0 V and is relatively constant over the upper 75 percent of its operating range. When this drop is critical to circuit operation, detailed information on the specific diodes used should be obtained.

12.8 EFFECT OF LEAKAGE INDUCTANCE ON DC OUTPUT VOLTAGE

Under the assumption of a constant current flowing through a choke of infinite inductance, the current will switch abruptly among the branches of a rectifier circuit. The voltage drop across the leakage inductance during switching and the disposition of the stored energy in the leakage inductance affect regulation. Since the output voltage is related to average values, only the average value of the voltage across the leakage inductance is needed. The average value of the voltage across an inductance subjected to a square pulse of current is as follows:

$$\Delta e_{av} = \frac{1}{\tau} \int_0^\tau L \frac{di}{dt} \, dt$$

$$= \frac{L}{\tau} \int_0^\tau di \tag{12.3}$$

In Eq. (12.3) τ is the interval over which the average is taken. Let $\Delta\tau$ be the interval during which switching occurs. Then

$$\Delta e_{av} = \frac{L}{\tau} \left(\int_0^{\Delta\tau} di + \int_{\Delta\tau}^{\tau-\Delta\tau} di + \int_{\tau-\Delta\tau}^\tau di \right) \tag{12.4}$$

Substituting the known values of current yields

$$\Delta e_{av} = \frac{L}{\tau} (I_{dc} - 0 + I_{dc} - I_{dc} + 0 - I_{dc}) \tag{12.5}$$

Only the intervals during which the current is changing contribute to the voltage drop across the inductor, and the voltage developed by the current rise cancels the voltage developed by the current decay. Note that the integral is dependent on the limits of integration and the current values at those limits, not on the rate of change of current. For the interval 0 to $\Delta\tau$

$$\Delta e_{av} = \frac{LI_{dc}}{\tau} \tag{12.6}$$

The voltage drop represented by Eq. (12.6) is associated with the energy $1/2\ I_{dc}^2 L$ stored in the inductance. In a lossless circuit, the negative voltage $-LI_{dc}/\tau$ associated with the decay of current would cancel the positive voltage and the stored energy would be recovered. In practical circuits the stored energy is not recovered. Consider the full-wave single-phase circuit of Fig. 12.4. In this circuit, CR_1 is initially conducting, being forward-biased by the positive transformer voltage e_1, and CR_2 is cut off, being negatively biased by the negative transformer voltage e_2. Voltages e_1 and e_2 now change their polarities. Voltage e_2 reaches a positive value above E_{dc}. CR_2 is at that moment forward-biased, and current i_2 begins to flow. A voltage drop develops across L_{L2} whose average value is given by Eq. (12.6). While this is occurring, a negative bias is developing on CR_1 while forward current i_1 is still flowing. Current i_1 begins to decrease. A transient voltage develops across L_{L1} whose amplitude is determined by the rate of decay of current i_1 and the value of the leakage inductance. This transient voltage tends to maintain a positive bias on CR_1. During this interval when both i_1 and i_2 are flowing, neither rectifier supports voltage, and a short circuit exists across the transformer winding. The resulting short-circuit current is limited only by the transformer winding resistances, the leakage inductance, and the rectifier resistance. The path of this short-circuit current is shown in Fig. 12.4. The short-circuit cur-

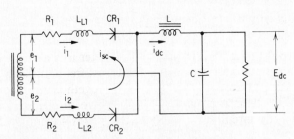

FIG. 12.4 Single-phase full-wave choke input rectifier circuit showing path of commutating current.

rent, which flows until i_1 reaches zero, dissipates in the transformer winding resistance and in the rectifier resistance the energy that was stored in leakage inductance L_{L1}. Rectifier CR_1, in the process of turning off, may present during this interval a relatively high resistance, thereby being forced to absorb much of the energy from the leakage inductance. Thus the rectifier is subject to considerable abuse during this switching process, called *commutation* by analogy with rotating machinery.

Because the energy stored in the leakage inductance is not recovered, the third term in Eq. (12.4) does not affect the average voltage drop. The first and only contributing term of that equation is evaluated in Eq. (12.6). The interval τ over which the voltage is averaged is the period during which one current pulse flows through the leakage inductance. This time is a function of the power line frequency and the type of circuit. Table 12.4 gives the formulas for leakage inductance drop for some common rectifier circuits. These formulas give the total voltage drop for all branches of multiple-branch circuits. Note that L_L in the table is the leakage inductance of one branch circuit. For example, in a three-phase wye-connected secondary, L_L is the line-to-neutral leakage inductance of one leg of the wye.

TABLE 12.4 Formulas for Calculating Voltage Drop across Leakage Inductance in Choke Input Filters

Circuit	Voltage Drop*
Single-phase half-wave	fL_LI_{dc}
Single-phase full-wave	$2fL_LI_{dc}$
Three-phase half-wave wye	$3fL_LI_{dc}$
Three-phase full-wave wye	$6fL_LI_{dc}$

*In the above formulas, f is in hertz, L_L is the leakage inductance of one secondary branch in henrys, and I_{dc} is the dc current in the branch in amperes.

12.9 CAPACITOR INPUT FILTER CIRCUITS

A capacitor input filter is a ripple-reducing filter in which a shunt capacitor forms the input circuit element. The most frequently used capacitor input filter circuits are shown in Fig. 12.1 together with some of their properties. Capacitor input filter circuits have poorer regulation than choke input filter circuits. The volt-ampere rating of the rectifier transformer for capacitor input filter circuits is higher than for an equivalent choke input filter. In spite of this, power supplies with capacitor input filters are smaller and lighter than those with choke input filters. Capacitor input filters are very sensitive to the source impedance. This source impedance is generally the winding resistance and leakage inductance of the rectifier transformer. The winding resistance includes the secondary winding resistance and the primary winding resistance referred to the secondary. In rectifier circuits with multiple branches, the impedance of one branch is separated and used to determine the filter properties. In the following sections the behavior of capacitor input filters is determined by means of various simplifying assumptions. None of these assumptions is completely rigorous. If the most appropriate set of assumptions is chosen for the particular application, the analysis will usually give sufficiently accurate results. It should be recognized that the substantial regulation that exists with capacitor input filters prevents them from being precision sources of dc voltage. Included in capacitor input filters are the various types of voltage-multiplier circuits. Their properties may be determined by the same methods as for other capacitor input circuits.

12.10 ANALYSIS ASSUMING INFINITE CAPACITANCE AND ZERO LEAKAGE INDUCTANCE

The simplest approximate analysis of capacitor input filters assumes that the input capacitance is infinite and the leakage inductance is zero. These assumptions are reasonable when the product $2\pi f C R_L$ is greater than 100 and the ratio $2\pi f L_L/R_w$ is less than 0.1. In these terms, f is the line frequency in hertz, C is the capacitance in farads, R_L is the load resistance in ohms, L_L is the leakage inductance of one rectifier branch in henrys referred to the secondary, and R_w is the total winding resistance in ohms. R_w includes all other circuit resistance, if significant. The assumption of infinite capacitance allows the load resistance and capacitance to be represented by a battery whose voltage is equal to the dc output voltage. Zero ripple voltage is inferred. The analysis of an isolated diode circuit, Fig. 12.5a, constituting a single-phase half-wave circuit, may be applied to multiple- and full-wave circuits. The process of applying this analysis is typical of that used for less restrictive assumptions.

Figure 12.5b is the equivalent circuit for the single diode circuit of Fig. 12.5a. Figure 12.5c is the waveform of the current i that flows in that equivalent circuit. This current starts to flow when e reaches a value equal to E_{dc}. This permits i to be described mathematically as follows:

$$i = \frac{e - E_{dc}}{R_w}$$

$$= \frac{E_{pk}\sin\omega t - E_{dc}}{R_w}\Bigg|_{\theta/\omega}^{(\pi-\theta)/\omega} \tag{12.7}$$

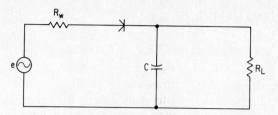

FIG. 12.5a Isolated diode segment of typical rectifier circuit.

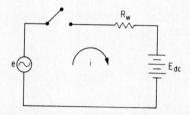

FIG. 12.5b Approximate equivalent circuit of isolated diode segment.

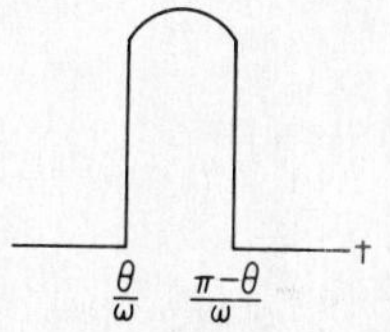

FIG. 12.5c Waveform of isolated diode current pulse in equivalent circuit of Fig. 12.5b.

where $\theta = \sin^{-1} \dfrac{E_{dc}}{E_{pk}}$ (12.8)

$E_{dc} = I_{dc}R_L$ (12.9)

Substituting Eqs. (12.8) and (12.9) in Eq. (12.7) and rearranging produces the following:

$$\frac{i}{I_{dc}} = \frac{R_L}{R_w} \left(\frac{\sin \omega t}{\sin \theta} - 1 \right) \Bigg|_{\theta/\omega}^{(\pi-\theta)/\omega} \qquad (12.10)$$

The average value of i for a single-phase half-wave circuit will be equal to I_{dc}. Therefore, if Eq. (12.10) is integrated between the limits of θ/ω and $(\pi-\theta)/\omega$, then averaged over a full cycle of line frequency, the result will be equal to 1:

$$f \int_0^{1/f} \frac{i}{I_{dc}} \, dt = 1$$

$$= \frac{R_L f}{R_w} \int_{\theta/\omega}^{(\pi-\theta)/\omega} \left(\frac{\sin \omega t}{\sin \theta} - 1 \right) dt \qquad (12.11)$$

The result of this integration is

$$\frac{R_w}{R_L} = \frac{2 \cot \theta - \pi + 2\theta}{2\pi} \qquad (12.12)$$

In Eq. (12.9) it was assumed that all the load current came from a single diode and that there were no multiplying circuits. To accommodate these different conditions, a factor n is introduced as follows:

$$E_{dc} = nI_{dp}R_L \qquad (12.13)$$

in which I_{dp} is the direct current resulting from 1 current pulse per cycle. Because of Eq. (12.13), Eq. (12.12) becomes:

$$\frac{R_w}{nR_L} = \frac{2 \cot \theta - \pi + 2\theta}{2\pi} \qquad (12.14)$$

In Eq. (12.13) n is equal to 2 for a single-phase full-wave circuit, since 2 current pulses per cycle from two different diode circuits are used to produce E_{dc}. A voltage-doubler circuit may be regarded as two diode circuits in series. Viewed in this manner, n would be equal to 1 for a voltage-doubler circuit if it were not for the fact that the actual R_L is twice as great as the resistance across one of the diode circuits. Therefore, n is equal to one-half for a voltage-doubler circuit. By similar reasoning, values for n for commonly used rectifier circuits have been determined and listed in Table 12.5. A plot of Eq. (12.14) is given in Fig. 12.6.

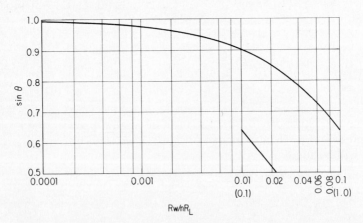

FIG. 12.6 Normalized source resistance over load resistance vs. sine of starting angle θ for infinite filter capacitor rectifier circuits, Eq. (12.14).

TABLE 12.5 Current-Multiplying Factors for Commonly Used Capacitor Input Rectifier Circuits*

Circuit	n	$(I_{rms})_{diode}/I_{hp}$	$(I_{rms})_{winding}/I_{hp}$
Voltage-doubler	½	1	$\sqrt{2}$
Multiplier (with M multiplying factor)	$\dfrac{1}{M}$	1	$\sqrt{2}$
Single-phase half-wave	1	1	1
Single-phase full-wave center-tapped	2	1	1
Single-phase full-wave bridge	2	1	$\sqrt{2}$
Three-phase half-wave wye-connected secondary winding	3	1	1
Three-phase full-wave wye-connected secondary winding	6	$\sqrt{2}$	2
Three-phase full-wave delta-connected secondary winding	6	$\sqrt{2}$	$\dfrac{2}{\sqrt{3}}$

*Legend

I_{dc} = output current in amperes.

I_{dp} = direct current in amperes resulting from 1 current pulse per cycle of line frequency.

$(I_{rms})_{diode}$ = rms current in amperes in one diode.

I_{hp} = rms value in amperes of 1 current pulse per cycle of line frequency.

$(I_{rms})_{winding}$ = rms value in amperes of the current in the transformer secondary winding.

$n = E_{dc}/I_{dp}R_L = I_{dc}/I_{dp}$ where R_L is the dc load resistance in ohms and E_{dc} is the dc output voltage across the load in volts.

The ratio of the rms current in 1 pulse per cycle, symbol I_{hp}, to the dc current for a single pulse per cycle, symbol I_{dp}, may be determined by applying the definition of rms current to Eq. (12.10) and using the values of n in Table 12.5:

$$\frac{I_{hp}}{I_{dp}} = \frac{nR_L}{R_w} \sqrt{f \int_{\theta/\omega}^{(\pi-\theta)/\omega} \left(\frac{\sin \omega t}{\sin \theta} - 1 \right)^2 dt} \qquad (12.15)$$

Performing the integration in Eq. (12.15) yields

$$\frac{I_{hp}}{I_{dp}} = \frac{nR_L}{R_w} \sqrt{\frac{\pi - 2\theta + 2\sin\theta\cos\theta}{4\pi\sin^2\theta} - \frac{2\cot\theta}{\pi} + \frac{1}{2} - \frac{\theta}{\pi}} \qquad (12.16)$$

The radical term in Eq. (12.16) requires separate evaluation:

$$f(\theta) = \sqrt{\frac{\pi - 2\theta + 2\sin\theta\cos\theta}{4\pi\sin^2\theta} - \frac{2\cot\theta}{\pi} + \frac{1}{2}\frac{\theta}{\pi}} \quad (12.17)$$

Eq. (12.17) is plotted in Fig. 12.7.

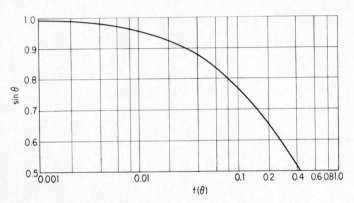

FIG. 12.7   Root mean square current function $f(\theta)$ vs. sine of starting angle θ for infinite filter capacitor rectifier circuits, Eq. (12.17).

The rms current in one pulse per cycle is related to the rms current in the diode and in the transformer winding by additional factors that account for multiple pulses per cycle occurring in the diode and transformer winding in some rectifier circuits. These factors are given in Table 12.5. This table also gives the ratio of dc load current to I_{dp}.

To apply the information in this section, the dc load voltage E_{dc}, dc load current I_{dc}, and the rectifier circuit used must be known, and the winding or source resistance R_w must be known or estimated. The load resistance is calculated from the known dc load voltage and current. The value of n is determined from Table 12.5. Sin θ is determined from Fig. 12.6. From sin θ the required value of peak ac secondary voltage is determined. From Fig. 12.7 $f(\theta)$ is determined. From Table 12.5 the factor relating dc load current to I_{dp} is obtained, and I_{dp} is calculated. From Eq. (12.16) I_{hp} is calculated using the values of $f(\theta)$ and I_{dp} just obtained. The rms current in the winding is determined from I_{hp} by using the factor in Table 12.5. The peak ac voltage on the secondary and the rms current in the secondary winding are data needed to design the transformer. After a trial design has been executed based on an estimated value of R_w, the process may be repeated with a more accurately known value of R_w. The rms currents in the primaries of transformers in which the secondaries have asymmetrical currents may be determined from Table 12.2, since the ratio of rms primary current to secondary current is the same for both choke and capacitor input rectifier circuits.

The conduction angle of 1 current pulse is $\pi - 2\theta$. For all the current pulses to have the same conduction angle, the following relationship must exist:

$$\pi - 2\theta \leqq 2\pi \frac{I_{dp}}{I_{dc}} \tag{12.18}$$

If an unusually large value of R_w/R_L yields a value of θ which violates Eq. (12.18), the method of this section is not valid. When this occurs, the transformer design should be examined to see that the temperature rise and regulation are acceptable.

Illustrative problem

A dc load requires 100 V dc at 1 A dc. The total source resistance is 2 Ω. Determine the rms no-load voltage on the secondary winding of the transformer and the rms current in the secondary winding for (1) a single-phase full-wave bridge circuit and (2) a three-phase full-wave bridge circuit with a wye-connected secondary transformer winding. Compare the volt-ampere ratings of both transformers to the power delivered to the load.

Solution

1. Single-phase full-wave bridge circuit

$$R_L = \frac{E_{dc}}{I_{dc}} = \frac{100}{1} = 100 \ \Omega$$

From Table 12.5:

$$n = 2 \qquad \frac{R_w}{nR_L} = \frac{2}{2 \times 100} = 0.01$$

From Fig. 12.6:

$$\sin \theta = 0.905 = \frac{E_{dc}}{E_{pk}}$$

$$E_{pk} = \frac{100}{0.905} = 110.5; \qquad E_{rms} = \frac{110.5}{\sqrt{2}} = 78.1 \ V$$

From Table 12.5:

$$I_{dp} = \frac{I_{dc}}{2} = 0.5 \text{ A}$$

From Fig. 12.7:

$$f(\theta) = 0.0285$$

From Eq. (12.16):

$$I_{hp} = \frac{0.5 \times 2 \times 100}{2} \times 0.0285 = 1.425$$

From Table 12.5:

$$I_{rms,winding} = \sqrt{2}I_{hp} = 1.425\sqrt{2} = 2.01 \text{ A}$$

2. **Three-phase full-wave bridge with wye secondary winding**

$$R_L = \frac{E_{dc}}{I_{dc}} = \frac{100}{1} = 100 \ \Omega$$

From Table 12.5:

$$n = 6 \qquad \frac{R_w}{nR_L} = \frac{2}{6 \times 100} = 0.00333$$

From Fig. 12.6:

$$\sin \theta = 0.952 = \frac{E_{dc}}{E_{pk}} \qquad E_{pk} = \frac{100}{0.952} = 105.04$$

$$E_{rms,leg} = \frac{105.04}{\sqrt{3}\sqrt{2}} = 42.88 \text{ V}$$

From Table 12.5:

$$I_{dp} = \frac{I_{dc}}{6} = \frac{1}{6} = 0.1667$$

From Fig. 12.7:

$$f(\theta) = 0.0115$$

From Eq. (12.16):

$$I_{hp} = \frac{6 \times 100 \times 0.1667}{2} \times 0.0115 = 0.575$$

From Table 12.5:

$$I_{rms,winding} = 2I_{hp} = 2 \times 0.575 = 1.15 \text{ A}$$

3. Comparison of volt-ampere ratings to dc power
Single-phase

$$78.1 \times 2.01 = 157 \text{ VA}$$

Three-phase

$$3 \times 1.15 \times 42.88 = 148 \text{ VA}$$

dc power

$$100 \times 1 = 100 \text{ W}$$

These figures illustrate the striking effect of capacitor input filters on the required volt-ampere rating of rectifier transformers.

12.11 ANALYSIS WITH FINITE CAPACITANCE AND ZERO LEAKAGE INDUCTANCE

A refinement of the analysis in the section above accounts for finite values of capacitance in the filter circuit. This analysis, a classic one by O. H. Schade, published in 1943, is valid when the leakage inductance meets the same criterion as that for the infinite capacitance analysis, $2\pi f L_L/R_w$ < 0.1. Limited to single-phase circuits, the design procedure is widely used in small and medium-sized power supplies.

The results are presented in the form of graphs in Figs. 12.8 through 12.12. The first three graphs give the normalized dc output voltage. The fourth graph gives the normalized rms diode current, and the fifth graph gives the normalized ripple voltage. All these quantities are plotted as functions of $2\pi f C R_L$ where f is the line frequency in hertz, C is the input capacitance of the filter in farads, and R_L is the load resistance in ohms. R_w/R_L is a parameter in which R_w is the total source resistance in ohms of a single rectifier branch.

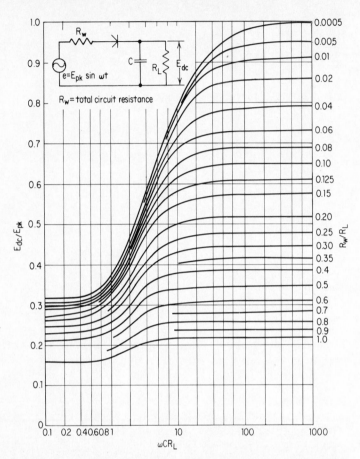

FIG. 12.8 Normalized dc output voltage vs. ωCR_L in a single-phase half-wave rectifier circuit with capacitor input filter. *Note:* ω is $2\pi f$ where f is the line frequency in hertz; C is in farads; R_L and R_w are in ohms; the leakage inductance is assumed to be negligible. *(O. H. Schade, Analysis of Rectifier Circuit Operation, Proceedings of the IRE, July 1943. Copyright IRE. Reproduced by permission of IEEE, successor to IRE.)*

Figure 12.9, which gives the normalized dc output voltage for a single-phase full-wave center-tapped transformer rectifier circuit, may be used also for a single-phase bridge circuit (Fig. 12.1c). The peak voltage on the secondary of the transformer in the bridge circuit is equal to the peak voltage from the center tap to either end of the secondary of the transformer used in a full-wave center-tapped circuit providing the same dc output voltage. In either circuit R_w is the total source resistance of one rectifier branch in that circuit.

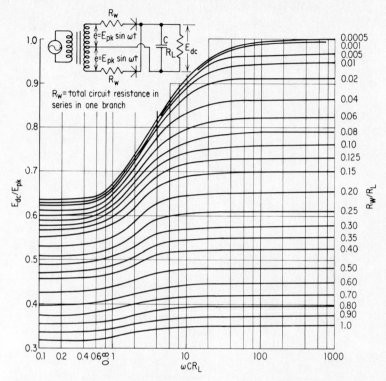

FIG. 12.9 Normalized dc output voltage vs. ωCR_L in a single-phase full-wave rectifier circuit with capacitor input filter. *Note:* ω is $2\pi f$ where f is the line frequency in hertz; C is in farads; R_L and R_w are in ohms; the leakage inductance is assumed to be negligible. *(O. H. Schade, Analysis of Rectifier Circuit Operation, Proceedings of the IRE, July 1943. Copyright IRE. Reproduced by permission of IEEE, successor to IRE.)*

Figure 12.10 gives the normalized dc output voltage for a full-wave voltage-doubler circuit. This figure may also be used for a cascade voltage doubler (Fig. 12.1*f*), provided the input capacitor has twice the capacitance of the output capacitor. The capacitance to use with the graph is the larger or input capacitance.

Figure 12.11 gives the normalized rms diode current for all the circuits analyzed. For single-phase circuits the rms diode current is identical with I_{hp}, the rms value of 1 current pulse per cycle. See the section above. The values of n and the factors relating current pulses to load and winding currents are given in Table 12.5.

The use of finite values of input capacitance yields finite values for ripple voltage. This is shown in Fig. 12.12.

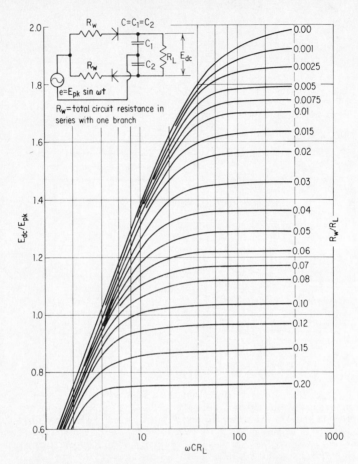

FIG. 12.10 Normalized dc output voltage vs. ωCR_L in a full-wave voltage-doubler rectifier circuit with capacitor input filter. *Note:* ω is $2\pi f$ where f is the line frequency in hertz; C is in farads; R_L and R_w are in ohms; the leakage inductance is assumed to be negligible. *(O. H. Schade, Analysis of Rectifier Circuit Operation, Proceedings of the IRE, July 1943. Copyright IRE. Reproduced by permission of IEEE, successor to IRE.)*

Illustrative problem

A dc load requires 100 V dc at 1 A dc. The total source resistance is 2 Ω. The input capacitance is 1000 mF. The line frequency is 60 Hz. Determine the rms no-load voltage on the secondary winding of the transformer, the rms current in the secondary winding, and the rms ripple voltage for a single-phase bridge circuit (Fig. 12.1c). Determine the volt-ampere rating of the transformer.

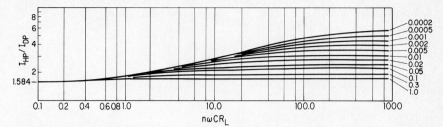

FIG. 12.11 Normalized rms diode current vs. $n\omega CR_L$ for single-phase rectifier circuits with capacitor input filters. *Note:* ω is $2\pi f$ where f is the line frequency in hertz; C is in farads; R_L and R_w are in ohms; $n = 1$ for half-wave circuits; $n = 2$ for full-wave circuits; $n = \frac{1}{2}$ for voltage-doubler circuits; the leakage inductance is assumed to be negligible. *(O. H. Schade, Analysis of Rectifier Circuit Operation, Proceedings of the IRE, July 1943. Copyright IRE. Reproduced by permission of IEEE, successor to IRE.)*

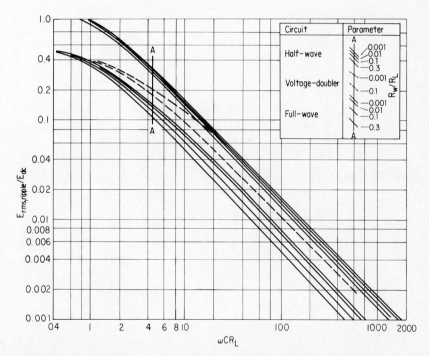

FIG. 12.12 Normalized rms ripple voltage vs. ωCR_L in single-phase rectifier circuits with capacitor input filters. *Note:* ω is $2\pi f$ where f is the line frequency in hertz; C is in farads, R_L and R_w are in ohms; the leakage inductance is assumed to be negligible. *(O. H. Schade, Analysis of Rectifier Circuit Operation, Proceedings of the IRE, July 1943. Copyright IRE. Reproduced by permission of IEEE, successor to IRE.)*

Solution

$$R_L = \frac{100}{1} = 100 \ \Omega$$

$$\frac{R_w}{R_L} = \frac{2}{100} = 0.02$$

$$\omega C R_L = 2\pi 60 \times 1000 \times 10^{-6} \times 100 = 37.7$$

From Fig. 12.9:

$$\frac{E_{dc}}{E_{pk}} = 0.918 \qquad E_{pk} = \frac{100}{0.918} = 108.9$$

$$E_{rms} = \frac{108.9}{\sqrt{2}} = 77.03 \ V$$

From Table 12.5:

$$n = 2 \qquad \frac{I_{dc}}{I_{dp}} = 2 \qquad \frac{I_{rms,winding}}{I_{hp}} = \sqrt{2}$$

$$\frac{R_w}{nR_L} = \frac{0.02}{2} = 0.01$$

$$n\omega C R_L = 2 \times 37.7 = 75.4$$

From Fig. 12.11:

$$\frac{I_{hp}}{I_{dp}} = 2.8 \qquad I_{dp} = \frac{I_{dc}}{2}$$

$$I_{hp} = \frac{2.8}{2} I_{dc} = 1.4 \times 1 = 1.4$$

$$I_{rms,winding} = \sqrt{2} \ I_{hp} = 1.4\sqrt{2} = 1.98 \ A$$

Using the full-wave curves in Fig. 12.12:

$$\frac{E_{rms}}{E_{dc}} = 0.018$$

$$E_{rms,ripple} = 100 \times 0.018 = 1.8 \ V$$

Transformer volt-ampere rating:

$$1.98 \times 77.03 = 152.5 \text{ VA}$$

Compare these results with the illustrative problem in the section above.

12.12 ANALYSIS WITH FINITE LEAKAGE INDUCTANCE AND INFINITE CAPACITANCE

When the leakage inductance greatly exceeds the criterion in the sections above, that inductance must be considered in determining the transformer secondary voltage. This section describes a method of determining the transformer secondary voltage considering leakage inductance when the input filter capacitance is very large. The rms currents may be determined from the above sections which do not consider leakage inductance. Inaccuracies from neglecting the leakage inductance when determining the rms current will be on the conservative side.

Figure 12.13a shows the circuit for a single-diode element with leakage inductance, and Fig. 12.13b shows the approximate equivalent circuit when the capacitance is assumed to be infinite. The equation relating the voltage and current in the equivalent circuit of Fig. 12.13b is

$$E_{pk} \sin (\omega t + \theta) - E_{dc} = L_L \frac{di}{dt} + iR_w \qquad (12.19)$$

The solution to Eq. (12.19) is

$$\frac{i}{I_{dp}} = \frac{nR_L}{R_w} \left\{ \left[\frac{\sin (\omega t + \theta - \Phi)}{\sin \theta \sqrt{1 + \omega^2 T^2}} - 1 \right] \right.$$
$$\left. - \left[\frac{\sin (\theta - \Phi)}{\sin \theta \sqrt{1 + \omega^2 T^2}} - 1 \right] \exp - \frac{t}{T} \right\} \qquad (12.20)$$

The terms in Eq. (12.20) are defined as follows:

i	is the instantaneous current in amperes
I_{dp}	is the dc current resulting from 1 current pulse per cycle of line frequency
n	is a factor relating I_{dp} and I_{dc} to multiple diode circuits; values for this factor are given in Table 12.5

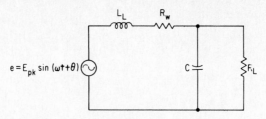

FIG. 12.13*a* Isolated diode segment of typical rectifier circuit with leakage inductance.

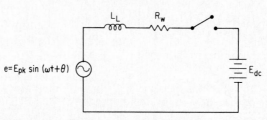

FIG. 12.13*b* Approximate equivalent circuit of isolated diode segment with leakage inductance.

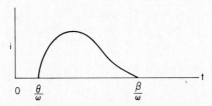

FIG. 12.13*c* Approximate waveform of single diode current pulse in equivalent circuit of Fig. 12.13*b*.

R_L is the load resistance in ohms

R_w is the total source resistance in ohms

$\omega = 2\pi f$ where f is the line frequency in hertz

$\Phi = \tan^{-1} \omega L_L/R_w$ in which L_L is the leakage inductance in henrys

$\theta = \sin^{-1} E_{dc}/E_{pk}$

$T = L_L/R_w$

The approximate shape of the current pulse of Eq. (12.20) is shown in Fig. 12.13*c*.

When Eq. (12.20) is integrated over the time during which current flows and averaged over a full cycle, the result is equal to 1. This operation results in the following expression:

$$\frac{R_w}{nR_L} = \frac{1}{2\pi} \left\{ \frac{\cos(\theta - \Phi) - \cos(\theta - \Phi + \beta)}{\sin\theta\sqrt{1 + \omega^2 T^2}} - \beta \right.$$

$$\left. - \omega T \left[\frac{\sin(\theta - \Phi)}{\sin\theta\sqrt{1 + \omega^2 T^2}} - 1 \right] \left[1 - \exp - \frac{\beta}{\omega T} \right] \right\} \tag{12.21}$$

In Eq. (12.21), β is the phase angle at which the current reaches zero. If β is substituted for ωt in Eq. (12.20), i is equal to zero. The result is

$$\sin(\theta - \Phi + \beta) = \sin\theta\sqrt{1 + \omega^2 T^2}$$

$$+ [\sin(\theta - \Phi) - \sin\theta\sqrt{1 + \omega^2 T^2}] \exp - \frac{\beta}{\omega T} \tag{12.22}$$

Equation (12.22) may be solved for β by successive approximation using various values of θ and ωT. Corresponding values of β, θ, and ωT obtained in this manner may then be substituted in Eq. (12.21) to obtain values of R_w/nR_L. The results of this process are plotted in Fig. 12.14. The curve in this figure for which $\omega L_L/R_w = 0$ is identical to the curve in Fig. 12.6.

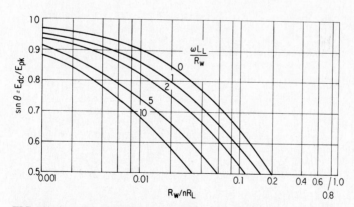

FIG. 12.14 Normalized source resistance over load resistance vs. sine of starting angle θ for infinite capacitor rectifier circuits with normalized leakage reactance as parameter. *Note:* ω is in radians per second; L_L is in henrys; R is in ohms; $\sin\theta$ represents the percentage of theoretical maximum for voltage multiplier circuits. See Table 12.5 for n.

To apply the information in this section, the dc load voltage E_{dc}, dc load current I_{dc}, and the rectifier circuit must be known. The source resistance R_w and leakage reactance ωL_L must be known or estimated. The load resistance is determined from the known load voltage and current. The value of n is determined from Table 12.5. The ratio $\omega L_L/R_w$ will be significant for high-power and high-voltage units. The larger the power, the larger this ratio is likely to be. It is usually necessary to refine the estimate of this ratio after a trial design has been executed. With the above quantities Fig. 12.14 may be entered for determining E_{dc}/E_{pk}. Currents may be determined by methods described in the sections above.

Illustrative problem

A dc load requires 100 V dc at 1 A dc. The total source resistance and leakage inductance referred to the secondary of the transformer are 2 Ω and 0.01 H, respectively. The line frequency is 60 Hz. Determine the rms no-load voltage on the secondary winding of the transformer for (1) a single-phase full-wave bridge circuit and (2) a three-phase full-wave bridge circuit with a wye-connected secondary transformer winding.

Solution

1. Single-phase full-wave bridge circuit

$$R_L = \frac{E_{dc}}{I_{dc}} = \frac{100}{1} = 100 \ \Omega$$

From Table 12.5:

$$n = 2 \qquad \frac{R_w}{nR_L} = \frac{2}{2 \times 100} = 0.01$$

$$\frac{\omega L_L}{R_w} = \frac{2\pi 60 \times 0.01}{2} = 1.88$$

From Fig. 12.14:

$$\frac{E_{dc}}{E_{pk}} = 0.83 \qquad E_{pk} = \frac{100}{0.83} = 120.5$$

$$E_{rms} = \frac{E_{pk}}{\sqrt{2}} = \frac{120.5}{\sqrt{2}} = 85.2 \ \text{V}$$

2. Three-phase full-wave bridge with wye-connected secondary

$$R_L = \frac{E_{dc}}{I_{dc}} = \frac{100}{1} = 100 \ \Omega$$

From Table 12.5:

$$n = 6 \qquad \frac{R_w}{nR_L} = \frac{2}{6 \times 100} = 0.0033$$

$$\frac{\omega L_L}{R_w} = \frac{2\pi 60 \times 0.01}{2} = 1.88$$

From Fig. 12.14:

$$\frac{E_{dc}}{E_{pk}} = 0.90 \qquad E_{pk} = \frac{100}{0.9} = 111.1$$

$$E_{rms,leg} = \frac{E_{pk}}{\sqrt{3}\sqrt{2}} = \frac{111.1}{\sqrt{6}} = 45.4 \text{ V}$$

Currents for this problem were determined in the illustrative problem of sections above. Compare voltages obtained in that problem with those obtained here.

CHAPTER 13

Inverter Transformers

Power inversion is one of several possible steps in a power conversion process. An inverter is a device for converting dc power to ac power by means of fast-acting switches. Power inversion makes possible the use of high-frequency transformers for changing voltages. In this application transformers must meet specialized requirements.

Power conversion has advanced recently because of the increased interest in high-voltage power transmission and solar electrical power generators. These operations require power conversion to make them compatible with existing ac power distribution systems. Other pressing needs for power conversion exist in ships, aircraft, and land vehicles where the prime electrical power is unsuitable for certain equipment which they carry. The great demand for inverters is in equipment where there is a need for minimum size and weight. Here the use of high-frequency switching results in a spectacular size and weight reduction over conventional equipment operating at power frequences.

The increased need for power conversion has been served by the development of superior magnetic materials and semiconductor devices with greatly improved switching properties. The switching circuitry used in inverters places no theoretical ceiling on efficiency as exists in proportional control circuits.

13.1 INVERTER CIRCUITS

Flyback Inverter

Among the most widely used inverter circuits is the flyback inverter; a simplified circuit of this kind is shown in Fig. 13.1. In this circuit a tran-

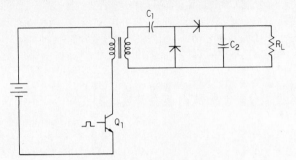

FIG. 13.1 Simplified flyback inverter circuit.

sistor switch Q_1 causes a dc voltage to appear across the transformer in approximately rectangular pulses. The operation of this circuit can be studied by means of the equivalent circuit of Fig. 13.2. During the interval when the switch Q_1 is closed, the ideal transformer T_1 in the equivalent circuit transforms this rectangular pulse to the load. The absence of an appreciable generator impedance prevents the shunt inductance L_e from altering the shape of the pulse as described in Sec. 5.3. When the switch opens, the current through the switch drops abruptly to zero. If other paths for the current which flows through the shunt inductance were not provided, a singularity would occur at the instant the switch opens. However, with resistance R_L and distributed capacitance C_D in parallel with the inductance, the singularity does not occur. This circuit has been analyzed in Sec. 5.4 as the trailing edge response of a pulse transformer. In this analysis the response of the parallel circuit has been determined for various initial conditions of current in the shunt inductance and charge on the capacitor. These responses are plotted in Figs. 5.9 through 5.12. These figures show the expected result that the larger the initial current in the inductance is, the larger will be the backswing voltage of polarity opposite the applied pulse. Also shown is the expected result that the higher the load resistance, the larger will be the backswing

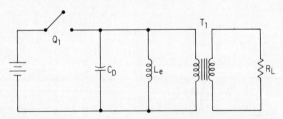

FIG. 13.2 Approximate equivalent circuit of flyback inverter circuit.

voltage and more sustained will be the subsequent oscillation. It is possible to develop across the primary of the transformer a backswing voltage higher than the forward voltage. The secondary of the transformer will reproduce the voltage envelope including both the forward and backswing voltages. The energy available to the load during the backswing interval is the energy contained in the magnetic field of the shunt inductance of the transformer when the switch opens.

The current that flows in the shunt inductance in response to a rectangular voltage pulse can be determined by integrating the defining relationship for inductance:

$$e = L \frac{di}{dt}$$

$$I_{pk} = \int_0^\tau \frac{E_{dc}}{L}\, dt$$

$$= \frac{E_{dc}}{L}\, \tau \tag{13.1}$$

where L is in henrys, E_{dc} is in volts, I_{pk} is the current in amperes flowing in the shunt inductance when the switch opens, and τ is the interval in seconds during which the switch is closed. The energy stored in the shunt inductance at the moment the switch opens can be obtained by substituting Eq. (13.1) in Eq. (1.21)

$$J = \frac{1}{2}\, I^2 L$$

$$= \frac{1}{2} \left(\frac{E_{dc}}{L}\, \tau \right)^2 L$$

$$= \frac{E_{dc}^2 \tau^2}{2L} \tag{13.2}$$

where J is in joules. Equation (13.2) shows that the energy storage may be controlled by varying the on time of the switch. The equation also shows that the energy storage is reduced by increasing the shunt inductance. The stored energy is available to the load during the interval when the switch is open.

If the load is a doubler circuit as shown in Fig. 13.1, the output capacitor C_2 will charge to the peak-to-peak value of the transformer secondary voltage less the efficiency factor for the doubling circuit. The determination of the transformer turns ratio needed with a fixed dc input voltage and a required dc output voltage draws upon experience. A typical peak-to-peak voltage across the primary is 2.5 times the dc input voltage.

It is possible to establish the amplitude of the backswing voltage and possibly improve efficiency by means of the clamping circuit shown in Fig. 13.3. In this circuit a center-tapped primary winding is used. The source voltage is connected to the center tap. One end of the primary winding is controlled by the transistor switch. The other end of the primary winding is connected to ground through a diode which is forward-biased when the backswing voltage exceeds the source voltage. Current then flows in the diode, preventing any further increase in backswing voltage. The energy removed from the magnetic field by the diode current is returned to the source, preserving efficiency. Shifting the position of the tap will change the amplitude of the output voltage at which diode current starts to flow. The voltage across the clamping winding during diode conduction will be the source voltage. The backswing voltage appearing on the switching leg of the primary winding will be the source voltage multiplied by the turns ratio of the two legs of the primary winding. The peak collector voltage will be the sum of the source voltage and peak backswing voltage appearing on the switching winding. Establishing the peak collector voltage accurately provides safer operation for the switch.

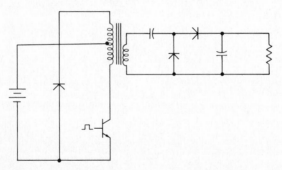

FIG. 13.3 Simplified flyback inverter circuit with clamping diode.

The on time of the transistor may be varied to adjust the output voltage. The on time plus the duration of the backswing voltage must be equal to or less than the period of the desired operating frequency. Transistor dissipation occurs principally during the switching intervals. This dissipation can be reduced by closing the switch when the voltage across it is low. The use of optimum switching points requires coordination of shunt inductance, natural frequency of the transformer, transistor on time, operating frequency, and output power required.

The flyback inverter circuit allows direct current to flow in the primary winding of the transformer. The current flowing in the clamping diode shown in Fig. 13.3 adds to the dc magnetization. An air gap is usually required in the core when dc magnetization is present.

Push-Pull Inverter

The push-pull inverter employs more than one switch to apply the source voltage to the transformer primary on both half-cycles. Three typical circuits are shown in Figs. 13.4 to 13.6. The circuit of Fig. 13.4 requires a transformer with a center-tapped primary. The half-bridge circuit of Fig. 13.5 requires two storage capacitors, and the voltage developed across the primary of the transformer is only half of the dc source voltage. The circuit of Fig. 13.6 applies the full dc source voltage to the primary but requires four switches instead of two. The copper utilization in the trans-

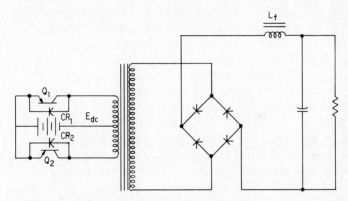

FIG. 13.4   Simplified push-pull inverter circuit.

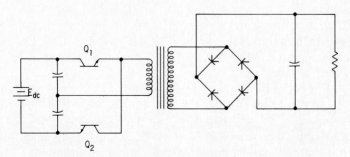

FIG. 13.5   Simplified half-bridge inverter circuit.

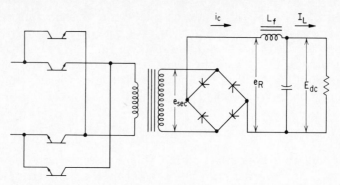

FIG. 13.6 Simplified full-bridge inverter circuit with flywheel inductor.

former of Fig. 13.4 is poorer than in the other two circuits since current flows in each half of the primary for half the time or less, making the ratio of primary rms current to load current high. These differences do not alter the basic similarity of the three circuits. Consider the circuit of Fig. 13.4. Transistor switches Q_1 and Q_2 are switched so that the source voltage E_{dc} is alternately applied to the two halves of the center-tapped winding so that $\pm E_{dc}$ is developed from center tap to either end of the primary. The on time of each transistor is one-half the period of the switching frequency or less. Efficiency of these circuits is determined in part by the leakage inductance of the transformer which lengthens the switching interval. The flow of load current through the leakage inductance stores energy which is dissipated at each switching sequence mostly in the off-going switch. The need for low leakage inductance is apparent.

When the on time of each transistor is one-half the period of the switching frequency, the current supplied to the transformer is transformed to the load without the use of energy storage techniques. If it is desired to control a rectified output voltage by reducing the transistor on time intervals, a flywheel inductor may be used to maintain the voltage during the intervals when neither transistor is conducting. The circuit of Fig. 13.6 contains such an inductor. The transformer secondary voltage, rectified voltage, and inductor current are shown in Fig. 13.7. The average load current I_L is related to the maximum and minimum currents in the inductor as follows:

$$I_L = \tfrac{1}{2}(I_{max} + I_{min}) \tag{13.3}$$

By letting I_{min} equal zero in Eq. (13.3) and substituting in Eq. (13.1), the minimum value of L_f needed to sustain the output voltage is

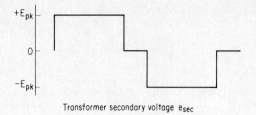

Transformer secondary voltage e_{sec}

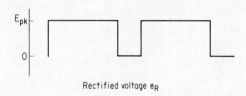

Rectified voltage e_R

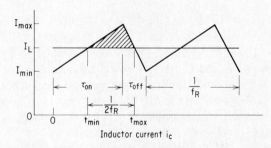

Inductor current i_c

FIG. 13.7 Voltage and current waveforms in circuit of Fig. 13.6.

obtained:

$$L_{f\text{min}} = \frac{E_{dc}\tau_{off}}{2I_L} \tag{13.4}$$

The dc output voltage is given by

$$E_{dc} = E_{pk}\frac{\tau_{on}}{\tau_{on} + \tau_{off}}$$

$$= E_{pk}\tau_{on}f_R \tag{13.5}$$

where f_R is the ripple frequency in hertz and τ is in seconds.

The maximum current in the choke will be

$$I_{max} = I_L + \frac{(E_{pk} - E_{dc})\tau_{on}}{2L} \tag{13.6}$$

The minimum voltage on the capacitor will occur at t_{min} and the maximum voltage will occur at t_{max} shown in Fig. 13.7. The peak-to-peak ripple voltage may be determined from

$$E_{pp} = \frac{1}{C} \int_{t_{min}}^{t_{max}} i \, dt$$

where the value of the integral is the shaded triangular area in Fig. 13.7. The value of that integral and the peak-to-peak ripple voltage is found by use of Eqs. (13.5) and (13.6) to be

$$E_{pp} = \frac{1}{8LCf_R^2} (E_{pk} - E_{dc}) \frac{E_{dc}}{E_{pk}} \tag{13.7}$$

Equations (13.4), (13.5), and (13.7) may be used to determine the minimum values of L and C when the ripple frequency, allowable peak-to-peak ripple voltage, and the dc output voltage are known. An inductance value above the minimum reduces the rms current in the transformer and inductance. Values of inductance and capacitance above the minimum reduce the ripple voltage.

The rms value of the inductor current is

$$I_{rms} = \sqrt{\tfrac{1}{3}(I_{max}^2 + I_{max}I_{min} + I_{min}^2)} \tag{13.8}$$

The maximum rms current will occur when I_{min} is zero. The minimum rms current will occur when the ripple current is zero, that is, I_{max} and I_{min} are both equal to I_L. The ratio of maximum to minimum rms current is $2 : \sqrt{3}$.

13.2 INVERTER TRANSFORMER OPERATION AND DESIGN

While their purpose is to transform electric power, inverter transformers closely resemble pulse and wideband transformers in that their proper functioning depends upon the parameters in the complete equivalent circuit. The switching circuitry applies rectangular waveforms to the inverter transformer, differentiating it from the power transformer which operates over a narrow band of sine wave frequencies. However, the frequency spectrum of the waveform is not often of direct interest in inverter transformers. The parameters of the equivalent circuit are examined in terms of energy storage, efficiency, damping, and overshoot characteristics, using some of the techniques developed for pulse transformers. Satisfactory inverter transformer operation depends upon interaction with both the driving and load circuits. These circuits must be considered in the design of the inverter transformer, and the design must be proven by operating the transformer in them. Inverter transformer operation and design are based on general transformer theory. (See Chaps. 1 through 5.) The various equivalent circuits apply. Magnetic material practice is similar to that for other types of transformers. The same mechanical considerations are required for such matters as heat transfer, insulation systems, and coil construction. (See Chaps. 7 through 9.)

Input waveshape and switching frequency are basic data in an inverter transformer specification. These values may not be known for the reason that the inverter transformer helps to determine them. Transformers designed from estimated values may require change after breadboard evaluation. Input waveshape, switching frequency, and voltage amplitude are used to determine flux density.

Table 1.1 gives formulas for calculating flux density from some commonly occurring waveshapes. If the waveshape does not conform to one of these functions, the flux density may be determined by applying Faraday's law directly. In applying Farraday's law, care is needed to choose the correct interval of integration and the correct constants of integration. When direct currents flow in the windings, the resulting dc flux may be determined by the methods described in Sec. 1.6. Direct current magnetization usually requires the introduction of an air gap in the magnetic circuit. In addition to preventing dc saturation of the core, an air gap improves the linearity of the magnetic circuit and reduces the remanent flux, an unwanted property of ferromagnetism resulting from the double-valued nature of the hysteresis loop.

The flyback inverter circuit in Fig. 13.3 is illustrative of the problems involved in determining flux density. In this circuit the operation of the transistor switch causes a rectangular voltage pulse to appear across the winding. This is a unidirectional pulse. The resulting flux will be a maximum at the end of the conduction interval. The flux density developed by this voltage pulse can be determined by applying function 5 of Table 1.1. This formula gives the flux increment produced by the voltage pulse. It does not give the initial value of the flux, the dc component of flux if any exists, and it does not provide any information about resetting before the next voltage pulse. To answer these questions, it is necessary to examine the operation of the circuit during the transistor off interval. In making this examination, it is convenient to use function 9 of Table 1.1 in which the flux density is shown to be proportional to the magnetizing current. (Load currents whose ampere-turns in the primary and secondary windings are equal do not generate flux.) With this relationship, it is possible to follow the behavior of the flux in the core from the magnetizing current.

During the conduction interval the magnetizing current and flux in the core will increase linearly in accordance with Eq. (13.1) as illustrated in Fig. 13.8. At the instant of turn off, energy is stored in the core by the peak magnetizing current and as given in Eq. (13.2). From the instant of turn off, the transistor winding is open and can no longer provide a path for the transfer of energy. Alternate paths are provided by the load and clamping diode windings. The energy term in Eq. (13.2) is proportional to the square of the current. Since the inductance is proportional to the square of the turns in the winding, the energy is proportional to the square of the ampere-turns. This relationship will give the available currents in the alternate windings. The energy available is proportional to the square of the sum of the ampere-turns in both alternate windings. That sum is equal to the peak magnetizing ampere-turns in the transistor winding. Upon turn off a rapid oscillation is initiated, causing the voltage on the windings to reverse and bias the clamping diode in the forward direction. The ensuing clamping diode and load currents do not represent transformer action but the removal of energy from the magnetic field by both windings simultaneously. The sum of these ampere-turns decreases linearly in accordance with the constraint imposed by the clamping diode circuit. These currents do not increase the flux but are associated with the reduction of flux in the core. If the switching intervals and ring period are coordinated, the ampere-turns will decrease linearly until they reach zero. This is illustrated in Fig. 13.8a. There is no reversal in the direction of flux.

There is another component of load current which flows during the conductance interval. The dc component of this current pulse flowing in the transistor winding adds to the magnetization of the core because it is not balanced by secondary ampere-turns which have no dc component. The dc component of load current is not directly measurable. The dc supply current contains the reverse flow from the clamping diode. The collector current contains the dc component of magnetizing current. This problem can be resolved by using the dc component of collector current to determine the dc flux and eliminating the dc component of flux from the calculation when function 5 of Table 1.1 is used. The dc component of flux may be determined from the dc component of collector current by using the method described in Sec. 1.6, or a simpler approximate method shown below may be used.

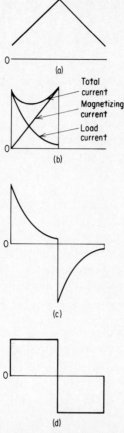

FIG. 13.8 Waveforms in flyback circuit of Fig. 13.3. (*a*) Magnetizing ampere-turns and flux in transformer core; (*b*) collector current; (*c*) transformer secondary current; (*d*) induced voltage.

To make an approximate calculation of the dc flux density, assume that all the dc magnetizing force will appear across the air gap. Now apply that assumption to Eq. (1.17) to obtain the following relationships:

$$\text{mmf}_{dc} = \phi_{dc}\mathcal{R}_a$$

$$0.4\pi NI_{dc} = \phi_{dc}\frac{l_a}{\mu_a A}$$

$$0.4\pi NI_{dc} = B_{dc}l_a$$

$$B_{dc} = \frac{0.4\pi NI_{dc}}{l_a} \qquad (13.9)$$

The sum of the dc and ac flux densities will be

$$B_T = \frac{0.4\pi N I_{dc}}{l_a} + \frac{E_{dc}\tau_{on} \times 10^8}{2NA} \tag{13.10}$$

where E_{dc} = supply voltage, V
$\quad A$ = cross-sectional area of core, cm^2
$\quad l_a$ = length of air gap, cm
$\quad N$ = number of turns in transistor winding,
$\quad I_{dc}$ = dc component of collector current, A
$\quad \tau_{on}$ = transistor on time, s

The ac term in Eq. (13.10) differs from function 5 of Table 1.1 in that the switching frequency has been replaced by the transistor conduction time and the constant in the denominator has been changed from 4 to 2. These differences account for the possibility that the on and off intervals may be unequal. If the two intervals are equal, the two expressions are equivalent. The waveform of the collector current is shown in Fig. 13.8*b*. The secondary current and induced-voltage waveforms are shown in Fig. 13.8*c* and *d*, respectively.

The dc component of collector current may be determined approximately from the maximum power required by the load by estimating the circuit efficiency, which will be approximately 85 percent:

$$I_{dc} = \frac{P_L}{0.85E_{dc}} \tag{13.11}$$

In Eq. (13.11) power returned to the supply via the clamping circuit has been neglected.

The determination of current, voltage, and flux relationships is needed to establish the properties of the magnetic circuit. The core material is selected mostly on the basis of switching frequency and availability. The properties of the core material selected determine the maximum flux density at which the core may be operated. This value and the current, voltage, and flux relationships determine the turn-core product. The load voltage and current determine the window area needed for the windings. At this point a search can be made for a core of the desired material with suitable window and core cross-sectional areas. In determining the suitability of a core, turns, operating flux density, core area, window area, and current density are adjusted to make a reasonable choice. When this tentative selection is made, the parameters of the equivalent circuit may be determined. Shunt inductance is calculated by means of Eqs. (1.19) and (1.20). When dc magnetization is present, the minimum air gap necessary to prevent saturation of the core places an upper limit on the shunt inductance achievable. This inductance should be sufficiently large to limit the magnetizing current to a value that will not burden the semiconductor switches and the power source. If a lower value of shunt inductance is needed for energy storage, the air gap may be increased.

Winding resistances, leakage inductance, and distributed capacitance are calculated by methods described in Secs. 10.5 to 10.7. The temperature rise consideration places an upper limit on winding resistances. In inverters where efficiency is such an important consideration, winding resistances are often kept low to reduce the copper loss. Excessive reduction of resistance will adversely increase size, cost, and the distributed parameters in the equivalent circuit. Leakage inductance is directly proportional to the coil buildup and the square of the turns. It is inversely proportional to the winding length. With the required turns–core area product established as a constant by Faraday's law, the leakage inductance becomes a direct function of the ratio of turns to core area. Thus for a core with a given core capacity, the leakage inductance is a direct function of the ratio of window area to core area and the ratio of window length to window height. Leakage inductance can be reduced by interleaving windings and by using two coils on a single U core. Reducing the leakage inductance frequently adversely affects other parameters. Distributed capacitance is determined by the geometric capacitance between opposing electrodes and by the voltages that exist between the various electrode pairs. Increasing the number of layers will decrease distributed capacitance. Universal windings may be used for the same purpose. Reduction of distributed capacitance is usually made at the expense of leakage inductance. The resonant circuit formed by the distributed capacitance and the shunt inductance causes overshoots and ringing. At high switching frequencies the dielectric loss associated with distributed capacitance must sometimes be considered. See Sec. 8.7. High-voltage inverters are particularly vulnerable to this problem. The design and construction of an inverter transformer requires a careful balance among all the competing parameters to obtain a workable compromise.

Illustrative problem

Design an inverter transformer for the inverter circuit shown in Fig. 13.4. The source voltage to the inverter will vary between 120 and 160 V. The dc output will be 2000 V at 50 mA. Pulse width modulation will be used to regulate the output voltage against source voltage variations. The flyback inductor L_f will be used to maintain the output voltage during intervals when neither transistor conducts.

The inductor has an inductance of 0.3 H. The operating frequency will be 10 kHz.

Solution

The step-up ratio of half the primary winding to the secondary winding should provide rated output voltage at the minimum source voltage and 50 percent conduction time on each transistor. If circuit drops are neglected, the ratio of the secondary voltage to the primary end to tap voltage will be

$$\frac{2000}{120} = 16.67$$

The on time of the transistors must be reduced to limit the average output voltage to its rated value when the source rises above its minimum value. The minimum on time of one transistor during a single period is

$$\tau_{on} = \frac{E_{min}}{E_{max}} \frac{1}{2f}$$

$$= \frac{120}{160} \frac{1}{2 \times 10^4}$$

$$= 37.5 \times 10^{-6} \, s$$

The secondary winding current consists of that portion of the inductor current shown in Fig. 13.7 during which either transistor is conducting and the current is rising. The free-wheeling or falling portion of the current flows through the bridge rectifiers. The secondary winding current will have its maximum rms value when the transistor on time is a minimum and the line voltage is highest. That current is determined from the duty cycle and Eq. (13.8). To use that equation, the maximum and minimum currents are needed and may be obtained from Eq. (13.6).

$$I_{max} = I_L + \frac{E_{in} - E_{out}}{2L} \tau_{on}$$

where τ_{on} is the minimum on time for one transistor. E_{in} for that condition will equal E_{out} multiplied by the ratio of the minimum to the maximum input voltage leading to

$$I_{max} = I_L + \frac{(E_{max}/E_{min} - 1)E_{out}}{2L} \tau_{on}$$

$$= 0.05 + \frac{(160/120 - 1)2000}{2 \times 0.3} \times 37.5 \times 10^{-6}$$

$$= 0.092 \text{ A}$$

The initial current in the secondary, called I_{min} in Eq. (13.8), may be obtained from Eq. (13.3):

$$I_L = \frac{1}{2}(I_{max} + I_{min}) \tag{13.3}$$

$$I_{min} = 2I_L - I_{max}$$

$$= 2 \times 0.05 - 0.092$$

$$= 0.008 \text{ A}$$

Substitution of values for I_{max} and I_{min} in Eq. (13.9) and application of the duty cycle gives the rms current in the secondary:

$$I_{rms} = \sqrt{\tfrac{1}{3}(I_{max}^2 + I_{max}I_{min} + I_{min}^2)(\text{duty cycle})}$$

$$= \sqrt{\tfrac{1}{3}[(0.092)^2 + (0.092)(0.008) + (0.008)^2](0.75)}$$

$$= 0.048 \text{ A}$$

An rms current less than the output current does not violate energy conservation because the transformer secondary voltage is higher than the output voltage.

The primary rms current is determined from the voltage ratio of the transformer and the duty cycle of the current in the primary, half of that in the secondary because of the alternating conducting intervals of the two transistors. The primary rms current is

$$(I_{rms})_{pr} = 0.048 \times 16.7\sqrt{0.5}$$

$$= 0.567 \text{ A}$$

With the voltages and rms currents established for both primary and secondary, it is possible to proceed with the core selection and coil design. A ferrite core is preferable for the operating frequency

WDG VOLTS	TURNS	WIRE SIZE DIA T/IN	OM Ω/KFT LB/KFT	AMPS OM/AMP	MARGIN VOLTS/MIL	T/L WDG, LGTH	LAYER INS V/MIL	WRAP V/MIL	BUILD UP CL .015 WF .040	Δ	MLT/FT	R ___°C	I²R	CU WT	MLT P=___
1	258.	H27.	202.	.567	.156	43.	.003K	6L	.098	.055	.179	2.37	.76		
160-0-	T12.9	.0164	51.4			6.		.005K	.015	.113	46.				
160		57.9	.634	356.		.75	67V/M	67V/M	.030	.030					
2	215.4	H37.	20.2	.048	.187	115.	.004K	8L	.108	.198	.271	299.	.69		
2667		.0057	512.			19.	4.7V/M	.005K	.072	.180	.584				
	168.		.0655	420.		.688			.040	.040					
3							93V/LAYER			.418					
							186V/LAYER PAIR								
4													1.45		
5															
6															

X1.3 (TEMP. FACTOR)

P_{cu} = 1.9 WATTS

TOTAL .418

CORE 50-0409 _____ WINDOW 1.125 X .75

MTL 24B (STACKPOLE)

WF 1.062 LG

AREA .46 DIA _____ IN .97 CM²

VOL 12.96CM³

WATTS/CM³ .06

CORE LOSS 0.8 WATTS

VA/LB

VA EX

NOTES MLT = (DIA. + 2Δ₁ + Δ₂) π/12

FIG. 13.9 Calculations for illustrative problem in Sec. 13.2.

FIG. 13.9   (*continued*)

Flux Density:

$$B = \frac{E \times 10^5}{4fNA} \text{ kG} = \frac{120 \times 10^5}{4 \times 10^4 \times 129 \times 0.97} = 2.4 \text{ kG}$$

Temperature Rise:

$$Sc \ 1.062\pi(0.46 + 2 \times 0.418) =$$

$$2\pi \left[\left(\frac{0.46}{2} + 0.418 \right)^2 - \left(\frac{0.46}{2} \right)^2 \right] - 0.46 \times 0.418 \times 2 = \begin{array}{c} 4.32 \\ \underline{1.92} \\ 6.24 \text{ in}^2 \end{array}$$

$$\Delta T = K \left(\frac{P_{cu}}{Sc} \right)^{0.8} = 100 \left(\frac{1.9}{6.24} \right)^{0.8} = 39°C$$

Leakage Inductance:

$$L_L = \frac{4\pi N^2 P}{D} \left(\frac{h_1 + h_3}{3} + h_2 \right) \times 10^{-9} \text{ H}$$

$$= \frac{4\pi(2145)^2 \times 6.58}{1.83} \left(\frac{0.287 + 0.457}{3} + 0.076 \right) \times 10^{-9} = 0.067 \text{ H}$$

Reactive Power in Leakage Inductance:

$$\tfrac{1}{2}I^2L = \tfrac{1}{2}(0.092)^2 \times 0.067 = 283 \times 10^{-6} \text{ J}$$

$$283 \times 10^{-6} \times 20,000 = 5.66 \text{ W}$$

Distributed Capacitance:

$$C_G = \frac{0.225A\epsilon}{d} \text{ pF} \qquad A = 0.27 \times 12 \times 0.688 = 2.24 \qquad d = 0.004 \qquad \epsilon = 3.5$$

$$= \frac{0.225 \times 2.24 \times 3.5}{0.004} = 441 \text{ pF}$$

$$C_{\text{eff}} = \frac{4C_G}{3} \left(\frac{N_L - 1}{N_L^2} \right) = \frac{4 \times 441}{3} \left(\frac{19 - 1}{19^2} \right) = 29 \text{ pF}$$

High-Frequency Cutoff:

$$f = \frac{1}{2\pi \sqrt{L_L C_{\text{eff}}}} = \frac{1}{2\pi \sqrt{0.067 \times 29 \times 10^{-12}}} = 114 \times 10^3 \text{ Hz}$$

Primary Inductance:

$$A_L = 1.280 \qquad \text{from core manufacturer's data}$$

$$1.280 \left(\frac{129}{1000} \right)^2 = 0.0213 \text{ H}$$

Peak Exciting Current:

$$I_{\text{pk}} = \frac{E\tau}{L} = \frac{120(1/20,000)}{0.0213} = 0.282 \text{ A}$$

specified. For the power level required, an adequate selection of sizes and shapes is available. A U-U type core was chosen because this configuration permits the use of layer-wound coils and good lead access for the high-voltage secondary. Stackpole core 50-0409 in Ceramag 24B material was found to be a reasonable choice. This is a high-permeability low-frequency ferrite widely used for inverters. The detailed calculations for the coil are given in Fig. 13.9. The methods used in this calculation are given in Sec. 10.7. The estimate of temperature rise is based on Eq. (9.20). The coefficients for this equation given in Table 9.3 do not apply to ferrite cores. The value used of 100 is a conservative estimate based upon a comparison between watts dissipated per unit volume in the ferrite core and the watts dissipated per unit volume in silicon steel cores at power frequencies which the table does cover.

Leakage inductance is calculated by using the formula given in Fig. 10.1a. Distributed capacitance is calculated by the methods described in Sec. 10.6. The flux density is determined from function 2 of Table 1.1. The open-circuit inductance is determined by applying Eq. (1.20). Ferrite core manufacturers often provide permeability information in the form of an inductance per 1000 turns for the core. This value then includes the necessary core geometry data. The actual inductance is calculated by multiplying the inductance per 1000 turns by the square of the ratio of the actual number of turns to 1000. The core loss is determined from data of loss per unit volume supplied by the manufacturer for various frequencies and flux densities. The exciting current is obtained by applying Eq. (13.1). The leakage inductance and the load current are used to calculate the energy storage in the leakage inductance at the end of each conduction interval. By multiplying this energy term by the number of conduction intervals per second (twice the operating frequency), a power term is obtained. This term is an approximation of the switching losses caused by the leakage inductance.

The high-frequency cutoff is calculated from the leakage inductance and the distributed capacitance using Eq. (4.13). The high-frequency cutoff provides a qualitative evaluation of the leakage inductance and the distributed capacitance. The high-frequency cutoff should be ideally more than 10 times the operating frequency. The load capacitance should be included in the distributed capacitance.

The appreciable peak exciting current that exists at the moment either transistor turns off represents energy storage which will cause ringing during nonconducting intervals. To prevent the ring voltage from becoming excessive, clamping diodes, CR_1 and CR_2, are used to limit the voltage across Q_1 and Q_2 to twice the supply voltage. The bridge rectifier in the secondary rectifies the backswing voltage, effectively increasing the width of the preceding pulse. Current flowing in CR_1 and CR_2 returns energy stored in the shunt inductance to the source. By these mechanisms some of the energy stored in the shunt inductance is recovered. Copper and iron losses in this design are reasonable. Because of the low cost, simplicity, and reliability of this design, prototype construction and breadboard evaluation would be justified.

CHAPTER 14

Static Magnetic Devices Other Than Transformers

This section is a survey of some of the commonly used iron-cored devices which resemble transformers. Although some of these devices are quite complex, many of the construction features and coil design techniques are the same. They are usually designed and manufactured by transformer facilities.

14.1 CURRENT-LIMITING REACTORS

Current-limiting reactors are used in circuits where temporary overloads and short circuits must be accommodated without damage. The inductance of the reactor must be large enough to limit short-circuit current to a required value. At the same time the inductance must be low enough to meet the requirements of minimum and maximum voltage changes with normal current variations. The upper and lower limits of inductance are established by these two requirements. The normal load reactance and resistance and the residual circuit reactance and resistance under overload must be considered when establishing the inductance limits. The reactor will have an rms current rating as well as a maximum current rating. The rms rating will be based on the duty cycle of the maximum current and the normal current. RMS currents are discussed in Sec. 1.7. If the maximum current is rarely developed, then adiabatic heating can be used as a guide to establish the maximum capability of the reactor. The *adiabatic rating* is an energy term expressed as the product of the current squared in amperes and time in seconds. When use of the reactor is kept within this adiabatic rating, the temperature rise of the reactor will be limited to a safe value. This is a one-time rating. The reactor cannot safely dissipate additional heat until it has had time to cool. Adiabatic rating is discussed in Sec. 9.3. During the interval in which excessive current flows, a higher than normal voltage exists across the reactor. The reactor must be capable of absorbing this voltage without saturating. Most reactors with a sufficient turn-area product to absorb this voltage have an inductance in excess of the maximum allowable until air gaps are introduced in the core to reduce the permeability. Thus the size of a current-limiting reactor is usually determined by its voltage, frequency, maximum current, and rms current. The size is roughly equal to a transformer of half the volt-ampere rating of the reactor with the same frequency rating. Current-limiting reactors are single-frequency devices.

14.2   WIDEBAND INDUCTORS

Wideband inductors have properties and limitations analogous to wideband transformers. Most wideband inductors are used in frequency-selective circuits in combination with capacitances. The ideal inductance for such applications is invariant with frequency, voltage, current, and temperature and has low winding resistance, low core loss, and negligible distributed capacitance. These demands can only be approximated over a limited range by iron-cored devices.

Because the application is critical, an immediate problem with wideband inductors is the tolerance on the nominal inductance and the method of measurement. See Sec. 18.2 on inductance measurements. Economics requires that the allowable tolerance on the inductance not be absorbed by test ambiguities, but be allocated to production variations. The use of air gaps reduces but does not eliminate the sensitivity of the inductance to variations in iron permeability. Variations in the air gap are a source of error in maintaining the inductance value with accuracy. There are manufacturing problems in controlling the air gap with precision. Adhesives are used on the mating faces of the core and the gap spacer material. The adhesive and the gap spacer material must be compatible. Dimensional changes in either the adhesive or the gap spacer material during processing after setting the spacer and testing will cause changes in the inductance. Both adhesive and spacer material should have good dimensional stability. Impregnating materials entering the gap spacer region alter the effective gap upon setting and curing. The gap spacer material should be impervious to the impregnant. Impregnants have greater dielectric losses and constants than the air they replace. The apparent inductance at high frequencies is affected by the dielectric constant of the impregnant because the distributed capacitance is increased.

The increased losses reduce the Q of the inductor. The impregnants commonly used in low-frequency devices are generally unsuitable for critical wideband inductors. Impregnants that have good high-frequency properties may lack the good mechanical and thermal properties of the low-frequency impregnants. Compromises in desired characteristics are required with the selection of the impregnant.

Distributed capacitance gives the inductor a natural parallel resonant frequency. Above this frequency the component has capacitive reactance. At frequencies below but in the vicinity of the natural resonant frequency, the inductance changes rapidly with frequency, particularly if the Q of the inductor is high. Thus for applications requiring a constant inductance, the natural resonant frequency must be well above the highest operating frequency. See Sec. 10.6 for a discussion of distributed capacitance.

The Q of an inductor is an important property. The familiar definition of Q is the ratio of inductive reactance to series resistance. For iron-cored reactors the series resistance must include a resistance equivalent to the core loss as well as the winding resistance. In Eq. (6.18) it is shown that the eddy current loss is proportional to the square of both the flux density and frequency. Although the eddy current loss is proportional to the square of the frequency for a constant flux density, for a given device and constant induced voltage, the flux density-frequency product is constant with frequency, thus keeping the eddy current loss approximately constant with frequency. The hysteresis loss found in Eq. (6.6) is proportional to the first power of the frequency and the area of the hysteresis loop. Since for a constant induced voltage the flux density decreases with frequency, the area of the hysteresis loop will decrease and the shape of the loop will change. Steinmetz developed the following empirical equation to relate hysteresis loss to flux density and frequency:

$$P_H = \eta f B^n \times 10^{-7} \qquad (14.1)$$

In this expression f is in hertz, B is in gauss, and P_H is in watts per cubic centimer. η is a function of the core material and n, while dependent upon B, has a value of about 1.6 over the range of 1000 to 12,000 G. η has a value of about 0.00046 for high-grade magnetic silicon steel. From Eq. (14.1) it is seen that for a given device and constant induced voltage, the hysteresis loss will decrease with frequency. As the frequency increases well above its lowest operating value, the hysteresis loss decreases to the point where the eddy current loss predominates. This justifies representing the core loss of an inductor as a resistance invariant with frequency in parallel with the inductance as shown in Fig. 14.1. The equivalent circuit in this figure applies when the frequency is well below the natural resonant frequency of the inductor. If the parallel resistance is at least 3

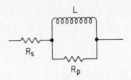

FIG. 14.1 Low-frequency equivalent circuit of an iron-cored inductor.

times greater than the inductive reactance, then the equivalent series resistance is approximately

$$R = R_s + \frac{\omega^2 L^2}{R_p} \tag{14.2}$$

in which R is in ohms and L is in henrys. Q, a dimensionless quantity, can be represented as follows:

$$Q = \frac{\omega L}{R}$$

$$= \frac{1}{R_s/\omega L + \omega L/R_p} \tag{14.3}$$

By differentiating Eq. (14.3) with respect to f and equating to zero, the following criteria for maximum Q are obtained:

$$Q_{\max} = \frac{1}{2} \sqrt{\frac{R_p}{R_s}}$$

$$= \frac{\pi f_{\max} L}{R_s} \tag{14.4}$$

in which $f_{\max}$ is the frequency in hertz at which the Q is a maximum. That frequency is given by

$$f_{\max} = \sqrt{\frac{R_p R_s}{2\pi L}} \tag{14.5}$$

The Q at any frequency f as a function of $Q_{\max}$ and $f_{\max}$ is

$$Q_f = \frac{2Q_{\max}}{f/f_{\max} + f_{\max}/f} \tag{14.6}$$

To determine Q at any frequency f, it is necessary to know only the maximum Q and the frequency at which the Q is a maximum. The Q approaches zero as the frequency approaches either zero or infinity.

For a given core size with the core window filled, the resistance varies approximately as the square of the turns because any reduction in wire size causes a proportionate increase in resistance per unit length of the conductor and a proportionate increase in the number of turns that may be placed in the window. The inductance is also proportional to the square of the number of turns, making the ratio of inductance to winding resistance almost constant. From Eq. (14.4) this makes Q_{max} invariant with L and the number of turns for a given core. The maximum Q is a function of the core loss but not the permeability. Particularly it is not a function of the air gap. A change in permeability will change the frequency at which the maximum Q occurs, and by so doing will change the Q at other frequencies, but the maximum Q remains constant. In this analysis the variable is frequency. Resistance is constant.

Eddy current loss, which constitutes most of the core loss, is inversely proportional to the resistivity of the core. High resistivity is a property usually obtained at the sacrifice of permeability. The higher the permeability, the lower will be the frequency at which the Q is a maximum. For this reason it is difficult to obtain high Q at low frequencies.

14.3   RESONANT-CHARGING REACTORS

A resonant-charging reactor is a pulse generator component with the dual role of, first, limiting the flow of direct current during the short interval when the pulse switch is closed and, second, charging a capacitor during the interval between pulses with the energy to be expended during the pulse. Figure 14.2 shows a simplified circuit of a pulse generator circuit used in radar modulators. E_{dc} is the dc power source. L_c is the resonant-charging reactor. CR_1 is a holding diode. Z_1 is an unterminated artificial transmission line called a *pulse-forming network,* consisting of a number of L sections of inductance and capacitance. V_1 is a thyratron or other fast-acting switch. T_1 is a pulse transformer. R_L is the load. A charged transmission line, when suddenly connected to its characteristic impedance, delivers a rectangular pulse of amplitude equal to half the voltage to which it was charged with a duration equal to the round-trip transit time of the transmission line. The charged pulse-forming network is discharged by closing switch V_1. Switch V_1 stays closed only for the duration of the pulse, during which time the pulse is applied to the load through transformer T_1 which matches the impedance of the load to the pulse-forming network. During the interval when V_1 is closed, a short is placed

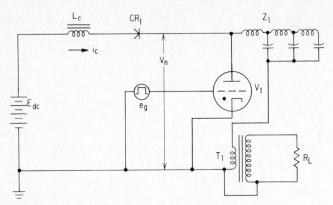

FIG. 14.2 Simplified schematic of line-type pulse generator.

across the dc supply in series with the resonant-charging reactor L_c. Because this time interval is short, the large inductance L_c is able to prevent an appreciable increase in current from E_{dc}. The transformer T_1 has a relatively low inductance compared with the charging reactor. During the pulse interval only the transformer inductance can support voltage and transform it to the load R_L. V_1 opens after the pulse-forming network is completely discharged. Current then begins to flow from E_{dc}, charging the capacitance in Z_1. Since the charging rate is relatively slow, the inductances of the transformer and the pulse-forming network do not function. The charging inductance L_c and the capacitance of the pulse-forming network form a resonant circuit which charges the capacitance sinusoidally to a voltage slightly less than twice the voltage E_{dc}. The charging current reaches zero as the voltage across the network reaches its peak. Past this point the voltage on the network would begin to drop, and the current would continue through zero to become negative, discharging the capacitor, if it were not for the presence of CR_1 which prevents the flow of current in the reverse direction. The resonant charging is thus halted at the point where the charge on the pulse-forming network is maximum. The line remains charged until the switch V_1 closes again in response to a signal from e_g and the cycle is repeated. The current- and voltage-charging waveforms are shown in Fig. 14.3.

Several relationships are useful in establishing requirements for the magnetics in a line-type pulse generator. The round-trip delay time of the pulse-forming network is

$$\tau = 2\sqrt{L_N C_N} \tag{14.7}$$

In this expression τ is in seconds, L_N, the total network inductance, is in henrys, and C_N, the total network capacitance, is in farads.

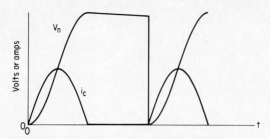

FIG. 14.3 Waveforms of charging current and voltage in line-type pulse generator shown in Fig. 14.2.

The characteristic impedance of the pulse-forming network in ohms is

$$Z_N = \sqrt{\frac{L_N}{C_N}} \tag{14.8}$$

From Eqs. (14.7) and (14.8) the expression for network capacitance is obtained:

$$C_N = \frac{\tau}{2Z_N} \tag{14.9}$$

The resonant-charging frequency, in hertz, is

$$f_R = \frac{1}{2\pi\sqrt{L_c C_N}} \tag{14.10}$$

The interval between pulses to achieve full voltage in the network must be at least half of the period of the resonant-charging frequency. Because of the presence of the holding diode, the interpulse interval may exceed this threshold value.

The charging current is given by

$$i_c = \frac{E_{dc}}{\sqrt{L_c/C_N}} \sin\left(\frac{1}{\sqrt{L_c C_N}} t\right) \tag{14.11}$$

One terminal of the resonant-charging reactor is at constant dc potential. This potential is typically thousands of volts. The other terminal of the charging reactor swings between zero and a peak voltage of twice the dc potential. The peak induced voltage applied to the reactor is E_{dc}.

From Eq. (14.11) the peak current in the reactor is

$$I_{pk} = \frac{E_{dc}}{\sqrt{L_c/C_N}} \qquad (14.12)$$

The direct current is

$$I_{dc} = \frac{I_{pk}}{1.57}\left(\frac{f_{prf}}{2f_R}\right) \qquad (14.13)$$

in which f_{prf} is the pulse repetition frequency in hertz.

The rms current in the reactor is

$$I_{rms} = \frac{I_{pk}}{\sqrt{2}}\sqrt{\frac{f_{prf}}{2f_R}} \qquad (14.14)$$

To establish the specifications for a resonant-charging reactor, the pulse width, pulse repetition interval, and pulse-forming network characteristic impedance must be known, from which C_N is calculated using Eq. (14.9). The resonant-charging half-period must be equal to or less than the pulse repetition interval. The rms current in the choke is minimum when the pulse repetition interval is equal to one-half the resonant-charging period. Making one-half the nominal resonant-charging period slightly less than the pulse repetition interval permits a plus tolerance on the inductance of the charging reactor, as can be seen from Eq. (14.10). In this equation are substituted the chosen value of resonant-charging frequency and the calculated value of C_N to determine the nominal value of the inductance in L_c. The holding diode permits the inductance to drop below the nominal value. The peak current from Eq. (14.12) and the maximum inductance are substituted in formula 9 of Table 1.1 to calculate the maximum flux density. The rms current in the reactor is calculated from Eq. (14.14). Since the peak current is used to calculate maximum flux density, it is not necessary to consider direct current.

The charging-reactor inductance should be essentially invariant with current. If the core of the inductor saturates, an unacceptable bistable condition can develop in which the network voltage on alternate charging cycles is of reduced amplitude. A typical specification may state that the inductance with zero current shall be no more than 5 percent higher than the inductance with rated direct current.

The special requirements of resonant-charging reactors occasion design problems that must be carefully addressed. Most charging reactors are high-voltage devices. The high voltage is both externally applied and induced. The usual choice for coil construction is a single-coil assembly layer-wound. The insulation for the start lead and start layer is less capable of withstanding dielectric stress than the finish layer and lead. The start lead is therefore chosen as the dc terminal. Nonetheless, the start region must be insulated for the high dc potential. The start lead exit is particularly troublesome. One way to cope with this problem is to use a winding form larger than the core so that an insulating pad may be placed between the winding form and the core at the point of exit of the start lead. The winding form wall, whose thickness is normally selected for mechanical strength, may need to be supplemented with additional insulation under the first layer.

Because of the asymmetrical magnetization of the core, the core loss in charging reactors may be greater than that predicted by data published by core manufacturers for symmetrical sine wave magnetization. The indeterminate nature of core loss and the need for inductance stability suggest operation at conservative flux densities. The most frequent choice for charging choke cores is the C core, usually two around a single coil. The strip thickness is typically 4 mils to control the losses in the customary hundreds to several thousand hertz used as the resonant-charging frequency. The flux density is usually kept at 10 kG or less. The wire size selection is based on the rms current. Pulse generators frequently operate at several repetition rates and pulse widths. It is necessary to rate the charging reactor on the basis of the most severe condition for all requirements even if all the most severe requirements do not occur simultaneously during any single operating mode. The use of a conservative maximum flux density in formula 9 of Table 1.1 will give a large turns-core area product. To obtain the correct inductance with this large turns-area product, a large air gap is needed in the core, an effect which will aid in making the inductance invariant with current.

Charging reactors are frequently hermetically sealed components, either oil-filled or potted with solid dielectric materials. They may often be recognized by their two high-voltage terminals of different size. This gives visual emphasis to the fact that terminal orientation must be observed.

14.4   SATURABLE REACTORS

A saturable reactor is an inductance placed in series with an ac voltage source and a load. By controlling the saturation of the core in the inductor, the current delivered to the load is controlled. Saturable reactors used in self-saturating circuits are sometimes called *magnetic amplifiers*. They are not amplifiers in the sense that they do not faithfully reproduce an input signal at an increased power level. *Saturable reactor* is a more accurate and descriptive term. These devices have been largely replaced by silicon controlled rectifiers (SCRs), which are smaller and less expensive and provide greater gain. There are some advantages to saturable reactors. They can be designed for almost any voltage-, current-, and power-handling requirement. They are tolerant of temporary overloads, and they are less inclined than SCRs to produce high-frequency transients which are troublesome to sensitive and high-voltage circuits. In addition to the winding(s) which carry the load current called *anode winding(s)*, saturable reactors have a control winding, current through which determines the saturation of the core. Inherent in this arrangement are two additional limitations of the saturable reactor: voltage from the ac power source induces voltage in the control winding by transformer action, and the response time is long. Various artifices are available to keep these limitations within manageable proportions.

The voltages, currents, and fluxes in the core of a saturable reactor are related by the magnetization curve of the core. The operation of saturable reactors is dependent upon the nonlinear nature of the magnetization curve. The lack of a convenient analytic function to describe the magnetization curve complicates the design and analysis of saturable reactors. Approximate analysis is accomplished by a combination of graphical and analytical techniques. This may include dividing the operating cycle of the reactor into a series of intervals, within each of which operation is nearly linear.

The operation of a saturable reactor and some of the simplifications used for analysis are illustrated by the series-connected reactor shown in Fig. 14.4. Two cores are used in this reactor. Each core has an anode winding of N_G turns. A single-control winding of N_c turns links both cores. The two anode windings are series-connected so that the ac fluxes from the two cores pass through the control winding in opposite directions. The two anode windings are in series with the ac voltage and the load

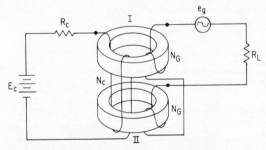

FIG. 14.4 Diagram of a series saturable reactor showing control and anode circuits.

resistor R_L. This resistance includes the resistance of the windings as well as the load. The control circuit contains a dc voltage E_c and a resistance R_c, which represents the total resistance in the control circuit. This resistance is assumed to be very small. With no control signal applied, the net change in flux through the control winding is zero, making the voltage induced in the control winding zero. The magnetization curve is assumed to be as shown in Fig. 14.5. The permeability is infinite except during saturation when it is zero. At the start of a positive half-cycle of ac voltage, core I is at flux density B_o and core II is saturated at flux density B_s. As the ac voltage increases, the flux density in core II decreases below B_s and the flux density in core I rises above B_o in accordance with Faraday's law, Eq. (1.13). Because of the high permeability of the cores, all the ac voltage appears across the cores and none across the load. The assumed shape of the magnetization curve requires that the magnetizing force be zero when the flux density is below saturation. Thus during this interval when neither core is saturated, the ampere-turns in the anode winding of each core must equal the ampere-turns in the control winding:

$$N_G i_G = N_c i_c \tag{14.15}$$

Note that the currents are instantaneous values. Since the anode current is negligible during this interval by virtue of the high permeability of the cores, the ampere-turns in both anode windings and the control winding may be considered to be zero. This situation continues until B_I reaches B_s, at which time B_{II} has reached the value of B_o. This and other significant time-variant functions are shown in Fig. 14.6. Upon reaching B_s the flux density in core I can increase no further. The voltage across

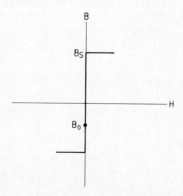

FIG. 14.5 Idealized magnetization curve used to analyze the series saturable reactor of Fig. 14.4.

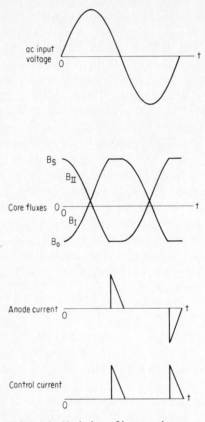

FIG. 14.6 Variation of input voltage, core fluxes, and currents with time in series reactor of Fig. 14.4.

the anode winding of core I drops to zero. A net rate of change of flux through the control winding different from zero now exists. The very low control circuit impedance presents an effective short circuit to the control winding, preventing this change in flux in core II. Since the flux in both cores is constrained from changing, the voltage across the two windings drops to zero. The ac supply voltage is immediately transferred to R_L. Core II is still in the unsaturated state, and so the ampere-turn relationship of Eq. (14.15) still holds. This results in the control current shown in Fig. 14.6. The average value of this current is the dc control current. From Eq. (14.15) it follows that the average ampere-turns of the control and anode windings are equal:

$$N_G I_G = N_c I_c \tag{14.16}$$

The control and anode windings of the unsaturated core act as a current transformer in which the conduction interval of the anode current is determined by the requirement that the average ampere-turns of the anode winding must equal the average ampere-turns of the control winding. This interval ends when the anode current and supply voltage reach zero, initiating a similar sequence with core II. In this second sequence the control current pulse is of the same polarity as the first sequence. The control current consists of an average component and a fundamental frequency equal to twice the supply frequency with higher-order even harmonics. With both the control and gate currents expressed in average values, Eq. (14.16) represents the transfer characteristic of the saturable reactor.

The gate windings may be connected in parallel instead of series, as shown in Fig. 14.7. In the series-connected saturable reactor with neither core saturated, the source voltage is equally divided between the two anode windings. Each core absorbs one-half of the volt-time integral of the supply voltage. Each anode winding carries the full anode current. In the parallel-connected reactor each core absorbs the full volt-time integral of the supply voltage requiring twice as many turns as the series-connected reactor using the same core, but each anode winding carries only one-half of the anode current.

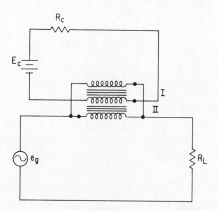

FIG. 14.7 Parallel-connected saturable reactor with control and anode circuits.

The current, voltage, and flux relationships in the parallel-connected reactor are similar to the series-connected reactor. At the start of a positive half-cycle of source voltage, core II is saturated and core I is at flux B_o. As the voltage increases, the flux in core I increases and the flux in core II decreases with the volt-time integral. When core I reaches saturation, the voltage across that core drops to zero. This clamps the flux in core II at B_o because of the parallel connection. The anode current is now limited only by the anode circuit resistance which includes the load resistance and the two anode winding resistances in parallel. Since the anode winding resistances are equal, the anode current divides equally between the two anode windings. The ampere-turns in the gate winding of core II, which remains unsaturated, must be balanced by the ampere-turns in the control winding:

$$N_c i_c = 2N_G(\tfrac{1}{2} i_g) \tag{14.17}$$

in which the lowercase letters represent instantaneous values. During the alternate sequence this relationship holds for the second core, so that the control ampere-turns, averaged over a full cycle, are equal to the total average anode ampere-turns, making Eq. (14.15) applicable to the parallel-connected reactor as well as the series-connected reactor.

The transfer characteristic which Eq. (14.15) represents is applicable to saturable reactors whose core material has a magnetization curve approaching that shown in Fig. 14.5. Some core materials depart widely from this ideal shape. The permeability is finite, and no sharp saturation point exists. From a saturable reactor made of such core material, it is possible to obtain data of average anode current vs. average ac voltage applied across the gate windings for various values of dc control current. From the dimensions of the magnetic circuit, it is possible to convert current and voltage to the magnetic units of flux density in gauss and magnetizing force in oersteds. A plot of anode oersteds versus gauss with control oersteds as parameter provides a family of generic curves for the core material independent of the windings used. The data are subject to variations in core properties. In converting anode amperes and volts to magnetic units, care must be taken to employ the correct turns and areas for the connection used in the saturable reactor. For example, if the series connection is used, the anode turns are the turns on one winding and the area is the area of the core enclosed by that winding. The voltage from which the flux density is calculated is one-half the applied voltage. If the parallel connection is used, the number of turns in either gate winding is used, but the current for calculating the anode magnetizing force will be one-half the total anode current. The voltage used to calculate flux density will be the entire applied voltage, and the number of turns will be the number of turns in either anode winding. A typical experimental curve using 4-mil silicon steel C cores is shown in Fig. 14.8. It is possible to use this curve to estimate transfer characteristics by plotting a load line across the curves whose slope is the negative of the load resistance expressed in

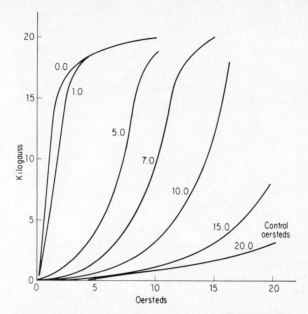

FIG. 14.8 Typical saturable reactor characteristic curve of ac flux density in kilogauss vs. ac magnetizing force in oersteds with dc control oersteds as parameter for 4-mil silicon steel C cores.

gauss per oersted instead of volts per ampere. The abscissa of the intersection of the load line with a curve of constant control in oersteds is the anode output in oersteds. The vertical intercept of the load line is the supply voltage expressed in gauss. The difference between the vertical intercept and the ordinate of the intersection of the load line with the constant control magnetizing force curve is the voltage expressed in kilogauss developed across the load for that control magnetizing force.

Saturable reactor performance can be greatly improved through the combined use of square hysterisis loop core material and the self-saturating circuit. A single reactor element in a self-saturating circuit is shown in Fig. 14.9. An idealized hysteresis loop showing minor loop-operating cycles is shown in Fig. 14.10a. Applied voltage, core flux density, and load voltage are shown as functions of time in Fig. 14.10b. In the analysis of this circuit, the control circuit impedance is assumed to be very high. The notion of balanced ampere-turns in control and gate windings is abandoned in favor of an assumed constant control current. When the control current is zero, the rectifier in the anode circuit will allow only positive magnetizing force to be applied to the core. The core will remain

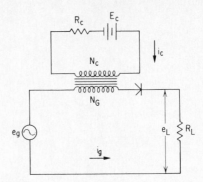

FIG. 14.9 Single-element self-saturating reactor circuit.

saturated except for the short intervals during which the flux moves from *b* to *c* and from *d* to *a* in Fig. 14.10*a*. An essentially half sine wave of voltage will appear across the load resistor R_L. To achieve control over the anode circuit, the control circuit must develop a magnetizing force in the core of opposite sense to that of the anode magnetizing current, a small but finite quantity which is inversely proportional to the unsaturated permeability of the core. A control current of the correct sense will generate a magnetizing force H_o, establishing an initial flux density B_o, as shown in Fig. 14.10*a*, during the interval when the applied voltage is negative and the rectifier prevents the flow of negative current. With the onset of positive applied voltage e_g, the flux density increases from the

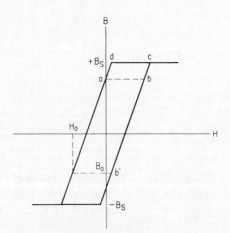

FIG. 14.10*a* Idealized oblong hysteresis curve showing minor operating loops.

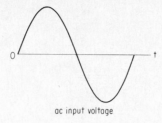

ac input voltage

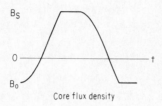

Core flux density

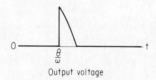

Output voltage

FIG. 14.10*b* Voltages and flux density in circuit of Fig. 14.9 as functions of time.

negative value B_o until $+B_S$ is reached. That point is determined from the following application of Faraday's law:

$$AN_G(B_S - B_o) = \int_0^{\theta/\omega} e_g \, dt \qquad (14.18)$$

The high permeability of the unsaturated core makes the impedance of the anode winding so much higher than the load impedance that all the applied voltage may be assumed to appear across the anode winding. The circuit will not function with an open-circuit load. The operating path of a controlled cycle is shown in Fig. 14.10*a* starting at the point H_o, B_o. The firing angle θ is predetermined by the value of B_o in accordance with Eq. (14.18). The input voltage e_g, the output voltage e_L, and the core flux density are shown in Fig. 14.10*b* as functions of time.

Individual reactor elements may be combined to obtain full-wave ac and dc outputs. Figure 14.11*a* and Fig. 14-11*b* illustrate two typical applications. When several elements are combined, performance differs from that predicted from the simplified analysis. Complicating factors force

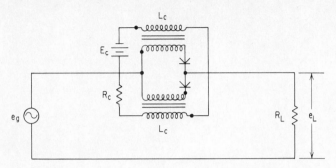

FIG. 14.11*a*   Self-saturating reactor circuit supplying single-polarity full-wave output.

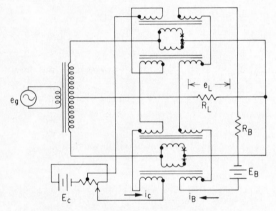

FIG. 14.11*b*   Self-saturating reactor circuit with bias windings supplying full-wave output of reversible polarity.

experimental verification by operation in equipment to determine acceptability of the reactor design.

A typical transfer characteristic of a self-saturating reactor circuit is shown in Fig. 14.12. The useful operating range is the high-gain, nearly vertical part of the characteristic. To establish a favorable operating point and to obtain an increase in output for an increase in signal input, an additional control winding, called a *bias winding,* is added to the reactor. The magnetizing force generated by the current in the bias winding opposes the magnetizing force generated by the anode winding. Correct choice of bias current will favorably position the operating point on the transfer characteristic for high gain. Bias and signal ampere-turns add algebraically to obtain net control ampere-turns.

Saturable reactors have slow response times compared with semiconductor devices and vacuum tubes. This, perhaps more than any other

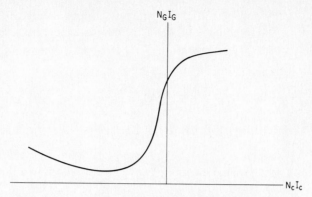

FIG. 14.12 Typical self-saturating reactor transfer characteristic.

property, diminishes their usefulness. An approximate knowledge of the response time is needed to help determine the suitability of saturable reactors for specific applications. The response time can be given as a time constant of the control circuit. For a step change in input, the output change reaches 63 percent of the final increment in one time constant. While this concept is derived from linear systems, which a saturable reactor obviously is not, the idea yields sufficiently accurate results for most purposes. In the circuit of Fig. 14.4 the time constant is

$$T = \frac{2L_c}{R_c} \tag{14.19}$$

where T = time, s
L_c = inductance of one control winding, H
R_c = total control circuit resistance, Ω

For the purpose of determining the time constant, inductance is defined as the average incremental change in flux linkages divided by the average incremental change in control current:

$$L_c = N\frac{\Delta\phi}{\Delta I_c} \tag{14.20}$$

From Faraday's law the average incremental change in voltage for an incremental change in flux is

$$\Delta E = N\,\Delta\phi\,2f$$

$$N\,\Delta\phi = \frac{\Delta E}{2f} \tag{14.21}$$

in which $1/2f$, the half-period of the sine wave input voltage, is the interval over which the average is taken. The incremental control and anode voltages are related as follows:

$$\Delta E_c = \frac{N_c}{N_G} \Delta E_G \tag{14.22}$$

Substitution of Eqs. (14.20) to (14.22) in Eq. (14.19) yields

$$T = \frac{\Delta E_G}{\Delta E_c} \frac{N_c}{N_G} \frac{1}{2f} \tag{14.23}$$

If the control and gate windings are converted to the same turns base, Eq. (14.23) becomes

$$T = \frac{A_v}{2f} \tag{14.24}$$

in which A_v is the voltage gain for a common turns base.

By use of similar reasoning, the time constant for the series-connected saturable reactor of Fig. 14.4 is

$$T = \frac{A_v}{4f} \tag{14.25}$$

Equations (14.24) and (14.25) give the not-surprising result that the response time is directly related to the supply frequency and the gain of the amplifier.

14.5 FERRORESONANT TRANSFORMERS

A ferroresonant transformer is a saturable device which regulates for changes in the magnitude of the input voltage. It differs from a conventional transformer by having a large leakage inductance, a saturating shunt inductance, and a large capacitor in parallel with the load. The equivalent circuit of a ferroresonant transformer is shown in Fig. 14.13. Inductance L_p is the transformer shunt inductance which saturates just below the low end of the input-voltage-regulating range. This inductance enters into ferroresonance with the capacitor and leakage inductance. Above the voltage at which ferroresonance occurs, the impedance of the parallel combination of L_s and C decreases with increasing voltage. Reg-

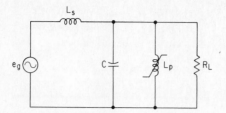

FIG. 14.13 Equivalent circuit of a ferroresonant transformer.

ulation takes place as a result because a larger voltage drop appears across L_s as the input voltage increases.

If a voltage at rated frequency is applied to an unloaded ferroresonant transformer, with the voltage being gradually increased from zero, the current will increase from zero, gradually at first, then rapidly with the onset of resonance. After peaking, the current falls rapidly to a minimum, after which it increases gradually again. The current behavior is illustrated in Fig. 14.14. The transformer will regulate for voltages above that at which the current peaks. An explanation for the current behavior can be made on the basis of an approximate linear analysis. The input impedance of the transformer is almost purely reactive. By reference to Fig. 14.13 that reactance can be seen to be

$$x_{in} = x_s + \frac{-x_c x_p}{x_p - x_c} \qquad (14.26)$$

As the voltage is increased, inductance L_p will vary from a relatively large value to zero. Table 14.1 gives significant values of x_{in} and x_p including those at which singularities exist. The magnitude of x_{in} may be plotted

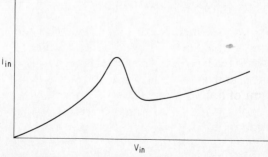

FIG. 14.14 Input current vs. input voltage of unloaded ferroresonant transformer.

TABLE 14.1 Significant Values of Input Reactance x_{in} vs Inductive Reactance x_p of Ferroresonant Transformers

x_{in}	x_p
x_s	0
∞	x_c
$+$	$x_c - \Delta x_c$
$-$	$x_c + \Delta x_c$
0	$x_s x_c$
	$x_s - x_c$
$x_s - x_c$	∞

as a reactance function in which the variable is x_p instead of frequency. That plot is shown in Fig. 14.15. The pole and zero of x_{in} coincide with the current minimum and maximum, respectively.

The average voltage across the inductance L_p tends to remain constant because of the integral form of Faraday's law in which the flux density term is constant and equal to the saturation flux density of the core. Since frequency also appears in the integration of Faraday's law, the average voltage across L_p will remain constant only when the frequency is constant. This makes ferroresonant transformers essentially single-frequency devices. The output voltage is almost directly proportional to the first power of the frequency.

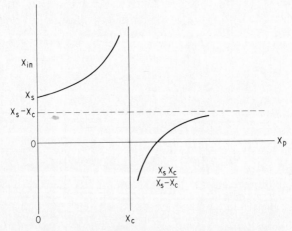

FIG. 14.15 Input reactance vs. saturating reactance of unloaded ferroresonant transformer.

Ferroresonant transformers regulate the average value of the output voltage. Core saturation introduces waveform distortion which varies with the input voltage. The rms value of the output voltage may therefore vary while the average value is held constant. Ferroresonant transformers are most useful with rectified loads in which the dc output voltage is more closely related to average than the rms value of the ferroresonant output voltage. The dc voltage across a constant load will vary about 1 percent for a 15 percent change in input voltage to the ferroresonant transformer. The device does not regulate for load changes. They are very similar to conventional transformers in regard to changes in output voltage with changes in load current.

The construction of a ferroresonant transformer is suggested in Fig. 14.16. Special ferroresonant EI laminations are generally used. Primary and secondary coils are placed on the center leg of the lamination separated by a magnetic shunt. The secondary coil consists of two windings, one for the capacitor and one for the load. Use of a separate winding for the capacitor allows freedom in selecting the capacitor voltage for optimum use of the capacitor dielectric. The capacitor volt-amperes are approximately twice those of the load. The magnetic shunt provides the path for the leakage flux of the leakage inductance L_s, shown in the equivalent circuit of Fig. 14.13. The resonant effect causes the voltage on the secondary to be about 1.2 times the voltage that would result from the turns ratio alone. This makes the maximum flux density in the secondary portion of the center leg higher than the primary portion. The difference between these two fluxes flows in the magnetic shunt. The secondary portion of the center leg will saturate before the primary portion. The outer legs are dimensioned so that the flux density there is lower than the center

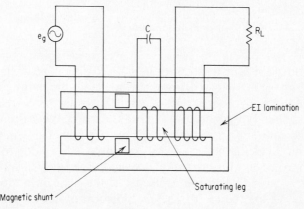

FIG. 14.16 Construction diagram of a ferroresonant transformer.

leg. With saturation confined to only a portion of the center leg, greater efficiency is obtained than if the lamination were proportioned as in a conventional transformer. The air gap in the shunt path is optimized to provide the correct value of leakage inductance.

Ferroresonant transformers have a load voltage-current characteristic as shown in Fig. 14.17. This curve shows that the device is self-protected against short circuits, a frequent and considerable advantage. Under rectified load with a capacitor input filter, the dc output voltage tends to peak when the current goes below about 5 percent of full load. Thus a given ferroresonant transformer will operate satisfactorily over a range of about 5 to 110 percent of full load. If the cross section of the magnetic shunt is made small so that it saturates with the onset of short circuit when flux through the shunt is maximum, the load voltage-current characteristic will take the form of the indicated curve in Fig. 14.17.

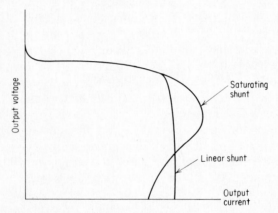

FIG. 14.17 Typical output voltage-current characteristics of ferroresonant transformer.

Ferroresonant transformers under light loads can be unstable. When the input voltage is increased from zero to rated, the input current may vary differently from the way it does when the voltage is decreased. If full voltage is suddenly switched on, the input current may assume either of two bistable values. This bistable condition is associated with the flow of subharmonic currents and may be accompanied by different audible tones from the core. This instability does not necessarily affect the output voltage. It is advisable to use a bleeder if the load on a ferroresonant transformer is expected to drop below 5 percent of full rating.

Ferroresonant transformers have the advantage of providing line regulation in applications where a transformer is necessary at an attractively small increment in cost over a conventional transformer. When a ferroresonant transformer is being considered, the limitations must be understood and weighed. Circuit evaluation of prototypes is needed, and production tolerances on electrical performance require consideration.

The nominal value of output voltage of a ferroresonant transformer is a function of the center leg area. The shape of the load voltage-current characteristic is a function of the magnetic shunt and air gap. Both of these items are subject to production variations. This makes ferroresonant transformers frequently troublesome in production when tight tolerances on electrical characteristics must be met.

CHAPTER 15

Magnetic Device Economics

15.1 GENERAL

Magnetic components present special problems in systems economics. The manufacturing methods and materials used in magnetic components are established. Improvements are made by a relatively slow evolution. The requirements and usage of magnetic components are varied and changing rapidly. The task of the equipment producers is to extract their varied and novel requirements from an established source at minimum cost.

The total cost of a production quantity of a component consists of the fixed cost and the unit costs times the quantity. The fixed cost is the engineering and development cost. The unit cost is the sum of the unit labor and unit purchased material costs required to produce the item. The evident engineering cost is the component design cost. Other significant fixed costs are the preparation of specifications, engineering liaison, prototype evaluation, and the correction of deficiencies. The unit costs are functions of the quantity to be produced. The unit cost of many purchased items will decrease with quantity. The unit labor cost consists of the amortized setup cost and the run cost. The amortized setup cost decreases with quantity. The run time will follow a learning curve in which the unit labor content decreases with each unit produced. The most significant cost elements in the total costs are largely determined by the quantity.

15.2 STANDARDIZATION

There is an inclination to view magnetic devices as components which may be selected from catalogs as resistors are. The users who find catalog items to which they adapt their circuits are fortunate, and the process should certainly be pursued. Too often the nature of magnetic components will cause this expectation to be met with expensive disappointment. The electrical requirements cover a continuous range of voltages, currents, frequencies, waveforms, source impedances, and load characteristics. Optimum performance of a magnetic device will be obtained only when it is operated over a limited range around its design center. Operation outside the optimum range is either inefficient, unsatisfactory, or destructive. Thus varied electrical requirements effectively preclude the extensive standardization necessary for catalog procurement.

The absence of standardization is the plague of magnetic component procurement. The most successful standards exist in the electric utility industry. It is instructive to examine some of the reasons for this success. The industry is mature and changes slowly. Associations that set industry standards labored for many years to achieve the present degree of success. The fiscal stability of the industry attracts the investment required to execute standardization. The environmental requirements, though difficult, are uniform. The problems of outdoor installation have long been solved. Distribution voltages and frequencies are standardized. Distribution networks are designed to be tolerant of discrete intervals in transformer kilovolt-ampere ratings. The conditions that led to standardization of distribution transformers are conspicuously absent in the equipment industry. It is fiscally volatile and rapidly changing. Environmental requirements vary. There are no electrical function standards and little hope of establishing any because of the continuously changing technology.

The mechanical requirements on magnetic components, while also quite varied, are more amenable to classification and standardization than the electrical requirements. Mechanical requirements tend to fall into a range of increasing severity. Operation under less than the most severe conditions, while inefficient, is neither unsatisfactory nor destructive. A single mechanical construction can accommodate a multitude of different electrical functions. Therefore, the opportunities for standardization of mechanical construction are far greater than they are for electrical functions. The mechanical requirements frequently are more significant in establishing the cost of magnetic components than are the electrical requirements. It therefore behooves the user to exploit whatever standardization exists in mechanical construction.

One set of mechanical standards of long standing is shown in Fig.

15.1. These standards, which apply to enclosure and mounting dimensions only, were developed by the U.S. Department of Defense. The arbitrary dimensional increments of these standards provide no intrinsic advantages. Their advantage is their ready availability from multiple sources. Purchasers of magnetic components gain the additional advantage, from specifying these standards, of avoiding proprietary constructions which limit multiple sourcing. The multiple sources which can supply enclosures to these standards use varied constructions within the limits imposed by these standards. It is possible to accommodate designs to these multiple sources given prior knowledge of construction variations.

There are other standards of interest to both user and supplier. Some of these standards are industry-wide and some will apply to only a single supplier. Magnet wire, laminations for small and medium-sized transformers, magnetic steels, wound cores, ferrite cup cores, and bobbins are standardized industry-wide. Established sources will have standards for cases, tanks, potting shells, mounting facilities, and terminations that are likely to be unique. Avoid single-source standards that will require expensive tooling to reproduce elsewhere.

Standardization reduces the length of time required to fill a requirement. Even if the item is not kept in stock, the time needed to produce it does not include engineering and development delays. Standardization has the technical advantage that the characteristics of the standard item are better known than those of an item that has never been made before. The cost advantage of standardization is related to quantity. If the usage is high and the standard industry-wide, the all-important factor of competitive pricing will work to the advantage of the user. Competitive bidding on special items with no previous price or production history is of doubtful value because of indeterminate hidden costs. These costs include delayed deliveries, inability of the supplier to meet specifications, poor quality, and user specification errors. It often happens that in these situations the lowest bid is not the least cost to the user. Competitive bidding on standardized items is invaluable even in small quantities. When the procurement quantity is small, standardization has its greatest advantage in pricing because there are no fixed costs to pay. The unit cost of the standard item will depend on the quantity and the supplier's appraisal of the prospect for future sales. This appraisal influences the decision on investment in production facilities. Most equipment is developed on a limited budget with uncertain estimates of the quantity to be produced. This atmosphere tends to influence decisions on parts selection in favor of standards. It is conceivable that if the equipment is later produced in substantial quantity, such decisions should be reversed. The

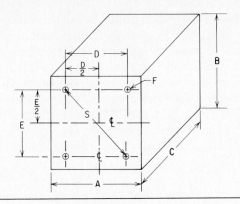

	Dimensions, in						
	Envelope			Mounting			
Case Symbol	A	B	C	D	E	S	F
AF	0.750	0.750	1.125			0.562	4—40x⅜
AG	1.000	1.000	1.375			0.750	4—40x⅜
AH	1.312	1.312	1.750			1.250	6—32x⅜
AJ	1.625	1.625	2.375	1.188	1.188	1.680	6—32x⅜
EA	1.938	1.812	2.750	1.375	1.250	1.858	6—32x⅜
EB	1.938	1.812	2.438	1.375	1.250	1.858	6—32x⅜
FA	2.312	2.062	3.125	1.688	1.438	2.217	6—32x⅜
FB	2.312	2.062	2.500	1.688	1.438	2.217	6—32x⅜
GA	2.750	2.375	3.812	2.125	1.750	2.753	6—32x⅜
GB	2.750	2.375	2.812	2.125	1.750	2.753	6—32x⅜
HA	3.062	2.625	4.250	2.297	1.859	2.955	8—32x⅜
HB	3.062	2.625	3.188	2.297	1.859	2.955	8—32x⅜
JA	3.562	3.062	4.875	2.625	2.125	3.377	8—32x⅜
JB	3.562	3.062	3.875	2.625	2.125	3.377	8—32x⅜
KA	3.938	3.375	5.250	3.000	2.438	3.866	10—32x½
KB	3.938	3.375	4.312	3.000	2.438	3.866	10—32x½
LA	4.312	3.688	5.562	3.312	2.688	4.266	10—32x½
LB	4.312	3.688	4.500	3.312	2.688	4.266	10—32x½
MA	4.688	4.000	6.000	3.688	3.000	4.754	¼—20x⅝
MB	4.688	4.000	4.938	3.688	3.000	4.754	¼—20x⅝
NA	5.062	4.312	6.812	4.062	3.312	5.243	¼—20x⅝
NB	5.062	4.312	5.500	4.062	3.312	5.243	¼—20x⅝
OA	5.500	4.500	6.750	3.750	3.000	4.802	¼—20x⅝

FIG. 15.1 Standardized envelope and mounting dimensions from U.S. Department of Defense Specification MIL-T-27D.

Inches	Millimeters	Inches	Millimeters
0.01	0.25	2.955	75.06
0.03	0.76	3.000	76.20
0.1	2.54	3.062	77.77
0.3	7.62	3.125	79.38
0.562	14.27	3.188	80.98
0.750	10.05	3.312	84.12
1.000	25.40	3.375	85.73
1.125	28.58	3.377	85.78
1.188	30.18	3.562	90.47
1.250	31.75	3.688	93.68
1.312	33.32	3.750	95.25
1.375	34.93	3.812	96.82
1.438	36.53	3.866	98.20
1.625	41.28	3.875	98.43
1.680	42.67	3.938	100.03
1.688	42.88	4.000	101.60
1.750	44.45	4.062	103.37
1.812	46.02	4.250	107.95
1.858	47.19	4.266	108.36
1.859	47.22	4.312	109.52
1.938	49.23	4.500	114.80
2.062	52.37	4.688	119.08
2.125	53.98	4.754	120.75
2.217	56.31	4.802	121.97
2.297	58.34	4.875	123.83
2.312	58.72	4.938	125.43
2.375	60.33	5.062	128.57
2.438	61.93	5.243	133.17
2.500	63.50	5.250	133.35
2.625	66.68	5.500	139.70
2.688	68.28	5.562	141.27
2.750	69.85	6.000	152.40
2.753	69.93	6.750	171.45
2.812	71.42	6.812	173.02

FIG. 15.1 (*continued*)

specially tailored part may be less expensive in quantity than the nearest available standard. It may also improve marginal performance resulting from the compromises made to accommodate the standard.

15.3 SIZE, WEIGHT, AND COST RELATIONSHIPS

Price is often considered to be proportional to size. For transformers this is a half-truth. The cost of routine power transformers which differ only in their volt-ampere rating will increase with increasing size. A high-temperature transformer designed for minimum size will cost more than a larger transformer with a lower temperature rise but otherwise similar ratings.

Transformers are bulky objects, and inevitably there is pressure to make them smaller. When minimum size is required, the cost will increase. Size reductions are best achieved by carefully specifying requirements and good design. Avoid size requirements that force the design beyond currently available manufacturing technique. The improvement in this technique is an expensive and time-consuming process which your single requirement will not be able to support. Beware of excessively limiting size in high-voltage devices. There is no substitute for adequate spacing around high-voltage electrodes. Endless problems with reliability can result from ill-advised size reductions. Some suggestions for obtaining minimum size at reasonable cost follow:

1. Deal with a competent source.
2. Determine your requirements with care.
3. Do not overspecify. Ask only for what you need.
4. Avoid unnecessary safety margins. Magnetic devices are rugged and may be reliably operated at their full ratings.
5. Take full advantage of limited duty cycles. On times may be related to the thermal time constant of the device. This allows temporary operation under very heavy loads.
6. Take advantage of the allowable temperature rise. Permit the use of high-temperature insulation.

15.4 ELECTRICAL REQUIREMENTS AND THEIR EFFECT ON SIZE, WEIGHT, AND COST

Current

The economics of magnetics design is concerned with true rms current. True rms current takes into account duty cycles, repeated momentary loads, and intermittent operation as well as waveshapes. The rms current determines the size of the conductors in the windings. The conductors are most often copper magnet wire. The size of round magnet wire is specified by American wire gauge (AWG) wire size. Round magnet wire is readily available in sizes from about AWG 48, the finest, to about AWG 6. The larger sizes of magnet wire, AWG 14 to AWG 2, are available as square wire. Square wire is also identified by gauge number signifying a square dimension approximately equal to the diameter of the round wire of the same gauge number. The square wire, which has a greater current-carrying capacity than the round wire of the same number, occupies the same core window area as the round wire. Rectangular magnet wire is available in larger cross-sectional areas, expanding the range beyond the largest available square wire. Magnet wire is sold by weight. The price is determined by the base price of copper plus adders which vary with the size of the wire and the type of insulation. The base price of copper is volatile, whereas the adders are relatively stable in price. The adders are a maximum for the finest gauges and decrease as the size increases. The adders for square wire are greater than for round wire of the same gauge number. The extra cost of square wire in the smaller sizes is an economic factor in design. The current-carrying capacity of a given size of magnet wire varies with the geometry of the device. However, for a given geometry and temperature rise, the rated current density will be constant and independent of the current.

The winding techniques used with magnet wire vary with size. Most suppliers are not equipped to wind the full range of sizes. Current specifications will help determine what sources can fill a particular requirement. Although the current-carrying capacity of magnet wire is not fixed according to size, the larger the rms current requirement, the larger in general will be the required wire size and the greater will be the size of the device. Reductions in current below the current-carrying capacity of the smallest wire size which the chosen source can wind will not affect any further reduction in the volume of that particular winding. Reductions in wire size are limited not only by winding capability but also by reliability considerations. The probability of opens developing is much greater in windings using the finest wire. This is a major cause of failure in small coils.

The rms current requirement affects the unit cost by contributing to the size and cost of the core and the enclosure, by determining the size and amount of wire required, and by affecting, through wire size requirements, the labor cost of winding. The labor cost of winding per turn is a minimum for the intermediate range of wire sizes and increases for either smaller or larger sizes. Winding labor content is complicated by the practice of multiple winding coils of fine and intermediate size wire and by the fact that the windings of finer wire tend to have more turns because of design requirements. Coils of fine wire require supplementary assembly operations, called *coil finishing,* during which durable terminations are provided for the fine magnet wire. Coil finishing adds to the labor cost of fine wire coil. The heaviest wire sizes are wound one coil at a time on slow-moving massive equipment, resulting in the greatest labor cost per turn.

Voltage

The turn area product, other factors being constant, is determined by the applied voltage as stated by Faraday's law. (See Sec. 1.3.) Decreasing the cross-sectional area of the core requires an increase in the size of the core window because of the added turns needed to maintain the turn area product constant. The net result is little change in the overall size of the core, so that the size of the core is approximately determined by the turn area product. Applied voltage for a given wire size has a direct bearing on core size and therefore cost. The properties of magnetic materials most significant to economical design are losses as a function of frequency and flux density, saturation flux density, and permeability. These properties vary not only with the alloy but also with the heat treat condition and the fabrication. In addition, the losses vary directly with the lamination thickness. Laminations of various shapes and sizes are punched from magnetic steels of several standard gauges. Both magnetic steels and laminations are priced by weight. The unit cost of steel sheet increases as the thickness decreases. The unit cost of laminations increases as the number of fabricating operations per unit weight increase, as occurs with smaller laminations and thinner gauges. The labor cost per unit weight for stacking also increases as the gauge becomes thinner and the laminations smaller. The highest-permeability materials are nickel-iron alloys. The raw-material cost of these alloys is very high compared with the more common silicon steel alloys. The requirement for high permeability is frequently combined with the requirement for low losses at high frequencies. It is thus common for cores of nickel-iron alloys to use the thinnest gauges. The cost per unit weight of such cores is an order of magnitude greater than the equivalent in low-frequency silicon steel material. Nickel-iron alloys have lower saturation flux densities than silicon steel, which can further increase cost. The highest-frequency performance is obtained with ferrite cores which are molded and fired by ceramic processes. The yield from these processes is low and becomes worse as the size of the cores increases. Thus large ferrite cores are epensive and limited in available geometries. The excellent high-frequency performance of ferrite cores is partly offset by their low-saturation flux density, which is the lowest of any of the commercial core materials.

Secondary voltages determine the number of turns required for their windings, the winding cost and core size increasing for larger numbers of turns. Voltages, either induced or applied, sufficiently high to invoke dielectric considerations introduce additional factors affecting cost. Spacing between electrodes within coils must be increased to keep insulation stress levels within ratings. The impregnation and other coil treatment processes become more costly. Enclosures become larger. The filling processes for enclosures become critical. Terminations become larger and more expensive. High voltage is likely to add hidden costs resulting from poor reliability.

Volt-Ampere Rating

For a fixed primary voltage and frequency, the primary winding will remain the same for varying secondary voltages and currents, provided the secondary volt-ampere rating remains constant. Excluding the considerations of high voltage, the increased cost of the secondary winding due to increased voltage will tend to be offset by a decrease in the cost due to decreasing current. Therefore, the cost of a transformer will tend to be a function of its volt-ampere rating independent of the voltage and current ratings of the load, other factors being constant.

Frequency

Frequency has two countervailing effects on cost. Faraday's law states that the product of frequency and area is constant, other factors remaining constant. The frequency here is the minimum operating frequency. Thus, the higher the minimum operating frequency is, the smaller the cross-sectional area of core required. Smaller cross-sectional core area permits smaller overall size and lower cost. For a given geometry an increase in frequency causes a decrease in flux density. However, to take advantage of an increase in minimum operating frequency, the flux density must be maintained near the same value by changing the geometry. Under these conditions, increasing the minimum operating frequency finally limits the usefulness of the core material by reason of the increase in core losses. This increase in core loss forces a change to a more efficient and more expensive core material. When the maximum operating frequency requires limiting the distributed capacitance and leakage inductance, additional costs are introduced because of the need for complex coil constructions.

Regulation

Regulation more stringent than that imposed indirectly by allowable insulation temperature and high-frequency response requires a decrease in winding resistances and a decrease in the number of turns in the windings. These changes are affected by using larger wire and greater cross-sectional core area, both increases raising the cost. A transformer with stringent regulation is under-utilized.

Source Impedance and Open-Circuit Inductance

Transformers driven by high-impedance sources have placed upon them either explicitly or otherwise a requirement for high open-circuit inductance and constraints on the distributed parameters. A source impedance becomes significant when it reaches the same order of magnitude as the load impedance referred to the primary. The control of distributed parameters, leakage inductance, and distributed capacitance involves the use of complex windings which increase labor cost and reduce the number of turns, increasing the cost of the core. Ferromagnetic inductance requirements on any magnetic component invoke peculiar problems. The turns in the windings are determined no longer by limiting the flux density or losses, but by the effective permeability of the magnetic circuit. This permeability is determined in part by the permeability of the core material and in part by the fabrication and assembly of the core. The vagaries of the magnetic circuit make it difficult to assess the effect of the variations in material permeability on the effective permeability. Published data on material permeability seldom cover the magnetizing conditions in actual use. The result is that indeterminate variations in material permeability and core assembly make control of effective permeability expensive.

15.5   EFFECT OF AMBIENT AND MECHANICAL REQUIREMENTS ON SIZE, WEIGHT, AND COST

Ambient Temperature

The size and weight of power-handling magnetic components is related to the maximum ambient temperature. The insulation temperature rise, which is proportional to the losses, added to the maximum ambient temperature must not exceed the safe operating maximum temperature for the insulation system used. If a device is operating at the maximum allowable insulation temperature, an increase in the ambient temperature will require an increase in the size of the device which simultaneously allows a reduction in the losses and an increase in its heat-dissipating ability. Conversely, a reduction in the ambient temperature will allow a reduction in the size of the device. Various insulation systems have different maximum safe operating temperatures. The high-temperature insulating systems which permit reduced size cost more. Various mechanical techniques are used to increase the heat dissipation of device enclosures. Fins are added to the case walls. Tubes are added to the walls of liquid-filled units to allow the liquid to flow by convection, bringing the hotter portions of the liquid in close contact with the outside surface and increasing the outside surface area. Convection flow is sometimes supplemented by pumps which force a more rapid circulation of the cooling liquid. Heat exchangers are sometimes placed in enclosures of liquid-filled units through which water or antifreeze coolant passes. All these techniques add to the cost of the unit.

The minimum ambient temperature may cause embrittlement and cracking of plastic encapsulants and potting compounds. Repeated exposure to temperature extremes as in thermal shock tests increases the tendency of plastic materials to crack. A requirement for absolute assurance against cracking at extremely low temperatures will force the use of gaseous or liquid instead of solid plastic dielectric materials. This will increase the cost of units in which the use of solid plastic materials would otherwise be the economical choice.

The ambient temperature extremes plus the temperature rise of the component determine the change in volume of liquid dielectrics. Provision for this change in volume must be made in the construction of enclosures. A substantial percentage of the volume of enclosure will be devoted to this provision. The cost depends upon additional requirements. If the component is to operate always in the same stationary position, the provision for expansion need be only a space above the liquid level. The least expensive construction will result if this space is allowed to be air-breathing. Alternatives to air-breathing are a sealed air space or a sealed dielectric gas space. In either case provision for increased pressure at the high-temperature extreme must be made. Sealed expansion volumes prevent exposure of the liquid dielectric to the atmosphere and sometimes allow operation in various positions and motion during operation, depending upon construction and high-voltage stresses. The expansion volume may be eliminated in favor of mechanical expansion devices, the most common of which is the metal bellows. Metal bellows cost more than expansion volumes and increase the total volume of the component, but they safely allow operation in any position and motion during operation.

Humidity

The insulating materials used in magnetic components are usually hygroscopic. Interstices in coils are further traps for moisture. Moisture reduces the effectiveness of insulation, causing circuit malfunctions or complete failure. There are three general constructions offering varying levels of protection against moisture. In order of increasing cost, they are impregnated open-frame, plastic-enclosed, and hermetically sealed construction. If the unit is to be installed where subsequent assembly provides protection against humidity, then it is unnecessary to provide additional protection on the transformer itself. If some humidity protection is required and space is critical, plastic encapsulation may be the best choice. Hermetically sealed metal cases provide the best protection and occupy the greatest volumes. It should be understood that hermetic sealing is not an absolute condition. Sensitive tests reveal small leaks even in the most carefully constructed enclosures. Sealed metal cases filled with liquid or solid dielectric potting materials are probably more than adequate for most applications. The determination of minimum acceptability should be made with care. Stringent arbitrary requirements increase the cost sometimes without any apparent benefit.

Shock and Vibration

Magnetic components, composed of large volumes of very dense materials, are especially vulnerable to failure from mechanical shock and vibration. These otherwise rugged devices will fail most often under shock and vibration due to inadequate mounting arrangements from which they break loose, damaging themselves and adjacent objects. The mountings must survive the abuse of shipment regardless of how benign the operating conditions may be. The severity of this abuse may be unrelated to the distance or the nature of the destination. Severe damage can occur on the user's own loading dock. Sound mechanical design thus is the economical choice. Severe operating conditions can impose additional requirements on mountings. Particularly troublesome are resonances in which mechanical movement is amplified. It is necessary to increase the natural resonant frequencies of the device above those to which the unit will be subjected. Provision for such conditions will increase cost.

Military Standards and Requirements

The stringent requirements placed on electrical and electronics equipment by the military originated with the complexities of maintenance logistics and the urgent need for reliability in the face of poor equipment service records. Funding for increasingly stringent requirements has generally been available.

The established document covering transformers and inductors for military equipment is U.S. Department of Defense Specification MIL-T-27. There also exist some special purpose specifications which are modeled on this original specification. First implemented in a less complex time, the specification had as an objective the standardization of requirements and the reduction of the number of procurement items. While vestiges of this original objective remain, the current emphasis is on the establishment and maintenance of reliability and quality standards. The facets of the specification may be classified as follows:

1. Nomenclature
2. Mechanical and electrical standards
3. Standardized electrical and environmental tests
4. Qualification procedures
5. Quality assurance provisions

The nomenclature in the specification is a standardized notation for describing and identifying magnetic devices covered within it. Its impact on product cost is minor in that it requires detailed performance information on the product.

The mechanical envelope standards in the specifications have been adopted and are generally useful. They are shown in Fig. 15.1. The electrical standards, an attempt to establish uniform electrical requirements, have not been widely accepted. See Sec. 15.2.

The standardized tests are compulsory methods for qualification and quality assurance. The standardized tests invoke a rigorous set of environmental conditions which are the heart of the specification. Meeting these requirements necessitates designing and constructing specifically for this purpose. Costs are substantially higher than for commercial equivalents.

The qualification and quality assurance provisions of the specification are the most controversial. Many persons using the specification for technical purposes are unaware of the administrative and legal complexities in these provisions. Briefly, they require that the environmental program be performed completely in order to be in compliance with the specification. The cost of this test program easily exceeds the cost of prototype design and functional evaluation. When budgeting for items upon which Specification MIL-T-27 is invoked, it is important to clarify whether environmental testing is to be performed and, if performed, what the liability will be for units made before the possibly unsuccessful results of the long test program are known.

There is a tendency for persons unfamiliar with the specification to invoke its requirements inappropriately. It is inevitable that a general specification will not be universally applicable to every individual requirement in complete detail. When such occasions arise, the situation should be negotiated. Failure to do so can result in futile expense and delays.

15.6   TOLERANCES AND COST

The word *tolerance* in industry usage means an allowable variation from a nominal value. The more descriptive origin of the word implies endurance. How much variation can you endure? The less you are able or willing to endure, the more you are going to pay. The concept of tolerance is a recognition that no single object is perfect and no two objects are identical. Tolerances are applied to attributes of imperfect objects signifying the allowable departure from perfection beyond which the objects will fail to perform their function. From this definition it is obvious that cost undergoes an exponential increase toward infinity as tightened tolerances force an approach to perfection. It is likewise obvious that minimum cost will result when tolerances are without limit. As a practical matter tolerances for low cost must be sufficiently large that further increases will result in only an insignificant reduction in cost.

Industrial tolerance requirements begin with the minimum performance limits of a final product. In the perspective of the design function, these limits are given information. The product consists of a set of subassemblies and detail parts which function together to meet the minimum performance limits of the final product. Each subassembly and detail part possesses attributes to which are assigned nominal values representing optimum conditions. Variations from these nominal values interact in complex ways to cause variations in the performance of the final product. Minimum cost is achieved by allowing the attributes of all details and subassemblies to vary as much as possible without exceeding the minimum performance limits of the final product. The selection and design of the details and subassemblies determine the effect of their variations on the final product. Design ingenuity provides for greater allowable variations.

The complex interplay of variations permits a myriad of options in the assignment of tolerances. It is a common mistake to allow excessively generous tolerances on some items, forcing the need for unreasonably tight tolerances on others. This pitfall is the more treacherous because of the large number of choices available. It is necessary to reduce the number of choices by first making a substantial number of independent decisions on economics. The key here is determining tolerances the increase of which will result in no further significant reduction in cost. This does not mean asking, "How tight can you hold it?" Many people, when asked this question, will consider it a challenge and give a reply which reflects their skill. The correct question is, "Will it cost substantially less if this tolerance is opened up?"

Some of the answers are implied in data available from suppliers. Tolerances given in catalogs, or supplied upon request for catalog items, are generally selected so that they may be held without economic penalty. Armed with a list of components with economical tolerances, the designer may then turn to a reduced number of items which require tolerance assignments. These remaining items may permit classification by manufacturing methods. It is reasonable to assign similar tolerances to values controlled by the same manufacturing method. When options exist among manufacturing methods, the obvious choice is the least expensive. Various manufacturing options tend to have a range of economical tolerances. Tolerances which force a more expensive option on an operation that can readily be performed by a less expensive process should be avoided.

Some quantities require only a single-sided tolerance which may be either a minimum or a maximum value. The designer should take advantage of this situation whenever possible.

Tolerance decisions are usually made on the basis of a worst-case analysis. In worst-case analysis the effect of the variation of all the attributes controlling a final result are evaluated. From the allowable variation of the final result, tolerances are assigned to individual attributes which keep the final result just within the allowable variation when all those attributes assume their more disadvantageous extreme values.

Once assigned, tolerances become administrative and legal as well as technical requirements. Control passes to others who are either forbidden or unable to exercise judgment in their enforcement. Administrative procedures perpetuate inadequately considered tolerances, frustrating the economics those procedures were designed to achieve. Part of the administrative procedure associated with tolerances often involves the measurement of an attribute under two different conditions, typically first by the supplier and then by the user. It is inevitable that identical results will not be obtained. When the supplier's measurement indicates that the attribute is within tolerance, and the user's measurement indicates that it is out of tolerance, a fruitless adversary relationship can develop. The designer specifying the tolerance in this situation is required to either defend the tolerance or relax it. This unhappy state of affairs is aggravated if the precision of the measurement is little better than the tolerance. The assignment of tolerances must of necessity recognize the precision available for measuring the attributes to which the tolerances apply. This precision is controlled by available instrumentation, calibration procedures, and the method used to make the measurements. Specifying a tolerance beyond the capability of available instrumentation is pointless because it will be impossible a priori to determine compliance with such a tolerance requirement.

The measurement of attributes sometimes requires the indirect approaches of simulation and the use of implicit rather than explicit quantities. Tolerances applied under such circumstances should be no tighter than is justified by the precision of the simulation or the degree to which the implicit requirements relate to the explicit requirements. For example, it is usually not practical to measure at the supplier's plant the output characteristics of a pulse modulation transformer operating in the actual circuit for which it is intended. Performance is sometimes specified in terms of equivalent resistive source and load impedances. Severe tolerances on the output-pulse shape characteristics measured under these conditions will add to the cost without necessarily assuring desired performance in actual operation. The same is true of implicit requirements on leakage inductance and distributed capacitance. The actual values of these quantities are of no interest except to the transformer designers who use them in their design procedures. Furthermore, the optimum values are known only approximately.

In assigning tolerances, consideration should be given to the precision with which design center values are known. The 1-s accuracy of a traveler's digital watch will not ensure keeping a distant appointment when the time zone of the destination is uncertain. Precision control of attributes whose nominal values are approximate is uneconomical and incongruous.

Some attributes vary with environmental conditions. Designers specifying tolerances need a knowledge of the stability of the attributes with which they are working. They must avoid tolerance bands which are narrower than the instability of the attributes to which they apply. The mechanical dimensions of many plastic materials vary with temperature and humidity. Their flexibility prevents the well-defined surfaces needed for precision measurement of length. It is not reasonable to assign to plastic parts the tight machine tolerances that are routinely applied to metal details.

Voltage Ratio

Voltage ratio describes the operation performed by transformers on the applied voltage. The full-load voltage ratio is determined by the no-load voltage ratio and the regulation. Regulation is the voltage drop caused by the load current flowing through the winding resistances and leakage inductance, usually expressed as a percentage of full-load voltage. The no-load voltage ratio is almost solely dependent upon the ratio of the turns in the respective windings across which the voltages are developed. No-load voltage ratio is usually free of spurious effects due to distributed parameters when measured at a frequency near the design center of the transformer. The turns in the transformer coil are established by a mechanical counter which is without error. Errors in turns result from misuse of the counter and faulty lead dressing. No-load voltage ratio can be held to ± 2 percent with no economic penalty. It can be held to 1 percent at a slight penalty. The cost increases rapidly when the tolerance is tightened beyond 1 percent. The achievable resolution in turns can be no greater than one turn, or in the case of certain constructions, one-half turn. The half-turn resolution is subject to an indeterminate error due to the fact that a mechanical half-turn does not link precisely half of the flux that a whole turn links. A single-turn or half-turn increment causes a smaller change in the voltage ratio when applied to the winding with the larger number of turns. This smaller change will generally provide a resolution finer than the precision required of the voltage ratio. The adjustment of voltage ratios in transformers with two or more secondary windings is constrained. This adjustment made to the finest possible resolution on one secondary winding establishes the turns on the primary winding. Adjustment of the voltage ratios on the remaining secondary windings can be made only by adjusting turns on those secondary windings. If those windings have few turns, the resulting coarse resolution may approach the allowable tolerance. Part of the tolerance is then absorbed by a design deficiency, an unwholesome production condition. Single-winding secondary transformers are preferable for precision low voltages.

Resistance

Variations in the resistance of windings are caused by variations in the resistance per unit length of the wire and variations in the tension on the wire during winding. The tension affects the total length of wire in the winding, and in the case of finer gauges tends to draw the wire. Winding resistances can be calculated during design to an accuracy of only about 10 or 20 percent. Manufacturing variations are between 10 and 20 percent. The measured values of resistance can therefore be as much as 50 percent different from the predicted value. Tight tolerances on resistance should not be enforced until the nominal values have been confirmed from measurements made on prototypes. Adjustments in the no-load voltage ratio are sometimes necessary to obtain the correct full-load voltage ratio when the measured resistances depart widely from the predicted values. After the nominal values have been confirmed, a tolerance of ± 20 percent can be held without economic penalty. There is some economic penalty for a ± 10 percent tolerance, especially on the finer wire sizes. There are no manufacturing adjustments available as corrective action to out-of-tolerance winding resistance. The coils must be scrapped, with a high probability that the same thing will happen again. Tight tolerances on winding resistance should be avoided. In many cases specifying maximum resistances only should be adequate.

Leakage Inductance

The number of turns in the winding and the coil geometry determine leakage inductance. Variations in leakage inductance are due to variations in the physical dimensions of the coils. Since these physical dimensions cannot be precisely held, neither can the leakage inductance. The nominal value of leakage inductance cannot be calculated with precision. Measurement and functional evaluation are often required to confirm the nominal value. Leakage inductance is frequently an implicit requirement of other tests such as regulation, pulse rise time, and short-circuit current. When it is specified explicitly, a maximum-only tolerance will satisfy most circuit requirements. When closer control is required, a variation of about 25 percent can be expected around a confirmed nominal value.

Full-Load Voltage Ratio

Variations in the full-load voltage ratio are the combined effect of variations in the no-load voltage ratio, the winding resistances, and the leakage inductance. Usually the full-load voltage can be held to within ± 5 percent without economic penalty. A sizable penalty will result for a tolerance tighter than ± 2 percent.

Open-Circuit Inductance

Inductance is an example of an attribute whose nominal value can be only approximately confirmed. The measurement of ferromagnetic inductance is subject to substantial variation in obtained values for different measurement methods and for varying magnetizing conditions in any single method. Maintenance of tight tolerances on inductance in production is troublesome and expensive. Manufacturing variations in inductance result from inability to set air gaps with precision, the variation of those air gaps during processing, and variations in the permeability of the ferromagnetic material. Many applications require a minimum inductance only. The higher the inductance is, the better many circuits work. Obviously it is a mistake to put an upper limit on inductance in such cases. When a minimum-only tolerance is placed on inductance, a safety margin is provided to accommodate manufacturing variations. Prototypes are likely to have inductances above the minimum. Care should be taken in circuit evaluation to avoid interpreting those inductances as the minimum allowable. When both upper and lower limits are required on inductance, full details on conditions and measurement methods must be included in the requirement. There is some economic penalty in specifying a tolerance of ± 10 percent on inductance as opposed to a minimum value. There is a substantial penalty for a ± 5 percent tolerance. The cost is very high for a tolerance tighter than ± 5 percent.

Exciting Current

Exciting current is closely related to open-circuit inductance but is more commonly specified, almost invariably as a maximum value. There is no economic penalty for a value chosen to account for normal manufacturing variations.

Capacitance

There are two capacitance requirements commonly placed on transformers. The first is the geometric or measurable capacitance of one winding to other windings and/or ground. Associated with the adverse effects of capacitive loading to ground, this requirement is usually specified as a maximum value. The second requirement is the effective capacitance across the winding. The effective capacitance differs from the geometric capacitance because the voltage distribution across the electrodes varies along the electrodes and differs from the voltage across the winding. Measured by indirect and less accurate means than the geometric capacitance to ground, this attribute is closely associated with the high-frequency performance of the transformer. When it is a factor in performance, both upper and lower limits must be set either explicitly or, more commonly, by specifying minimum high-frequency performance. Variations in both capacitance values are due to variations in the physical dimensions of the coils and the dielectric constants of the insulation between electrodes. The problem of control of variations is similar to that of leakage inductance. The design prediction of nominal values of capacitance is rather crude. Confirming these predicted values is essential before implementing tolerances. Once the nominal value is established, a tolerance of about ± 25 percent can be held without economic penalty.

Bandwidth

Bandwidth is a functional design requirement specified with minimum limits at upper and lower frequencies. Variations in bandwidth result from variations in the parameters discussed above. Achievable bandwidths depend on impedance and power levels and other operating conditions as well as transformer parameters.

Pulse Characteristics

Pulse characteristics depend on the equivalent circuit parameters the same as does bandwidth. Rise time is frequently an explicit requirement. A rise time tolerance of ± 20 percent under carefully specified conditions and after construction of a successful prototype can be held with little economic penalty. Pulse droop, inevitably a maximum allowable requirement, is the same as a minimum requirement on open-circuit inductance. Backswing and recovery are functional ways of specifying open-circuit inductance and effective distributed capacitance. The backswing and recovery times are usually maximum-allowable values.

Mechanical Attributes

The mechanical attributes to which tolerances are applied are mainly dimensions. Attributes such as weight, obviously a maximum, are design requirements which, once met, do not require further economic consideration. In contrast, mechanical dimensions are perpetually economic factors in manufacture. Dimensions may appear on a drawing as maximum, with bilateral tolerances, or for reference only. Many of the vital dimensions on a specification control drawing need be only maximum. A few critical dimensions require bilateral tolerances. Reference dimensions are nominal ones to which no direct tolerance control is applied. A dimension may be a reference dimension because it is already controlled indirectly by the tolerances on other dimensions, or because it provides general information which does not require control by tolerances, at least not from that drawing. Judicious use of reference dimensions is an effective way of conveying information without invoking tolerance economics.

The envelope dimensions are the overall dimensions of the device. Since the purpose of these dimensions is to ensure that the device will fit in the place intended, they may be maximum dimensions only. The envelope dimensions of open-type transformers usually include the dimension across the coil(s). Not closely controlled, this dimension should be specified as a generous maximum. The coil dimensions obtained on prototypes should not be regarded as maximum values. Envelope dimensions may be determined by impregnation and coating processes. These processes manifest poor dimensional control. Adequate allowance should be made in envelope dimensions determined by such processes. Mechanical perturbances frequently affect the silhouette of the device and consequently help determine whether the device will fit in place. Only features of concern in the installation should be controlled by tolerances. It is neither necessary nor advantageous to dimension the device in full mechanical detail on the specification control drawing. Mechanical perturbances should be specified as maximums whenever possible.

Mounting dimensions require alignment with mating dimensions in equipment. Bilateral tolerances are required here. Figure 15.1 shows a widely accepted mounting dimension system with tolerances. Enclosures larger than those shown in the figure frequently make use of slots which permit more generous tolerances. The mounting dimensions of open-type transformers often encompass an assembly composed of brackets and the lamination stack. The lamination stack dimension cannot be held to a close tolerance. Slots oriented to favor the stack dimension are customarily used for mounting transformers with this construction.

CHAPTER 16

Procurement Practice

The procurement of magnetic devices which are catalog items is a straightforward process. The list price and discount schedule are readily obtained and generally not subject to negotiation. Delivery times are short. Performance specifications and general quality levels are known. Such procurements can usually proceed in a routine fashion. The special items are the preoccupation of the procurement process.

16.1 QUOTATIONS

A satisfactory business atmosphere is prerequisite for quoting. Vendors' responses to requests for quotations on special items will be colored by past experience with the purchaser. Vendors are extensions of a purchaser's own facilities. It is important to the purchaser that the vendor function as efficiently as possible to the mutual benefit of both parties.

The work should be scheduled to allow reasonable time for the procurement process. Specifications should be economically obtainable. Communication channels for specification questions must be maintained. Purchaser's bureaucratic processes should not work hardships on vendors. The quality control machinery for incoming material should be capable of arbitrating between rejections for serious defects and trivial objections. Accessory agreements regarding testing and special services should be clearly defined. Vendor proposals for changes should be given consideration, and additional costs incurred as a result of changes should be fairly arbitrated.

16.1

The accuracy of responses to invitations to quote can be no more exact than the specification which accompanies the invitation. Subsequent changes in specifications can result in price increases. It is advantageous to prepare adequate specifications immediately. Various bidders will have different approaches which will be of varying usefulness to the purchaser. Evaluation of bids should not be on the basis of lowest bid alone but should consider the total probable cost.

Often bid invitations are for several quantities. The response of vendors to this will vary depending on the way they provide for start-up costs. Some vendors may assign these costs to overhead. Others may assign all the cost to the first piece. Some may list the start-up costs as a separate item. Stipulation by the purchaser of the pricing method to be used will ensure uniformity and preclude prices which may duplicate start-up costs.

16.2   PROTOTYPE PROCUREMENT

Prototype procurement is the occasion when provision for economical production procurement can be most effectively made. Specifications are provisional. Technical options are still open. Vendor and purchaser are both in tentative positions allowing negotiations and problem-solving to proceed freely.

Selecting Sources

Prototypes should be procured from sources which are capable of meeting production requirements. The design of a magnetic component is directed at the shop which will produce it. A shop devoted exclusively to prototypes can take liberties with manufacturing practice which may be uneconomical in production. Patronizing this type of facility may result in commitments to a prototype design that is not easily obtainable in production. The criteria for selecting sources for prototypes should be the same as those for selecting sources of production.

An early but not necessarily the most important consideration in selecting a source will be the previous procurement history between purchaser and vendor. This history should be weighed with the answers to the following questions:

Are previous successful procurements an adequate way of judging the current procurement?

Were previous unsuccessful procurements within the vendor's power to control?

Has the vendor's management emphasis shifted since the last procurement?

The vendor must be technically competent and have adequate facilities for production. Those facilities should include a versatile winding section, processing facilities for all the operations that may be needed for the current procurement, and a mechanical facility for the required case and tooling. The vendor's experience with similar items made for other customers should be evaluated.

Minimizing Cost

There is a dual consideration in the control of the cost of prototypes. The first is the out-of-pocket cost of obtaining and evaluating the prototypes themselves. The second is the preparation for economical production. When the prospect for substantial production exists, excessive concern about the immediate prototype cost should not be allowed to create conditions that will result in much greater costs in production. The real cost of prototypes is difficult to determine. Vendors regularly provide prototype service at below cost. The prototype orders which this practice accrues place the vendor in a better position for the follow-on production work. Of course, the real cost of the prototype is then buried in the production cost. This practice has the advantage to the purchaser of passing along to the vendor some of the investment risk of a new development, but it makes the cost accounting more difficult. In order to stabilize this situation, purchasers ask for prices on a range of production quantities as well as prototype quantities before placing orders. The apparent economical choice then is the source that quotes the minimum total cost for the most probable production quantity. Requests for such quotations should state where the fixed costs are to be inserted. Specifications can change during the prototype phase. By the time production orders are ready to be placed, the period for which the quotations were firm may have expired. The vendor may then raise the price if experience in the prototype phase indicates higher costs than anticipated.

Another cause of inaccuracy in prototype costs is the cost incurred by the purchaser. The obvious costs are specification preparation, liaison, and prototype evaluation. When a prototype does not work and must be returned to the vendor, the vendor may assume the additional cost of constructing new prototypes, but the labor cost and time delay experienced by the purchaser must be absorbed by the purchaser. The probability of this happening is lessened with a technically competent supplier.

The selection of the prototype vendor is of vital importance to cost. In spite of the best efforts to provide for multiple sourcing, the purchaser makes a commitment with this selection. There is the investment in prototype evaluation, a process that must be repeated for other sources. Technical personnel will be reluctant to undertake another evaluation because of the risks involved in substituting an unknown for a fully evaluated product and because of the time delay involved in the evaluation. Large-volume production requirements should be protected by developing two or more satisfactory sources during the prototype phase.

16.3 SPECIFICATION PROCEDURES

Procurement of a special component must begin with a specification. (See Secs. 18.1 and 18.2 on how to specify.) The circuit engineer, in preparing the specification, is faced with a difficult problem. A decision to use the component which forces circuit compromises may be tentative. The engineer is uncertain about what is needed and the minimum allowable performance requirements and is uncertain also about how to communicate needs by means of the specification to the supplier. The failure to promptly resolve these dilemmas is often cause for delay, misunderstanding, and unnecessary expense. It should be understood by all parties that the specification will initially be tentative. This understanding must be matched by the specifier's appreciation of invoking a process involving many other people who do not share the same knowledge or orientation. The specifier is obligated to distill needs into a residue with which the procurement process can cope. At the earliest opportunity the circuit engineer should communicate directly with the component designer, and these two must reach a common technical ground. Here the vendor selection begins to be of vital importance. The vendor's designer must aid in the distillation process, arriving finally at a set of requirements sufficiently coherent to justify prototype design and construction. The conclusion of this dialogue should be a written specification subject to further revision in response to comments from other prospective vendors.

During the development of electrical specifications, the mechanical design of the equipment will be proceeding. From this work will come insights into the mechanical requirements for the component, particularly the size. When size is a problem, close liaison with the vendor's designer should be maintained. Specification requirements should be reviewed and trade-offs evaluated. Other vendors should be consulted. The reconciled mechanical requirements will become a part of the component specification.

During this negotiation the possibility of multiple sourcing should always be kept in mind. The specification should avoid stipulations that tend to limit procurement to a single vendor. Significant verbal under-

standings with the favored vendor should be included in the specification. When the specification has reached a form acceptable to both purchaser and favored vendor, it should be sent to the favored vendor and
the other selected vendors for quoting and proposals. (See Sec. 16.2 on
selecting sources.)

In evaluating the proposals of various vendors, it is essential that all
the vendors work to the same set of ground rules. Suppose, for example,
that a preliminary transformer specification is sent to several vendors.
One may suspect from the context of the requirements that the duty cycle
which was not specified is low. The vendor calls with this question, and
the suspicion is confirmed. The vendor bases the quotation on a low duty
cycle, and the price and size will be less. The preliminary specification
should be adequate. Where it is found to be inadequate, it should be
amended and all vendors should be made aware of the changes. Meaningful comparisons among vendors' responses can be made only on the
basis of uniform specifications and conditions.

Electrical evaluation of prototypes is needed quickly. Proper circuit
performance needs confirmation so that work on the entire equipment
may proceed with confidence. There are uncertainties about whether the
prototype specification is correct and whether the vendor can meet the
specification. To expedite the evaluation process and reduce the liability
in the event of failure, prototypes are sometimes evaluated in the core
and coil stage. The core and coil is the functional part of the transformer
before mounting in a metal case or other protective facility. The core and
coil can generally be provided sooner and at less expense than the complete unit. If the core and coil is to operate at a high voltage, it may be
possible to evaluate it by placing it in a temporary container of oil. Some
sensitive components may present difficulties because the characteristics
of the core and coil shift when placed in the final configuration. Core and
coil evaluation can be included as part of a purchase contract.

16.4 ECONOMICS OF QUANTITY PROCUREMENT

The unit selling price of a manufactured item is generally based on an
estimate which follows a pattern similar to the following example:

Estimated unit labor cost, h	5.0
Average labor rate, $/h	$ 5.00
Unit labor cost, 5.0 × $5.00	$ 25.00
Overhead factor	2.2
Burdened labor cost, 2.2 × $25.00	$ 55.00

Estimated unit material cost	$ 60.00
Total factory cost	$115.00
General and administrative cost factor	0.85
Total cost, $115.00 divided by 0.85	$135.29
Profit factor	0.9
Selling price, $135.29 divided by 0.9	$150.33

The average labor rate, the overhead factor, and the general and administrative (G & A) cost factors will have been established, but on the basis of policy rather than precise knowledge. The true labor rate varies with time and work assignments. The established factory overhead and G & A factors are based upon recent past experience. The true factors are determined by the current volume of business being conducted as well as other variables. The labor and material costs are indeterminate factors. The material cost is bound by the design, the quantity of the item being purchased, and the purchasing practices of the manufacturer. Part of the effort made during the prototype phase is aimed at controlling the unit material cost. The objectives are to allow the use of materials which have the lowest intrinsic cost and to allow the selection of materials which permit maximum opportunity for combining requirements with the material needs of other items being manufactured. This pooling of requirements, usually handled by large-quantity purchases for stock, is a significant cost saving for the purchaser. Materials in this category will vary little in price over the quantity range under consideration for the single item because of the larger quantities in which they are being obtained. The fewer special materials required, the better the control will be on the total material cost and the less variation there will be in that total material cost with quantity. The labor cost is more indeterminate than the material cost. Labor cost may be estimated from the item's own cost-accounting history if previously manufactured, from the cost-accounting history of a similar item, or by estimating the labor by applying labor standards to the labor operations performed in manufacturing the item. There are limitations to any of these methods. Cost-accounting figures are only approximate. Conditions under which the cost history was obtained may be different from those for the current production. Typical variables in the conditions are the quantity, lot size, production rate, labor skill, methodizing, disruptions due to material shortages, and rejection rates. In using the cost history of a similar item, the above variables apply as well as the inaccuracy in applying a quantitative factor to the degree of similarity. Estimating by the use of labor standards is subject to the uncertainties of the accuracy, applicability, and achievability of the standards. For an important pricing effort, all three of these pricing methods may be used in arriving at a selling price.

When the cumulative average unit cost for one quantity is known or assumed, a learning curve is sometimes used to establish the labor cost for other quantities. It has been found that the unit labor cost of a manufactured item tends to decrease by a fixed ratio every time the quantity doubles. The generally accepted ratio for magnetic components is 0.9. Figure 16.1 shows a learning curve using the 0.9 learning rate with a standard of 1.0 for 1000 units. The learning factor for any quantity less than 1000 should be multiplied by the average unit labor cost for the 1000 quantity to obtain the average unit labor cost for the smaller quantity. The curve may also be used when the average unit labor cost for a quantity less than 1000 is used as the standard. This is done by dividing the average unit standard labor cost by the factor for the quantity for which the standard applies and multiplying the result by the learning factor for the quantity for which the average unit labor cost is needed. The learning curve is an approximation. Its claim to validity is based on averaging over many items for many quantities. It does not provide for changes in manufacturing methods, or other abrupt changes in the cost structure of specific items. It does possess one characteristic which is a matter of common experience. The rate of decrease in unit cost decreases as the quantity increases. Since the value of a manufactured item, both material and labor, is ultimately the result of labor effort, the learning curve can sometimes be used as a guide in estimating unit selling prices for other quantities when the unit selling price for one quantity is known.

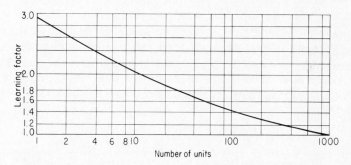

FIG. 16.1 Cumulative average learning curve for 0.9 learning rate.

In quantity procurement the all-important factor in minimizing cost is competitive pricing. Suppliers actively seek quantity orders and will make strenuous efforts to reduce costs in order to win the bidding competition. To exploit competitive bidding, the purchaser must be in a position to accept a bid from among the lowest bidders without jeopardizing interchangeability, quality, or delivery schedules. The first line of defense against these difficulties is the judicious selection of sources. (See Sec. 16.2 on selecting sources.) The second line of defense is a good procurement specification. By the time competitive bidding starts, the procurement specification should be near its final form, having been reviewed by at least two suppliers. Review by a second supplier helps prevent proprietary features of the first supplier from being frozen into the specification and provides a valuable second opinion on the reasonableness of the requirements. Serious differences of opinion in this matter among suppliers should be reconciled before bidding is concluded. Their constructive suggestions should be incorporated into the specifications. Suggestions for relaxation, particularly in tolerances, should be given the most serious consideration. The specification should be amended to include allowable relief. This specification reconciliation should continue into the production phase. It is counterproductive to deal with a specification which contains requirements found to be unnecessary, expensive, and never met and yet which remain because changes have been neglected. Less likely but more disastrous is the failure to formally add to the specification a requirement found to be necessary during production and provided by the supplier through informal agreements.

When the low bidder in a production procurement has not qualified the product, the purchaser is faced with a dilemma. The saving resulting from using the lowest bid is offset by the risks of being committed to an unproven source. There is concern about interchangeability. The purchaser will have the expense of evaluating a new prototype. There may be delays while prototypes are constructed and evaluated and performance problems resolved. Price reductions are often negotiated between the purchaser and the original qualified supplier as a way out of this dilemma, but developing the low bidder as a good second source places the purchaser in a stronger position for subsequent procurement. The risk of going to a second source can be reduced by splitting the order between the qualified source and the low bidder. Once the second source is qualified, the purchaser is free of the problem on subsequent procurements. procurements.

16.5 EVALUATING BIDS

Awarding a purchase contract to the lowest bidder is not necessarily the wisest action, and there is no obligation to do so except in special cases where required by contract or law. Even when it is a requirement, the requirement can be waived for good cause. However, the invitation to bid should have been preceded by an adequate screening of bidders. The critical auditor can question why a bidder was invited to bid if after the fact the bidder's ability to perform was challenged, but this is no reason to pursue a wrong course. Particularly when the lowest bidder cannot provide satisfactory production deliveries, the procurement is usually split.

Bids will tend to follow a distribution curve. To approximate such a curve, a large number of bids would be required. Most of the bids would be centered on an average value. Before using low bids outside the center cluster, an attempt should be made to determine the probable reasons for the bids being low. It may be that the low bidder has built the same or a very similar part before, making the start-up costs and risks less. The bidder may be willing to take greater risks in order to fill open production capacity. The low bidder may have a very efficient plant for the item in question. The bidder may have made errors in interpreting the specification. The specification should be reviewed with the supplier before making an award. The low bidder may have made arithmetic errors in preparing the bid. It is not in the long-range interest of the purchaser to capitalize on such errors. The low bidder may have an ingenious and less costly design approach. At the time of the specification review the supplier's design approach should be appraised. The bidder may plan to use less costly manufacturing processes, and this should raise concern about the quality of the product.

A great deal has been said about the relationship of price to quality. A commonly held view is that you get what you pay for, the implication being that if you insist on buying at the lowest price, you must be prepared to accept poor quality. While this notion has an element of truth in it, there are factors other than quality that contribute to cost. In a magnetic component quality means reliability, product uniformity, and excellence in manufacturing processes such as plating painting, impregnating, potting, soldering, coil winding, and coil finishing. A distinction should be made between design quality and manufacturing quality. The prototype phase of procurement should be directed toward assuring design quality. Using available materials and manufacturing processes, the designer tries to meet a set of requirements at minimum cost. If the designer does a poor job, the part will display what is often labeled poor manufacturing quality or poor workmanship. For example, the dimensions between centers of mounting holes of a component must be held to a reasonable tolerance in order to line up with the corresponding holes in a chassis. A commonly used construction in magnetic devices is inherently poor for this purpose. Special care in prototype assembly will ensure meeting needed tolerances, but in production this care will not be exercised. Many units will be out of tolerance. Obviously a workmanship problem, the purchaser may say. In truth this is a design defect. The way to cope with this problem is to use slots instead of round holes oriented in the direction in which it is difficult to hold a tight tolerance. If the problem exists in two directions, slots may also be used in the chassis oriented at right angles to the slots in the component. It is incumbent upon the purchaser to be willing to change the chassis if necessary and to maintain a receptive atmosphere so that proposals of this nature are encouraged and given full consideration when made. Once the drawings are changed and the tooling is in place, the cost of making a part with the slot is the same as that with round holes, but the functional quality will be superior.

The reasons for poor quality are complex. The design and specification requirements must be consistent with available manufacturing processes and materials. Sanguine efforts to force materials and processes beyond their capabilities are a design failure. Optimum design choice of materials and processes will result in lowest cost consistent with acceptable quality. Good design adds quality with no increase in cost. It can sometimes reduce cost and improve quality simultaneously. Extra cost to improve design quality is a wise investment for future production.

The materials used in the manufacture of magnetic components are available to all suppliers. Variations among suppliers in quality associated with materials arise most often from differences in the selection and application of these materials. Materials mistakenly chosen because of insignificant cost savings can have a disastrous effect on product quality. Costly materials are sometimes chosen because of a conviction that the quality will be improved when inexpensive materials would actually provide equal or better quality. The design can affect product quality independent of unit cost of the product.

Variations in manufacturing processes can affect both the price and quality of the product. It is here where the best correlation between cost and quality is found. It is usually impractical for a manufacturer to maintain multiple levels of quality in manufacturing processes at a single location. The quality obtained from manufacturing processes depends upon people using equipment. The available equipment is the result of investment decisions made by the supplier. Equipment is a relatively constant factor in the quality of the process. Variations in the quality are likely caused by the people using the equipment. In a production atmosphere these people receive general instructions which they follow regardless of the selling price of the product being manufactured. To vary these instructions for individual products in an attempt to correlate quality and cost would be hazardous. Thus the level of quality of processes tends to remain constant unless the general instructions are changed by managerial choice. When this is done, a different level of quality will result which, in turn, remains constant until changed by another managerial choice. There may be occasions when the process falls out of control inadvertently. A major function of the quality control system should be the detection and correction of these failures. It is advantageous for purchasers to be familiar with the manufacturing processes used in magnetic component manufacture and the general level of quality of those processes maintained by all bidders. This level of process quality is likely to be maintained independent of the price the supplier may charge for an individual product. With this knowledge the purchaser can make a judgment of whether the quality of the processes that will be used by the low bidder is likely to be satisfactory for the item being procured. The risk in making this judgment is reduced by the fact that the quality-control organizations of both supplier and purchaser are most effective in monitoring manufacturing processes, since variations in the data supplied by the inspection and test procedures used in quality control are mainly due to variations in manufacturing processes. Neither judgments on manufacturing processes nor quality control techniques are helpful in determining the adequacy of the design and material selection processes. This is dependent upon the skill and effort of the design staffs of both purchaser and supplier.

16.6 DELIVERY SCHEDULES AND COST

Delivery schedules can have a significant effect on cost. Most production deliveries are scheduled on an accelerating rate until the desired final rate is attained. This allows evaluations of the product by the customer before large quantities are completed and allows the supplier to troubleshoot the production operation before commitment to full capacity. Most magnetic components are produced in lots rather than continuous production lines. The use of discrete lots is forced by the assembly and processing methods commonly used. There is usually an optimum lot size for the individual product and shop. The schedule is most often one lot a month. The 1-month interval is compatible with the production flow of most shops. If the lot quantity is too small and the lots are scheduled frequently, then the setup costs associated with each lot become excessive. If the size of each lot is beyond the capacity of the shop to produce within the required interval, either the delivery will be late or the lot will be split. Splitting lots increases cost and multiplies the production control problems. Purchasers frequently wish to schedule deliveries to minimize the amount of cash tied up in inventory. This objective must be weighed against the need to schedule deliveries far enough in advance to provide for production capacity of the supplier and possible schedule slippage. If the main assembly line is closed down by a component shortage, the cost may be greater than the cost of maintaining generous inventories.

Production delivery requirements spread over an extended period present a cost problem during times of inflation. The supplier's labor costs are subject to revision perhaps twice a year. The costs of some of the materials used by magnetics suppliers are very volatile, changing at monthly intervals. The supplier may understandably resist efforts to establish firm prices for a period greater than 6 months. Multiple-source opportunities should be maintained by the purchaser so as not to be at the mercy of the supplier who uses inflation as an excuse for unjustified or excessive price increases in the middle of a procurement cycle.

16.7 CAPTIVE SHOPS

The history of equipment manufacturers that utilize magnetic components extensively often contains episodes in which the internal manufacture of magnetics was considered and occasionally undertaken. These episodes have their roots in procurement and technical problems sufficiently serious to affect equipment production and performance. The captive shop can usually solve these problems. Priorities in the captive shop are under the user's control. Deliveries can be scheduled as needed, if necessary without regard to production economics. A captive shop will develop valuable capabilities in design and product performance that will enhance the equipment in which it is installed. Specification conflicts and performance difficulties can be handled more readily than with an outside source. Advantageous changes can be made more readily when the component is constructed in-house. A captive shop is a great convenience to an equipment manufacturer.

Once established, a captive shop becomes a cost center. Cost comparisons for make or buy decisions are figured at the captive shop's factory cost (less G & A expense and profit) vs. the vendor's selling price. The G & A expense and profit are applied to either purchase price or factory cost of the item in determining its effect on the selling price of the equipment. (See Sec. 16.4.) In spite of this advantage the captive shop usually has trouble with excessive costs. Its raw labor rates will be determined by the rates for similar work in the rest of the equipment manufacturing plant. These rates are likely to be higher than those in a component supplier's plant. The overhead rates of the equipment manufacturer will usually be higher than the overhead rates of the component supplier. The purchasing practices of the equipment manufacturer will be different from those of the component supplier. The same materials will cost the component supplier less than the equipment manufacturer. The superior workmanship and quality of the captive shop, fundamental reasons for its existence, will cause its unit worker-hours to be higher than the component supplier's. These handicaps make the factory cost of the captive shop frequently higher than the selling price of the outside supplier, who will make a profit on some items which will offset the losses on others. The captive shop is generally not a profit center. It cannot average its gains and losses. Any saving it may achieve on an item benefits the project on which it is used. Cost overruns on other items must be borne by the project on which they are used. The advantages of the captive shop to the equipment manufacturer do not appear in the cost-accounting data. In order to survive fiscal scrutiny, the captive shop must provide such outstanding service that it is clearly worth the additional cost over outside procurement.

A captive shop will develop the same work-load problems as other production facilities. There will be times when the shop is underloaded. Work may be placed in the shop at a cost greater than with an outside vendor just to load the shop. The issue will be raised of whether the shop should attempt to sell its products to other companies. The labor and overhead rate may make competing difficult for the captive shop. The sales organization may not be organized to handle the marketing of components. The next step is to make the captive shop a separate subsidiary. When this is done, the advantages of the captive shop are usually lost.

In an attempt to obtain the best of both worlds, an equipment manufacturer may set up prototype facilities where breadboard and prototype units only are made. After the specifications and design are established, the units are purchased in quantity from outside sources. This practice generates many problems. The prototype shop, lacking a production capacity, does not consider the production problems in the initial design stage. Size and performance limits are established which outside sources, faced with production constraints, find uneconomical or impossible to meet. Magnetics designers must consider the capabilities of the shops for which they are designing. They will consider economical choices of materials and standards. Some of these standards will be industry standards, and some may be house standards. The equipment manufacturer's prototype shop will not have complete knowledge of the options available. The very useful exchange between magnetics and circuit designers during the design phase is eliminated. The magnetics designer during this exchange can make valuable contributions to the performance and economics of the final product. It often happens that units made in prototype shops do not conform to all the final mechanical and environmental requirements of the equipment. Size commitments made on the basis of such prototypes frequently must be changed, or the outside source is forced to resort to more expensive constructions. Meeting the complete set of requirements can force changes in performance specifications previously considered established. If prototype operations are considered necessary by the equipment manufacturer, they should be conducted in close cooperation with outside suppliers who can help avoid some of these pitfalls. The equipment manufacturer can obtain many of the advantages of the captive shop through the effective use of outside sources. Good planning, proper specification procedures, clear lines of communication, and astute procurement are the ingredients for success.

16.8   PROCUREMENT SCHEDULING

Prototypes

The need for a magnetic component typically arises in conjunction with prototype equipment for which there is a firm scheduled completion date. In order for this date to be met, the magnetic component procurement must be well planned. The plan should call for a series of events to take place in sequential order at specified times. The notation of Line of Balance Technology is useful for this purpose. Figure 16.2*a* and 16.2*b* shows a sample plan and plan key. The plan is drawn on a horizontal scale of working days prior to completion of the prototype equipment. The plan key identifies with consecutive numbers significant events in the magnetic component procurement cycle. Such a plan tailored for the individual organization and circumstances will give insight to the procurement process and facilitate surveillance.

Production

The need for efficient magnetic procurement in production makes careful planning essential. As in prototype procurement, a sequence of events must take place in a timely fashion to ensure on-schedule delivery of usable parts to the equipment production line. A good specification must be available to start procurement. Time must be allotted for sending the specification to prospective bidders and for their replies. Additional time will be required for the evaluation of their replies and possible specification revision. If the favored bidder does not have prototype approval, time must be allotted for this purpose. There is a risk that the prototype will not be acceptable. Multiple sourcing, one source which has submitted an approved prototype, is common in these circumstances. Then the production schedule must be planned with consideration of the production capacity of each source and the comparative costs. Care must be taken to obtain interchangeability of the products from every source. Conformance to the same specification by prototypes from different sources does not ensure interchangeability. There may be necessary attributes not covered by the specification which one prototype exhibits while a second prototype does not. Careful compatibility evaluation in the equipment is necessary. Test and inspection procedures must be established and

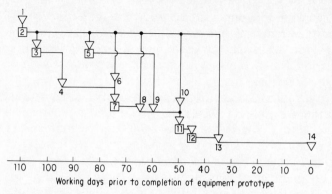

110 100 90 80 70 60 50 40 30 20 10 0

Working days prior to completion of equipment prototype

FIG. 16.2a Sample plan for prototype procurement. Numbers refer to activities and events listed in the plan key, Fig. 16.2b.

1. Determine need for magnetic component.
2. Establish and maintain technical liaison with magnetics designer.
3. Prepare specification.
4. Place order for prototype.
5. Order critical purchased items for prototype. (Vendor action.)
6. Receipt of core and coil for evaluation.
7. Evaluation of core and coil.
8. Approve core and coil.
9. Vendor receives critical purchased items for prototype.
10. Receive prototype.
11. Test prototype.
12. Install prototype magnetic component in prototype equipment.
13. Accept prototype, complete technical liaison.
14. Completion of prototype equipment.

FIG. 16-2b Numerical activity list for procurement plan of Fig. 16.2a.

accepted by contracting parties. Test data correlation must be obtained between source and user. Tooling may be required by the source before starting production. The time required to obtain this tooling must be considered when preparing the schedule. If procurement of the tooling requires a release by the user, that release should be scheduled and the release date met. Additional risks which this may involve must be weighed against providing the source with cause for delay. Magnetic components contain several purchased parts which are critical. They are frequently single-source items with lengthy procurement cycles. Typical of such items are terminals, cores, and bellows. Release for the purchase of some of these items can frequently precede release for full production. Understandings between source and user concerning the release of these items can reduce lead times.

The schedule will consist of a time when deliveries must start and a delivery rate thereafter. The interval between the moment when the decision is made to proceed with procurement and the moment required for the first production delivery is the period available to solve all the pre-production problems. This period must take into account the lead time needed between delivery of the first quantity to the equipment line and delivery of the first equipment. It must also provide for the time required for shipping, receiving, inspection, and handling. Under ideal production conditions deliveries will proceed routinely if preparations have been properly made. Ideal conditions may not prevail for a number of reasons. Vendors, in their enthusiasm to obtain business, may overestimate their capacity to produce. The item may fall victim to queue priorities. Magnetic component shops are organized by a lot system. Parts do not flow on a continuous assembly line; they are carried from one work station to another in lots, where they will wait their turn to be processed. Most of the lapsed time is spent in queues at the various work stations. Priorities are assigned at every work station. Shifting frequently, these priorities are assigned for a variety of reasons, some unrelated to delivery requirements. The priorities may be assigned in a manner to increase the workload in an underloaded work station. They may be assigned to improve billings near the end of a fiscal period. When a problem develops on an item at a busy work station, the item may be shunted aside from the queue until the workload lightens, a good reason for solving problems before production begins.

16.9 COORDINATION OF SPECIFICATION, DESIGN, AND USE

The design, procurement, and use of a magnetic component is a complex process. It involves several organizations and people, each with particular skills and viewpoints. Conflicting interests can frustrate the process. Planning and coordination are needed to blend these resources into a single mechanism to accomplish the objective of obtaining a usable device on schedule at lowest possible cost. Circuit engineers should specify what they need but avoid uneconomically conservative requirements. They should determine that what they specify is readily achievable in the component design. If it is not, they should be prepared to consider circuit changes that make the component requirements more reasonable. Component designers should be conversant with the circuit and understand the requirements in terms of the circuit. It is their job to convert the requirements from circuit to component language. They need an intimate knowledge of what is achievable by the component and must be prepared to discuss these possibilities with the circuit designer to arrive at economical mechanical requirements. Effective communication is extremely important. It is profligate for engineers to struggle to meet a specified characteristic which in reality is not needed, while essential requirements go unfulfilled. The final objective is the equipment which uses the component. Component specifications and design are only means to this end.

CHAPTER 17

Reliability and Safety

Magnetic devices are on middle ground with reputed reliability. Specification MIL-HDBK-217A assigned a failure rate of 2.46 failures per 10^6 h for a transformer with 105°C insulation operating at 105°C. This compares with a failure rate of 0.16 failure per 10^6 h for composition resistors operating at rated wattage at 70°C, and a failure rate of 4.0 failures per 10^6 h for paper capacitors operated at rated voltage and 85°C. These figures are statistical. Individual performance will vary depending on many factors. It is not surprising that resistors, being simple devices, have a lower failure rate than magnetic devices. There is no obvious reason why magnetic devices should be intrinsically more reliable than capacitors. Tradition and marketing expectations are partly responsible. Design stress levels, by tradition, are more conservative for magnetic devices than for capacitors. In the manufacture of capacitors, greater reliance is placed upon the control of the quality of materials and processes than in magnetic devices. User demands for small capacitors and tolerance of lower reliability in capacitors encourage this practice.

Magnetic devices conform to the "bathtub" failure rate curve illustrated in Fig. 17.1. This curve is characterized by a relatively high initial failure rate, sometimes referred to as *infant mortality*. Infant mortality is due to design and manufacturing defects which tend to reveal themselves quickly during operation. The failure rate decreases rapidly with time, approaching a low and constant value during most of the life of the component. This is the stress-related failure region in which components fail due to stresses in normal operation. As the components begin to wear out, the failure rate rises, as shown in the rising portion of the failure rate

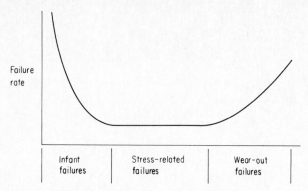

FIG. 17.1 Failure rate vs. time of a magnetic device.

curve. In magnetic components this is mostly attributed to insulation ageing. Some common failure mechanisms are described below.

17.1 COMMON FAILURE MECHANISMS

Dielectric Breakdown

Dielectric breakdown can occur in magnetic devices between electrodes at different potentials resulting in destruction of insulation and electrical failure. The potential can be induced by the transformer or it can be externally applied. Induced voltages can cause breakdown between portions of the winding in which the voltage is induced, or they can cause breakdown from a portion of the winding to ground or other windings. Externally applied voltages other than induced voltages can cause breakdowns between the winding to which the voltage is applied and other windings or ground. Breakdown is accompanied by a high current, causing overheating and carbonization along the breakdown path, permanently reducing the resistance between electrodes. Opens can develop in the windings in the breakdown region due to melting of the conductors from the concentrated heat. Breakdown is a highly localized condition. The carbon path can usually be seen if the device is carefully dissected. Identification of the breakdown path is important for determining corrective action.

Overheating

When the temperature of insulation becomes excessive, its resistance decreases, allowing the flow of leakage currents which add further to the heating and temperature. Insulation damage develops, culminating in a dielectric breakdown. Unlike dielectric breakdown, overheating is a diffuse condition. Internal examination of overheated coils will show discoloration over a large region. Breakdown will occur in this region at a point subject to especially high dielectric stress. Careful analysis is sometimes necessary to determine whether the failure is dielectric breakdown or overheating when specimens show evidence of both problems. Overheating is closely related to insulation ageing which is accelerated by high temperatures.

Corona Deterioration

Corona is an electrical discharge localized around one or both electrodes. Ionization occurs in the insulating medium, usually a gas, near an electrode. The ionization associated with corona is insufficient to establish a low-resistance path between electrodes allowing a high current to flow as in a dielectric breakdown. Random low-level currents rise and decay rapidly. The spectrum of this discharge includes the radio frequencies. Low levels of corona are first detected on sensitive equipment, such as high-gain AM receivers. Higher levels of corona can be detected by the unaided ear as a low hissing sound. High levels of corona in air can be seen as a bluish glow. If the corona is occurring in the presence of oxygen, it will generate ozone, which has the familiar pungent odor associated with electric discharges. Magnetic devices will operate with corona present. The equipment in which the magnetic device is installed will usually operate with the device exhibiting corona. In fact, the corona may go undetected until a premature failure occurs. Ionization causes complex chemical changes, destroying the insulation. Corona is usually not a problem until the potential between electrodes reaches several thousand volts. It is sometimes a problem in sensitive circuits because it raises the noise level. This phenomenon can occur at a level below that which affects insulation life. Circuits have been developed which detect corona. As qualitative determiners they work well, but they do not provide much useful information for determining acceptable passing limits. The problem is very complex. The place where the corona occurs and the type of insulation present alter the effect of corona on insulation life. Few very high-voltage circuits are completely quiet, yet many very high-voltage circuits are quite reliable. This is an area where much more work is still to be done. The fact remains that corona is one of the most troublesome problems affecting high-voltage equipment.

Open Windings

Opens develop in windings wound with very fine wire. This problem is distinct from opens resulting from the melting of conductors associated with breakdowns. Fine magnet wire is very fragile and becomes increasingly so during handling because of the tendency for copper to work-harden. The assembly processes of winding and lead dressing severely stress the wire. Subsequent thermal changes and vibration can fracture wire previously weakened by handling during manufacture, a common cause of premature failure. Lead dressing and termination techniques that avoid mechanical stress concentrations in fine wire are essential.

Shorted Turns

Dielectric stress exists between turns as a result of induced voltages. The only insulation between adjacent turns is the insulating film on the magnet wire. Normally the voltage between adjacent turns is low, and the insulating film is adequate to withstand the voltage. Shorts develop under temporary overvoltage conditions in regions where insulation is weak. Insulation defects can occur in the magnet wire film in manufacture. The film can also be damaged in winding. A particularly serious condition is a crossover in which a turn crosses over a previous turn instead of remaining in an even helix. The area of contact between turns in a crossover is subject to excessive pressure and dielectric stress. Crossovers are a common cause of shorted turns. Shorted turns can also occur between adjacent layers of layer-wound coils by puncture of layer insulation. The dielectric stress between layers exceeds the stress between adjacent turns. Under steady-state conditions the induced voltage across a winding distributes itself uniformly across the turns of the winding. Under transient conditions the voltage will tend to develop across the end turns of the winding, resulting in a dielectric stress many times greater than normal. Devices subject to this condition need special insulating provisions at the end turns, usually in the form of space between adjacent end turns, and increased layer insulation under the end layer.

Overstressing

Magnetic devices are frequently subject to overstress conditions. The effect of overstressing is not determinable. Common examples of overstressing are overloading or shorting of secondary windings pending the operation of primary circuit protection, dielectric overstressing as an acceptance test procedure, and high-voltage transients occurring during operation. These overstress conditions do not result in immediate failure, but the damage they cause may result in subsequent failure during normal operations. This situation is difficult to deal with because the complete operating history of the device is seldom known. Failure analysis and corrective action may have to rely on statistical data.

Wear-out Failures

End-of-life failures in magnetic devices occur as a result of a combination of the ageing effects of time, temperature, and dielectric stress on insulation. Additional factors contributing to wear-out are mechanical stresses from repeated thermal cycling and vibration of conductors due to the varying currents they carry and their presence in a magnetic field.

17.2   RELIABILITY ENHANCEMENT

Properly designed and constructed magnetic devices, i.e., those relatively free of infant failures, are long-lived. Life can be further increased by reducing the thermal and dielectric stresses which age the insulation. In an effort to predict insulation life, accelerated tests of many insulating materials have been run at elevated temperatures and the results extrapolated to lower temperatures. These tests provide guidelines for maximum safe operating temperatures and dielectric stress levels. The magnetic device will have good insulation life if the temperature of the copper is always well below the maximum safe operating temperature of the insulation and the dielectric stress is well below the maximum safe operating stress. Most data on lives of insulation systems in magnetic devices come from the electric utility industry, where insulation life is a matter of great economic significance. This industry is especially concerned about operating and maintenance costs. The life of the equipment used in this industry is measured in decades. In the equipment industry, obsolescence overtakes equipment before end of life for the magnetic components is reached to such an extent that data are lacking. Differences in design techniques and operating conditions make the application of experience data on transformers in the utility industry of doubtful use to the equipment industry. The prediction of insulation life remains highly conjectural. Premature failures are a much more pressing concern than end-of-life failures. The causes for these failures can be separated into design and manufacturing defects.

Design Defects

A common cause for early failure is the lack of compatibility between design and usage of the device. This can be due to the failure of the designer and user to communicate properly. It can also be due to unforeseen stresses which the device is not capable of withstanding. Or it can be due to the failure of the designer to provide for specified or implied conditions which the designer would usually be expected to consider.

When reliability problems develop in a new design, the relationship between design and usage must be investigated. Free communication is especially important here, but frequently it is difficult to maintain because financial liabilities may be involved and schedules disrupted at this stage to an extent that the parties involved are reluctant to accept responsibility. Nevertheless, if the problem is to be resolved, the facts must be established. It should not be very surprising that reliability problems arise in new equipment. This is the moment for design and specification errors to be discovered.

Dielectric failures can frequently be traced to design defects. Dielectric stresses are distributed through coils in complex ways in response to both steady-state and transient conditions. These stresses must be identified and accommodated. Transient conditions are troublesome because they are difficult to define and detect. When a pattern of dielectric failures develops, careful analysis of failed specimens is usually advantageous. The location and nature of the failures will help determine corrective action.

Failure resulting from corona deterioration can be a design or manufacturing problem or a combination of both. Corona develops in a region of high dielectric stress in an ionizable insulating medium. That medium is most often air. Corona can be eliminated by reducing the stress or by eliminating the ionizable medium. Air can be present by design as in the interstices of the coil, or it may be present as a result of defective processing in a nominally void-free insulation system. The choice of insulating system is a significant decision. The more susceptible an insulation system is to corona, the lower the design stress levels must be. The stress levels are reduced by increasing the distance between electrodes and increasing the radii of proximate electrode surfaces.

Insulation temperature is the premier design consideration in magnetics. The insulation temperature is the sum of the maximum ambient temperature of the equipment plus the temperature rises of the internal equipment plus the temperature rise of the magnetic device conductor. Since the temperature of the conductors is not uniform throughout the device, the hottest point is the place where the insulation is subject to the greatest thermal stress. That hottest temperature should not exceed the maximum safe operating temperature of the insulation. The average temperature rise of the device can be readily measured when loaded as specified. The difference between average and hot spot is usually estimated as about 10 to 15°C. Since the temperature rise of the device depends on the rms currents in the windings, knowledge of the actual load is essential. Problems arise from improperly defined temporary overload conditions or duty cycles. Inaccuracies in the equipment ambient and internal equipment temperature rise have a direct bearing on the insulation temperature of the device. The internal temperature rise of the equipment is usually not uniform. If the device is in an especially hot or poorly ventilated region, the insulation temperature may be higher than anticipated. The thermal time constant of the device is significant in its ability to withstand temporary overloads.

Manufacturing Defects

Opens in fine-wire windings are generally attributed to handling in manufacture. Winding shops are rated by the finest wire they can handle. A shop unaccustomed to handling fine wire is more likely to make handling errors. Winding machines, tension devices, and operators must have fine-wire capability for success. The greatest number of opens occur at or near terminations. Fine-wire lead dressing is a skill requiring patience and good finger dexterity. The general rules for lead dressing are to avoid mechanical stresses in the wire and avoid work-hardening by limited handling.

Poor impregnation and potting processes during manufacture are detrimental to product quality. The impregnation improves both the electrical and mechanical properties of the insulation system if done properly. Some processes eliminate voids and form moisture barriers, vital factors in reliability. When these processes are improperly done, premature failures can result. Many different processes and materials are available. The wrong choice of process and materials can lead to unreliable performance.

Coils are made with flexible materials. These materials do not provide the coils with sufficient dimensional stability to maintain close mechanical tolerances. In spite of this a considerable amount of mechanical precision is required in coil manufacture. End margins must be held above specified minimums. The overall size of the coil must be maintained below specified maximums in order to permit subsequent assembly without damage and to allow adequate clearances between opposing electrodes. Failure to maintain required mechanical dimensions is a frequent explanation for shorts to ground or between windings.

Termination failures are common. Any of the various ways for making terminations can pose reliability problems. Faulty terminations can result in opens or shorts, and they can be avenues for the entrance of moisture. Flexible leads as an intergral part of the device, while one of the least costly terminations, are vulnerable to fatigue failure, particularly at the point where they exit from the device. A broken lead here requires replacement of the unit. Any number of rigid terminations in which the leads are external to the device avoid this problem.

Oil leaks are a problem in the many magnetic devices operated in oil-filled containers. Oil has a large coefficient of thermal expansion; provision for this expansion is required. Large devices operating in a fixed position have expansion volumes which are air-breathing. Small devices, particularly those that must operate in any position, are generally hermetically sealed with mechanical features for accommodating the oil expansion. These features mean an increase in the internal pressure at elevated temperatures. The combination of pressure and capillary attraction makes it difficult to establish and maintain a seal against insulating oil. Many malfunctions increase the internal pressure, causing leaks. Leaks do not necessarily signal electrical failure, but hermetically sealed units which develop leaks are generally regarded as failed since an excessive loss of oil will lead eventually to electrical failure. Many reported failures are described as oil leaks because this is the obvious symptom. When such leaks occur in operation, they require investigation to differentiate between those due to failed seals in normal operation and those due to excessive pressure from abnormal conditions. Dielectric breakdowns in oil cause decomposition of the adjacent oil into gases. This action substantially increases the pressure in hermetically sealed containers, commonly resulting in oil leaks. Excessive oil temperatures will cause excessive pressures. Excessive temperature can be caused by a variety of malfunctions: overloading, excessive ambient temperature, and defective magnetics.

Hermetically sealed oil-filled units are tested for leaks during manufacture by heating them to a temperature which will develop slightly higher pressure than will be experienced in normal service. Performed on the first unit, this test checks the design. Performed on subsequent units, the test checks the materials and workmanship associated with the hermetic seal. Leaks occurring at this stage are troublesome and expensive to correct. The underlying causes for leaks should be determined and corrected. The substitution of makeshift repairs for effective long-term corrective action is a serious quality control lapse. The fact that a unit has passed this simple leak test does not mean that freedom from leaks is assured. Further cyclical stressing can cause leaks in units that pass the initial test.

17.3 SAFETY

Personnel Safety

The safety of personnel in the vicinity of high-voltage magnetic devices is mostly under the control of equipment designers. For the benefit of service personnel, high-voltage devices should have warning signs. It is prudent, and often required, to place shields around high-voltage terminals to prevent personnel from making inadvertent contact with them. A ground plane placed between high voltage and personnel or equipment establishes a guarded region. Devices packaged in grounded metal cases are in this category, provided the high-voltage terminals are also behind a ground plane. Some high-voltage magnetic devices are enclosed in plastic. The plastic functionally insulates the device, but it does not necessarily protect personnel who come in contact with the insulated surface. Capacitive coupling can create a hazardous condition. Identification and isolation of these surfaces is necessary for safety.

Polychlorinated biphenyls (PCBs) have been extensively used as liquid insulators. Being fire resistant, this material has been highly regarded for use in indoor transformers. It is nonbiodegradable. The discovery of its alleged carcinogenic property has resulted in its discontinuance. Some PCB-filled devices are still in the field, where they may pose a hazard if the material is released.

Circuit Safety

Circuits are planned to accommodate overloads and component failures with a minimum amount of related damage. Part of these plans includes the ability of the transformer to withstand temporary current and voltage overloads, often reflecting overload currents to protective devices. When the transformer has several secondaries, the operation of circuit protection in the primary becomes a problem. If the volt-ampere rating of one secondary winding represents only a small percentage of the total volt-amperes of the transformer, a current overload in that secondary will have a negligible effect on the total current in the primary. The primary circuit protection will be ineffective in protecting the low volt-ampere secondary circuit. Either circuit protection should be added to the secondary, or the low volt-ampere requirement should be on a separate transformer with separate primary circuit protection.

Fire and Explosion Safety

Magnetic devices with solid insulation often contain combustible materials. Some slight risk is added to the fire hazard because of the inclusion of these combustible materials. Where this is a serious problem, it is possible to introduce flame-retarding chemicals to reduce the hazard. Insulating oil is a much more serious hazard. Commonly used mineral oil has a flash point near 135°C (275°F). The oil is operated at a temperature of about 80°C (176°F). Electrical malfunctions can cause the temperature to reach the flash point. When the malfunction is accompanied by a leak, hot oil is set loose which may be ignited. An intense and destructive fire can result. More hazardous is the generation of explosive gases by the decomposition of oil during arcing. Hydrogen and acetylene are liberated in this process. When these gases leak outside the container, explosive mixtures with air can form which may be detonated by high-voltage electrodes in the vicinity. The use of good overload circuit protection can help reduce the probability of such a disaster. Units with large oil volumes are often equipped with thermal sensors that trip alarms if the coil temperature becomes excessive. Conservative design should be followed to reduce the likelihood of dielectric failure.

CHAPTER 18

Magnetic Device Specifications and Testing

18.1 SPECIFICATIONS, GENERAL

A magnetic component specification performs a dual role. First, it provides information from which a designer generates construction information for manufacturing a prototype. Second, after the completion of a satisfactory prototype, the production function of the specification is invoked, which is to provide criteria to maintain subsequent units of the same design similar to the prototype within functional and economical limits. A proper differentiation of these two functions is helpful in controlling costs and avoiding supplier-user misunderstandings. A specification is a document upon which both technical and business commitments are made. Few specifications completely define all the requirements. Some dependence is placed on the integrity and common sense of the supplier. An unwillingness to place any confidence in suppliers often leads to overspecifying, which can take several forms: asking for attributes that are not needed, asking for the same attribute in two different ways leading to restrictive and contradictory requirements, and calling for operating and test conditions in excess of need. The specification can aid in controlling costs by projecting a tone of reasonableness. Suppliers should not be frightened by the specification into inflating their estimated costs.

18.2   DESIGN SPECIFICATIONS

Electrical Specifications

There are two general approaches that can be taken with electrical design specifications. One is to specify the transformer parameters. In this approach values are given for such quantities as terminal voltage, exciting current, winding resistances, regulation, open-circuit inductance, and leakage inductance. The other approach is to provide information about the circuit in which the component will operate and how the circuit and component are expected to function. The parametric approach has the advantage of being simple and direct. It has the disadvantage that knowledge and understanding required of the specifier about the device characteristics may be limited or incorrect during the design stage. The functional approach requires that the magnetics designer be knowledgeable about component-circuit interfacing. The latter approach avoids artificial parametric requirements which, if met, do not assure satisfactory performance. The magnetics designer following the functional approach can more readily optimize the design through an understanding of what the device is required to do. It is advantageous for the magnetics designer to have adequate information about the circuit in which the component will operate. Circuit features which appear initially to be irrelevant may later prove to be significant to the design of the component. Discussed below are some circuit-component features that should be considered for incorporation in a design specification.

> *Input voltage.*   The waveform and the magnitude of the input voltage should be given. In specifying magnitude, it should be made clear what the measurement is: rms, peak, or average. The widespread use of nonsinusoidal waveforms has diluted the common understanding that all voltages are rms unless otherwise specified. Nominal sine wave voltages are often distorted. Processing circuitry may respond to other properties of the voltage than its rms value. The rms value of an imperfect sine wave can be an ambiguous specification.

> *Input frequency range.*   The minimum frequency determines the maximum flux density and is thus essential information. The maximum frequency determines indirectly the requirements for the distributed parameters of the device. Common usage for nonsinusoidal waveforms considers the frequency to mean the repetition frequency of the nonsinusoidal wave. Acceptable response to the nonsinusoidal wave is usually described as an allowable deterioration of the input waveform rather than a wideband frequency response to ensure fidelity.

Source impedance. Consideration of source impedance separates power transformers, which are designed for zero source impedance, from other types. Source impedance here refers to all types of control devices including switching devices. Control impedances place requirements on the open-circuit inductance, leakage inductance, and distributed capacitance. These parameters assume great significance when a wide band of frequencies must be passed. The type of control impedance and its characteristics should be included in the design specification.

Abnormal operating conditions. If it is required that the component survive abnormal operating conditions, those conditions must be specified. Typical abnormal conditions are temporary reduction in frequency, momentary overvoltages, and high currents which flow until protective devices function.

Direct currents. Unless provision is made for them by placing air gaps in the cores, direct currents readily cause saturation. Direct current in either a primary or secondary winding can cause asymmetrical magnetization and saturation of the core. A complete description of direct currents present is needed in the design specification.

Output voltage. When the load consists of equipment whose input voltages have been established, output voltage is easily specified. When the device feeds circuitry where the interest is in the output from that circuitry, the determination of transformer output voltage becomes difficult. In the rectifier circuit, for example, the impedance of the rectifier transformer helps determine the dc output voltage. The winding resistances and leakage inductance affect the waveform of the current. The user cannot accurately predict what the transformer terminal voltage should be without knowing these parameters. In choke input filter circuits the waveform is predictable and so the terminal voltage is determinable, but both the choke resistance and the transformer winding resistances contribute to the regulation of the circuit. More nearly optimum design is likely to be obtained by integrating the choke and the transformer designs rather than by specifying the two components independently with the total allowable regulation being arbitrarily divided between the two components. For design specifications, it is often better to specify the required dc voltage together with the rectifier circuit and anticipated voltage drops in the rectifiers than it is to specify the transformer terminal voltage.

Output current. The true rms output current is needed by the magnetics designer to select wire sizes. As with voltage, the rms current may be determined partly by the transformer impedance. For rectifier loads it may be better to specify the direct output current and let the component designer determine the rms current by an iterative process. For determining true rms current, the duty cycle, if less than 100 percent, should be given.

Inductance. Inductance is specified as a design requirement on filter inductors and similar devices. The nonlinear nature of ferromagnetic cores limits the precision of confirming measurements of inductance. The design specification for inductance should detail the operating conditions for the inductor, including direct current, ac voltages and frequency, and allowable changes as appropriate in inductance when these conditions change. It is generally advisable in transformer design specifications to provide functional requirements rather than specify inductance.

Operating voltages. External noninduced voltages are frequently applied to the windings of magnetic devices for which insulation must be provided. When these voltages are large, they become a major consideration in design, often forcing compromises on other design features. The manner in which noninduced voltages are specified varies. Substantial safety factors are sometimes included, and sometimes only the test voltage is specified. It is advisable to be unambiguous about this requirement. The nature of the voltage (dc, 60-Hz, sine wave, pulse, etc.) should be specified. The insulation requirements vary with the nature of the voltage. The actual operating voltage should be given because overvoltage test practice varies, and the correlation between overvoltage testing and reliability is poor.

Capacitance. The operation of pulse and high-frequency circuits is frequently affected by stray capacitances to ground such as exist in secondary transformer windings. When this occurs, special constructions are invoked by a capacitance to ground requirement. This requirement becomes a basic design consideration.

Frequency response. Output-voltage change with frequency is a circuit response problem to which the transformer is one among several contributors. The specification is usually given in the form of an equivalent Thévenin generator with generator impedance and a load impedance. The permissible variation in output voltage with frequency is given for a constant Thévenin generator voltage.

Electrostatic shielding. Electrostatic shielding between windings reduces capacitive coupling and sometimes helps to improve electrode configurations of high-voltage windings. It is relatively easy to install in many configurations. The use of electrostatic shielding is routine in equipment with conducted noise problems.

Winding polarity.   Correct winding polarities are necessary in such applications as dual primary windings that may be connected in either series or parallel and in feedback windings.

Mechanical Specifications

Mechanical requirements largely determine the cost and quality of components. Care is needed to avoid either unnecessarily adding to the cost by overspecifying or underspecifying, from which unacceptable product results. Some features normally included in mechanical design specifications are discussed below.

Ambient temperature range.   The maximum ambient temperature of the component, not the equipment, is one of the starting points in component design. This value, plus the component temperature rise, determines the insulation operating temperature. Size, weight, and cost are intimately related to the maximum ambient. The minimum ambient limits the material choices and the type of construction.

Size and weight.   The density of magnetic devices of a given general type is nearly fixed, and so weight and volume control are common objectives. In most devices, size reduction is limited by insulation operating temperature. Magnetic devices are limited in the proportions achievable as well as in volume. A severely restrictive dimension along one axis is not necessarily offset by a generous dimension along another axis. Mounting facilities and electrical terminations have a great effect on the component silhouette. The type of mounting and the location of terminations have important implications in the assembly of the component. It is advisable to remain flexible in this area until the matter is explored with the supplier. Magnetic components are sometimes called upon to perform extraneous mechanical functions, such as serving as a mounting for other components or structures. Such requirements can complicate the construction and tolerance problems for the component. Increases in the size and cost of the component may result.

Shock and vibration.   Magnetic devices are especially vulnerable to shock and vibration because they are among the heaviest components in equipment. Even when explicit shock and vibration requirements are absent, these devices must be able to survive the considerable abuse of handling and shipping. When shock and vibration requirements exist for equipment, it should be remembered that the shock and vibration applied to the equipment may be either amplified or dampened before reaching the component.

High humidity and moisture. Open coils are very vulnerable to moisture, and steel cores cannot be plated, giving open core and coils poor humidity and moisture resistance. The popular varnish treatment provides little more than sufficient moisture resistance to operate in room environments. Resistance to severe humidity conditions requires either plastic encapsulation or a metal enclosure. These protective features, which add to the cost, are regular parts of mechanical specifications. Moisture resistance is often specified in terms of the ability of the device to survive exposure to repeated cycling in an atmosphere of high humidity combined with rapid temperature changes. This test is described in military Specification MIL-STD-202.

Altitude. High altitude reduces convection cooling by air, reduces the dielectric strength of terminals exposed to the atmosphere, and lowers the corona threshold. Altitudes of 10,000 ft and higher are critical for magnetic components. When high-altitude operation is required, it is an essential part of the mechanical specification.

18.3 PRODUCTION SPECIFICATIONS

Production specifications set limits for uniformity and quality of product. They become legal as well as technical documents which are executed by persons who lack either the authority or ability to make technical judgments. The preparation of a production specification deserves meticulous attention. See Sec. 16.3 on specification procedures. Specifying attributes and setting limits of acceptance require a knowledge of the electrical and mechanical functioning of the device and also of the variables introduced by manufacturing processes and materials. Economics limits production testing and inspection to only those attributes whose monitoring is necessary to maintain uniformity and quality within acceptable limits. It is not necessary to repeat in production the design verification tests that are performed on prototypes. What is required is to verify similarity of other units to the acceptable prototype. The production test and inspection specifications for different designs for a single application may not be identical although the several designs may all be satisfactory. A factor to consider in choosing attributes for production testing is the existence of established test methods and equipment which require a minimum setup and no additional equipment. While the standardized conditions imposed by these tests may depart from operation conditions, they do provide comparisons between prototype and subsequent production.

Electrical Specifications

The magnetic properties of cores vary because of differences in magnetic materials and differences in the way manufacturing processes are carried out. Attributes which depend upon magnetic properties are prize candidates for production testing. These attributes include exciting current, open-circuit inductance, core loss, and the transfer characteristics of saturable reactors. Many applications require only that these attributes be below a required maximum or above a required minimum. Tolerances specified in this manner reduce costs substantially.

Voltage ratio is usually a production test. The voltage ratio closely approximates the turns ratio. Variations in turns ratio are due to operator error.

Winding resistances are of considerable value in monitoring uniformity of product and are frequently included in a production specification. To serve as a check on uniformity of product, the nominal value specified must be correct, and the tolerances around the nominal must be tighter than is usually required for functional purposes. Winding resistances are frequently an internal specification of the supplier but not a contractual obligation to the user. When it is made a contractual requirement, the nominal value must be determined with great care. Two different suppliers of the same article may have different nominal values of winding resistances.

Overvoltage specifications consist of induced-voltage tests and high-potential tests. Induced-voltage tests develop above-rated voltages between terminals of a single winding. In high-potential tests an external voltage is applied to the entire winding. Either or both of these tests are used as appropriate to stress the insulation system. Low-voltage devices are regularly tested at an induced voltage of twice the operating voltage and twice the frequency so that the exciting current is maintained at a safe value. High-potential testing of low-voltage units is usually performed at twice the operating voltage plus 1000 V rms power line frequency. Very high-voltage devices are often tested at less than twice the operating voltage. Overvoltage tests are of short duration. They confirm the ability of the tested device to withstand short-term overvoltages and will uncover gross insulation defects. There is little correlation between overvoltage testing and long-term reliable operation at rated voltages.

Insulation resistance is sometimes used as a production test for monitoring the quality of insulation. A general requirement in military Specification MIL-T-27 is that the initial insulation resistance of transformers and inductors be 10,000 MΩ minimum. Insulation resistance is a function of the geometry of the device and is directly proportional to the resistivity of the insulation. The geometry of the device is nearly constant for a given design. Variations in the insulation resistance of a given design will be largely the result of variations in the resistivity of the insulating materials and associated processes. The usefulness of insulation resistance as a quality control attribute lies in detecting these variations. To do this requires the establishment of minimum limits for individual designs. This can be done by taking data on a representative lot whose insulation integrity is considered acceptable because of other criteria.

Corona detection is often a production test on high-voltage components. The corona onset voltage is a production variable involving assembly, processing, and materials. Corona testing is troublesome because measuring techniques are imprecise and passing limits are difficult to establish. The troublesome nature of the test does not lessen its importance or alter the fact that corona is a major cause of poor reliability in high-voltage magnetic devices.

Leakage inductance and distributed capacitance are sometimes invoked as production tests in lieu of more complex operational tests. Although design procedures establish theoretical nominal values for these distributed parameters, those nominal values should not be used without confirmation by careful measurement on satisfactory prototypes. Since both design calculations and measurements are only approximate, close agreement between the two cannot be expected. In setting nominal values for production tests, prototype measurements performed by production test methods must be favored over design values.

Capacitance between windings and windings to ground are common production requirements for pulse- and high-frequency components. This is a lumped parameter which may be measured and predicted with greater accuracy than distributed capacitance. Maximum values of lumped capacitance are frequently based on circuit requirements and become a production test.

Mechanical Specifications

The principal instrument for conveying production specifications is the control drawing. By means of outline views, this drawing specifies installation dimensions, electrical terminations, and marking requirements. The control drawing gives electrical specifications either as an integral part of the drawing or by reference to a separate document. Also either as an integral part of the drawing or by reference to a separate document, the control drawing invokes mechanical and environmental requirements. Production mechanical requirements invoked in this manner include protective finishes, sealing test, and workmanship norms. The supplier needs an internal specification control drawing or equivalent documentation to administer the user's requirements. When the supplier specification control drawing is of appropriate format, it is frequently given to the user for information and comment. The user sometimes chooses to invoke this supplier drawing in the purchase contract. This places an obligation on the supplier to maintain strict control over drawing change procedure, obtaining prior concurrence by the user to changes which may affect quality or interchangeability. A user specification control drawing has the advantage of uniquely identifying the item and conveying to the supplier, in a form less likely to be misunderstood, the user's needs. It also channels the requirements into a systematic arrangement which helps persuade the specifiers to describe their needs realistically, correctly, and completely.

18.4 TESTING AND INSPECTION

Design Verification Testing

The fundamental requirement of a magnetic component is that it work in the place intended. All other requirements are derivative. Operation of a prototype component in its equipment offsets errors and supplements limitations of subsidiary requirements. It is an essential step in the evaluation process. Several problems present themselves in this process. The prototype must be evaluated on the basis of whether it represents the design center. If it does not, then appropriate changes may be required in the nominal values of subsidiary attributes which will shift the design center in the direction required. Sometimes it may be advisable to construct and evaluate another prototype with values closer to the required design center. Other components with which the magnetic device interacts also have attributes which vary. The attributes of the interacting components must be scrutinized for their effect on equipment performance. Ideally the interacting components should represent nominal performance. The derivative requirements should be reviewed at the time of the functional evaluation. This review should consider whether the derivative requirements are sufficient, redundant, or need quantitative revision.

Full-load voltage. A full-load voltage test is regularly performed as part of a prototype evaluation. In power transformers full-load voltage is equal to the no-load voltage less the regulation caused by winding resistances and leakage inductance. The term does not refer to the drop across the impedance of a high-impedance generator, as in wideband transformers. In that context the term *insertion loss* is customarily used to describe the drop through the transformer. The load used for the measurement must be correct in both magnitude and impedance angle. With rated voltage and frequency applied to the primary, the voltage across the loaded secondary is measured. If the test consists of measuring the absolute value of the secondary voltage, the primary and secondary voltages must be measured with equal care at the same instant to obtain an accurate measurement. If the phase shift between primary and secondary voltages is small, the voltage may be measured more accurately by using the voltage ratio bridge method as shown in Fig. 18.1. This method has the advantage of eliminating sensitivity to line voltage variations and substituting a more accurate and easily calibrated ratio arm for two voltmeters. The full-load voltage is equal to the nominal primary voltage multiplied by the ratio for which a null is obtained in the bridge circuit. The primary voltage does not have to be maintained with precision during the measurement. The quality of the null affects the precision obtainable in the ratio method. Phase shift, harmonic distortion, and noise can affect this quality. Frequently test conditions preclude the ratio method of measuring full-load voltage. The accuracy of the voltmeter method for making the measurement can sometimes be improved if the no-load voltage ratio of the transformer is known with precision, possibly by means of the ratio bridge measurement at low voltage. The transformer operated at no load is used to calibrate the two voltmeters by setting the primary voltage at nominal value as measured by the primary voltmeter. The difference between the no-load voltage predicted by the measured no-load voltage ratio and the no-load voltage measured directly is used as a calibration of the primary-secondary voltmeter combination when the transformer is under load. Transformer voltage ratios are capable of a good deal

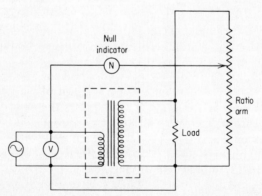

FIG. 18.1 Full-load-transformer ratio bridge circuit.

of precision, but the absolute value of the output has no greater precision that that with which the absolute value of the input voltage is maintained. Full-load voltage measurements must address this problem.

Frequency response.   Transformer frequency response is a circuit consideration requiring testing in either the actual circuit, a simulated circuit, or both. For design purposes, it is necessary to assign numerical values to both load and source impedances. It may be necessary to know whether the transformer meets specifications with these numerical values. This determination is best done in a simulated circuit. The change in output voltage across the simulated load with frequency is observed while maintaining a constant voltage across an impedance equal to the generator impedance and transformer input in series. Actually both load and generator are frequently dynamic devices in which the impedances vary, making evaluation in the actual circuit also necessary. Failure to perform in the actual circuit may be due to improperly chosen design values for generator and load impedances. See Chap. 4 for a discussion of wideband devices.

Square wave response.   Response to a square wave or rectangular pulse is determined with oscilloscope presentations. The response is the result of interactions between the transformer parameters and the load and generator circuits. These circuits are frequently nonlinear. Realistic determination of performance should be done in the actual circuit. See Chap. 5 for a discussion of the effect of transformer parameters on rectangular pulses and Fig. 5.1 for pulse shape notation. Measurements of time intervals and amplitudes with oscilloscopes tend to have subjective content and lack the precision obtainable with indicating instruments. The source and load circuits as well as the transformer are subject to variations which are difficult to control and measure. The actual devices in the circuit used to evaluate a prototype ideally should have close to nominal properties. This ideal is not always obtainable. Production variations of all the related components can result in production equipment performance being out of tolerance without a clear indication of which components are defective. The transformer parameters themselves can be monitored by careful measurements on an acceptable prototype. These measurements and the methods used to make them become the basis for establishing nominal values for production tolerances and production test methods. Photographs of oscilloscope traces are useful for documenting prototype performance. They help to eliminate some of the subjectivity of oscilloscope measurements and permit evaluation of performance without viewing the actual oscilloscope presentation in the laboratory.

Temperature rise. Temperature rise is most effectively determined by measuring the change in resistance of the windings from which the temperature of the windings may be computed. In the operating range of electrical equipment, copper undergoes a proportional change in resistance with temperature. The projected temperature for which copper resistance is zero is −234.5°C (−390.1°F). This behavior of copper may be expressed in the following relationship:

$$\frac{R_2}{R_1} = \frac{t_2 + 234.5}{t_1 + 234.5} \qquad (18.1)$$

In this expression, R_1 is the stabilized resistance at temperature t_1, and R_2 is the stabilized resistance at temperature t_2; R is in ohms and t is in degrees Celsius. Readings R_1 and t_1 are taken before the start of the temperature rise test. Reading R_2 is taken after thermal stability is reached under load conditions. This leaves t_2 unknown, which may be determined by substituting the known values in Eq. (18.1). The final copper temperature is t_2. If the initial component ambient temperature t_1 is maintained during the temperature rise test, then the temperature rise is $t_2 - t_1$. If the ambient temperature is not constant, the test is still valid provided thermal stability is achieved at the final ambient temperature. Thermal stability is demonstrated by successive resistance measurements with so little change that the effect on t_2 is negligible. The difference between the final and initial ambient temperatures is subtracted from t_2 in calculating the temperature rise if the final temperature is higher than the initial temperature, and it is added to t_2 if the final ambient temperature is less than the initial temperature.

Temperature rise determined by the change in resistance method gives the average temperature of the winding. The hot spot is usually between 5 and 15°C above the average temperature rise. The hot spot is the region of greatest thermal stress and will probably determine the life of the device. Section 9.1 discusses temperature gradient in windings.

The ratio of hot to cold winding resistance is typically about 1.2 or 1.3. Because the change in resistance is small, small-percentage errors in measuring resistance can cause large-percentage errors in calculated temperature rise. Care is required to obtain accurate measurements. A Wheatstone bridge for resistances over 1 Ω and a Kelvin bridge for smaller resistances, or instruments of comparable accuracy, are required. To prevent damage to these instruments, it is necessary to remove the voltage from the device while taking resistance readings. The resistance drops quickly after removal of voltage, especially in units of small thermal mass. It is advisable to switch voltage and instrumentation in a single operation. When the Kelvin bridge is used, two sets of connections must be switched. The setup should be arranged to avoid errors due to lead resistances and contact resistances.

The most realistic environment in which to place the device during the temperature rise test is the actual equipment in which it will be used. The mounting, the flow of convection currents, and adjacent warm objects contribute to the test result. If the actual load is used in the test, the load current should be monitored. The load current may differ from the specification deliberately to provide for worst-case conditions or inadvertently. Inadvertent differences between specified and measured load current should be reconciled. Temperature rise tests are sometimes performed under arbitrary or standard conditions. The results of such tests are not directly applicable to specific equipment. Temperature rise may be determined by placing sensors in the coil. This requires special constructions. Useful comparative data are sometimes obtained by measuring device surface temperature rise directly.

Environmental Tests

Humidity tests are performed in environmental chambers in which the moisture is controlled by wet- and dry-bulb thermometers that indicate humidity for only a small region in the chamber. Temperature differences in the chamber cause the humidity to vary in other parts of the chamber. As a result, condensation on test specimens is to be expected in high-humidity tests. Some tests require sudden reduction in temperatures, ensuring that condensation will occur. Likely failure modes are lowered insulation resistance in hygroscopic insulating materials, corrosion of metal surfaces, and high-voltage breakdown due to the accumulation of moisture on high-voltage creepage surfaces. A problem in humidity testing is the establishment of good pass-fail criteria. The requirement that the device operate during the humidity test is a severe one, particularly if high voltages are present. A less severe test requires that the specimen operate after the completion of the test and evaporation of surface moisture. An arbitrary minimum value of insulation resistance is sometimes used as a passing limit.

Mechanical Shock and Vibration

Mechanical shock tests are performed with machines that apply a sudden blow to a table upon which the specimen is mounted. The table is either dropped in a controlled and repeatable manner or struck by a large hammer dropped from a controlled height. The shock blows are applied along each of three mutually perpendicular axes. Shock tests have nominal acceleration and duration ratings, but the tests are defined in terms of the construction and operation of the test machines. Most specimens undergo reverberation during the shock test. Survivability is improved if this reverberation is minimized.

Vibration testing is performed with vibration tables upon which the specimen is mounted. As with shock testing, the vibration is performed along three principal axes. Vibration test frequencies vary from 10 to 2000 Hz. The maximum acceleration developed by the test is proportional to the amplitude and the square of the frequency. The test is usually specified as a maximum acceleration. The most troublesome frequencies for most magnetic components are between 10 and 55 Hz because that is the range in which the principal resonance usually occurs. Most vibration test specifications contain a requirement that the frequency be made to dwell at the resonant point of the device. At resonance the amplitude of the motion is magnified and failure is most likely to occur. The most vulnerable part of a magnetic component to shock and vibration is usually the mounting. The mass of the core and coil is usually attached to the equipment by means of screws or bolts and some type of sheet-metal construction. Failure under either shock or vibration most often occurs during resonance or reverberation due to fatigue failure in regions of stress concentration. Vulnerability to mechanical failure can be reduced by keeping the center of gravity as close to the mounting surface as possible, increasing the resonant frequency by increasing the rigidity of the device with thicker cross sections, gussets, and braces, and by limiting stress concentrations through the use of generous radii and adequate transition sections. Mechanical shock and vibration tests are described in detail in U.S. Department of Defense Standard MIL-STD-202.

Thermal shock consists of soaking the specimen at alternately hot and cold temperature extremes, the transition being made rapidly before the specimen temperature has had the opportunity to moderate. Plastic-encapsulated transformers are vulnerable to this test, particularly if the test includes turn-on operation at the low temperature after a period of soaking. Most plastic encapsulations will crack after repeated thermal shock testing at temperatures below $-40°C$ ($-40°F$).

Production Tests

No-load voltage ratio, a convenient and inexpensive production test, is best performed by a bridge measurement as shown in Fig. 18.2. The step-down ratio may be measured whether the usage is step-down or step-up. This avoids drawing the current to excite the ratio arm through the transformer, a possible source of error in small units. The test need not be performed at rated voltage. It is usually necessary to use a frequency near rated to avoid distributed parameter effects. Accurate testing of voltage ratio with a sine wave voltage may be made on transformers which transform nonsinusoidal voltages.

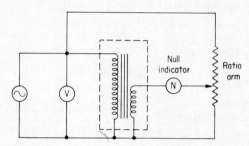

FIG. 18.2 No-load-transformer ratio bridge circuit.

DC resistance. The use of dc resistance as a control parameter requires a bridge or instrument of equivalent accuracy. Low resistances require a Kelvin bridge, which eliminates contact resistance as a source of error. Variations in dc resistance can be caused by using the wrong wire size, shorts, or errors in turns.

Exciting current. Exciting current is a useful test for monitoring production quality. Requiring only a voltmeter and an ammeter, it is simple to perform and does not require great precision. Precautions are needed with high-voltage step-up transformers. Processing must be advanced to the point where rated voltage may be safely developed, and appropriate terminals must be grounded. To be useful as a quality check, the exciting current limit must be set by manufacturing, not functional, considerations. Excessive exciting current is an indication of a variety of material and workmanship defects, among which are turns errors, shorted turns, defective cores, shorted electrostatic shields, and shorted mounting hardware.

Shunt inductance. Measurements of ferromagnetic inductance will vary with the conditions and method of measurement. The conditions include applied voltage, frequency, and direct current. Measuring methods include impedance, resonance, and bridge. Circuits for measuring inductance by these three methods are shown in Fig. 18.3. Each method will yield a slightly different value for inductance. The Owen bridge circuit shown is adapted for the simultaneous application of direct current. Requirements for inductance must be explicit regarding the conditions and method of measurement. Available test equipment frequently does not have the range needed to establish actual operating conditions. Approximations based on measurements of an acceptable prototype are common.

Capacitance. Capacitance of a winding to ground or between windings can be measured with a capacitance bridge as any other lumped

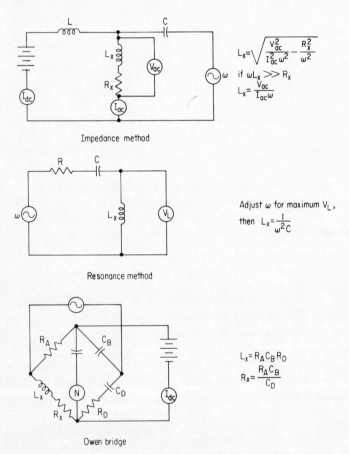

Impedance method

$$L_x = \sqrt{\frac{V_{ac}^2}{I_{ac}^2 \omega^2} - \frac{R_x^2}{\omega^2}}$$

if $\omega L_x \gg R_x$

$$L_x = \frac{V_{ac}}{I_{ac}\omega}$$

Resonance method

Adjust ω for maximum V_L,

then $L_x = \dfrac{1}{\omega^2 C}$

Owen bridge

$$L_x = R_A C_B R_D$$

$$R_x = \frac{R_A C_B}{C_D}$$

FIG. 18.3   Circuits for measuring inductance.

capacitance is measured. The measurement of distributed capacitance, being a function of voltage distribution, requires indirect methods. The open-circuit inductance may be measured by one of the methods discussed above. The frequency of parallel resonance between the open-circuit inductance and the distributed capacitance is determined. The distributed capacitance may be calculated from the known open-circuit inductance and the resonant frequency.

Leakage inductance. Determination of leakage inductance may be made by shorting the low-voltage winding and measuring the residual inductance in the high-voltage winding. If this is done by the impedance method, the winding resistances referred to the winding being measured should be considered. If the distributed capacitance is known, then the leakage inductance may be calculated from the resonant frequency of the high-voltage winding with the low-voltage winding shorted. The measurement of distributed parameters is approximate. If they are made part of the specification requirement, the method of measurement should be included.

Polarity. The output voltage of a transformer is either approximately in phase or out of phase with respect to the input. When the voltages on two terminals of different windings are in phase, those two terminals are said to have the same polarity. Polarity is determined by the no-load voltage ratio test because the bridge will balance only when the polarities are correct.

Corona testing. Corona testing is a controversial quality control procedure. See Sec. 8.5. Suppliers avoid corona tests because unpredictable production test results increase economic risk. Severe corona in an accessible site is detectable by the senses. It may be seen in a darkened room as a glow, it may be heard as a sizzling noise, or ozone produced by corona may be recognized by its distinctive odor. A number of electrical tests are available for detection of corona in less accessible sites, such as deep within the insulation system. A small battery-operated AM broadcast receiver, placed near a corona source, will produce an audible noise.

A method which attempts to acquire quantitative data on corona discharge is described in Specification MIL-T-27 and shown in Fig. 18.4. Two situations are shown, one where the high voltage is induced in the test specimen itself, and the other where an external voltage source is required. Either situation uses the same detecting circuit consisting of a series LC circuit tuned to approximately 71 kHz. The capacitor is insulated for the high-power frequency voltage. The selective property of the series circuit permits only a small fraction of that voltage to appear across the inductor, the lower end of

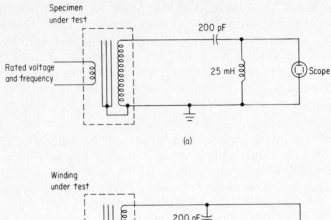

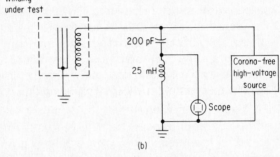

FIG. 18.4 Circuits for detecting corona. (*a*) High voltage induced in test specimen; (*b*) high voltage externally applied.

which is grounded. An oscilloscope monitors the voltage across the inductor. The electric noise generated by corona contains a wide band of frequencies. The resonant circuit separates a narrow band of those frequencies for display, helping to reduce ambient noise interference. The onset of corona will appear as a noise burst superimposed on the residual power frequency voltage, usually on the leading slope of the sine wave. The amplitude of this burst is an approximate measure of corona intensity. Specification MIL-T-27 has established as the threshold rejection an amplitude of this noise burst of 0.1 V peak to peak. The amplitude of the noise burst is a circuit function as well as a function of corona intensity. The destructive energy associated with corona discharge is proportional to the product of the number of ions formed and the voltage across the discharge site. The time integral of corona discharge current which has the dimensions of charge is proportional to the number of ions formed. This charge accumulates on the capacitance of the specimen, producing a voltage whose average value is proportional to charge. This average voltage, a measurable quantity, together with the specimen capacitance, provides a quantitative measure of corona intensity superior to peak voltage. Commercial corona-measuring test equipment based on this principle is available with calibrating voltages which are applied across the sample. This voltage and the known specimen capacitance provide calibration points in picocoulombs.

Overvoltage testing.   By increasing the frequency and applied voltage proportionally above ratings, an induced voltage test will maintain the same flux density that the rated voltage and frequency do. In the performance of this test, the voltage and current are monitored. The criterion for acceptance is no evidence of breakdown. Onset of breakdown may be indicated by an intermittent current or a sudden increase in current. It is not necessary for the component to be corona-free at double the voltage. In the performance of this test, correct grounding conventions must be observed. Many high-voltage windings are insulated for high voltage at one end only, the other end being maintained near ground by external connections. When testing such windings, the low end must be grounded, which usually means connecting to the core. The induced voltage test may be performed at twice the rated frequency or higher. It is convenient in many test facilities to perform the induced voltage test on 60-Hz transformers at 400 Hz. The current at 400 Hz is often much higher than would be expected from core excitation requirements. The high current is usually due to the distributed capacitance of the windings. It should not be mistaken for evidence of breakdown.

High-potential tests consist of applying external voltages to entire windings. To avoid the development of electrostatic voltages above insulation ratings, all windings not under test are grounded. Since the winding under test is everywhere at the same potential, the magnitude of the high-potential test must be based upon the lowest operating voltage between any portion of the winding and other windings or ground. Low-voltage and high-ac-voltage devices are usually tested with ac high potentials. High-dc-voltage devices are usually tested with dc high potentials. The test is of short duration, usually 5 s and seldom over a minute. The threshold of failure is usually readily detectable with dc high potentials. The current is quite small until the onset of failure. It rises rapidly as the breakdown voltage is approached, and it is usually intermittent. The threshold of failure with ac high potentials is less obvious and is occasionally the subject of controversy. Capacitive currents of appreciable magnitude frequently flow during ac high-potential testing. The method of determining failure must differentiate between capacitive and breakdown current. Breakdown current is characterized by a sudden increase and is intermittent. Capacitive current is steady and increases proportionally with voltage. Alternating current high-potential testers which indicate failure by means of a limiting magnitude of current will not make the required differentiation. It is necessary to reduce the sensitivity of the failure indicator in these equipments when testing devices with high capacitive currents. High-potential tests use voltages far above rated. It is not necessary for the devices under test to be corona-free at the test voltage.

Mechanical inspection.   Mechanical performance is equal in importance to electrical performance, and inspection to determine compliance with mechanical requirements is a necessary part of quality assurance. Mounting dimensions are critical. The probability of out-of-tolerance dimensions is more likely if fabrication is by assembly operation than by tooling or fixtures. Coil dimensions are not held with the same precision as metal constructions. When a coil forms a part of the device silhouette, close monitoring of dimensions is required. Encapsulation by the dipping process provides poor dimensional control. The encapsulation should be inspected for cracks and fissures which admit moisture. Electrical connections and mounting surfaces should be free of encapsulant. Workmanship inspection should include examination of machine threads on terminals, mounting facilities, or other fittings, and observation of painted and plated surfaces. Markings should be correct and legible. Ceramic parts and ferrite cores should be free of cracks. Checks on hermetic sealing are a regular part of production inspection. It is particularly important to check hermetically sealed oil-filled units for leaks. This is usually done by baking at elevated temperatures. See Sec. 17.2. Failure of the leak test is common in oil-filled construction.Rework is an almost routine matter. The method of reworking and retesting can have an impact on subsequent oil-tight integrity and electrical performance. Hermetically sealed units that are filled with solid potting materials are sometimes given a bubble test which consists of placing the component in a bath of hot water. A continuous stream of bubbles issuing from the unit is an indication of a leak.

Bibliography

Allied Chemical Corp.: *Sulfur Hexafluoride for Gaseous Insulation,* Bulletin TB-85603, undated.

Arguimbau, L. B.: "Losses in Audio-Frequency Coils," *The General Radio Experimenter,* vol. XI, no. 6, November 1936.

Brown, A. I., and S. M. Marco: *Introduction to Heat Transfer,* McGraw-Hill, New York, 1958.

Buchsbaum, W. H.: *Complete Handbook of Practical Electronic Reference Data,* Prentice-Hall, Englewood Cliffs, N.J., 1978.

E. I. du Pont de Nemours & Co., Inc.: *Kapton Polyimide Film, Summary of Properties,* Bulletin E 42727, June 1981.

E. I. du Pont de Nemours & Co., Inc.: *Mylar Polyester Film, Electrical Properties,* Bulletin M-4C, undated.

E. I. du Pont de Nemours & Co., Inc.: *Mylar Polyester Film, Physical-Thermal Properties,* Bulletin M-2D, undated.

E. I. du Pont de Nemours & Co., Inc.: *Properties and Performance of Nomex Aramid Paper, Type 410,* Bulletin NX-16, October 1981.

Fink, D. G.: *Electronic Engineers' Handbook,* McGraw-Hill, New York, 1982.

Finzi, L. A., and R. R. Jackson: "The Operation of Magnetic Amplifiers with Various Types of Load," *AIEE Transactions,* vol. 73, pp. 270–288, July 1954.

Finzi, L. A., and G. F. Pittman: "Comparison of Methods of Analysis of Magnetic Amplifiers," *Proceedings of the National Electronics Conference,* vol. 8, pp. 144–157, January 1953.

Forster, G. A., L. J. Stratton, and H. L. Garbarino: "Research and Development of New Design Method for Power Transformers," final report under U.S. government Contract No. DA-36-039 SC-52656, Armour Research Foundation, Chicago, March 1956.

Glasoe, G. N., and J. V. Lebacqz: "Pulse Generators," vol. 5, *Radiation Laboratory Series,* McGraw-Hill, New York, 1948.

Goldman, S.: *Transformation Calculus and Electrical Transients,* Prentice-Hall, Englewood Cliffs, N.J., New York, 1949.

Grossner, N. R.: *Transformers for Electronic Circuits,* McGraw-Hill, New York, 1983.

Halleck, M. C.: "Calculation of Corona-starting Voltage in Air-Solid Dielectric Systems," *AIEE Transactions,* April 1956.

Harder, E. L., and W. F. Horton: "Response Time of Magnetic Amplifiers," AIEE Technical Paper 50-177, May 1950.

Jakob, M., and G. A. Hawkins: *Elements of Heat Transfer and Insulation,* Wiley, New York, 1957.

Kilham, L. F., and R. R. Ursch: "Transformer Miniaturization Using Fluorochemical Liquids and Conduction Techniques," *Proceedings of the IRE,* vol. 44, no. 4, April 1956.

Knight, A. R., and G. H. Fett: *Introduction to Circuit Analysis,* Harper, New York, 1943.

Lee, R.: *Electronic Transformers and Circuits,* Wiley, New York, 1947.

McLyman, W. T.: *Magnetic Core Selection for Transformers and Inductors,* Dekker, New York, 1982.

————: *Transformer and Inductor Design Handbook,* Dekker, New York, 1978.

Massachusetts Institute of Technology, Electrical Engineering Staff: *Magnetic Circuits and Transformers,* Wiley, New York, 1943.

Matisoff, B. S.: *Handbook of Electronics Manufacturing Engineering,* Van Nostrand Reinhold, 1978.

Millman, J., and C. C. Halkias: *Electronic Devices and Circuits,* McGraw-Hill, New York, 1967.

NVF Company: *Forbon Vulcanized Fibre,* Bulletin 2.5M-4762, February 1977.

Reference Data for Radio Engineers, 6th ed., Sams, Indianapolis, 1975.

Schade, O. H.: "Analysis of Rectifier Circuit Operation," *Proceedings of the IRE,* vol. 31, pp. 341–361, July 1943.

Schaefer, J.: *Rectifier Circuits: Theory and Design,* Wiley, New York, 1965.

Shell Oil Company: *Diala Oil AX Electrical Insulating Oils,* Product Information Bulletin SOC:39-77, undated.

Simoni, L.: "A General Approach to the Endurance of Electrical Insulation under Temperature and Voltage," *IEEE Transactions on Electrical Insulation,* vol. EI-16, no. 4, pp. 277–289, August 1981.

Society of the Plastics Industry: *Plastics Engineering Handbook,* Reinhold, New York, 1954.

Terman, F. E.: *Radio Engineers' Handbook,* McGraw-Hill, New York, 1955.

U.S. Military Specification MIL-T-27D, *Transformers and Inductors, General Specification for,* 1968.

Volk, M. C., J. W. Lefforge, and R. Stetson: *Electrical Encapsulation,* Reinhold, New York, 1962.

Wood, P.: *Switching Power Converters,* Van Nostrand Reinhold, New York, 1981.

Index

ABOUT THE AUTHOR

William M. Flanagan has been involved for 37 years in the design, manufacture, and application of almost every conceivable type of transformer. His orientation has been toward solving practical industry problems where cost and scheduling must be considered as well as meeting the highest standards of quality and performance. He received a B.S. degree in liberal arts from Hampden-Sydney College and a B.S. degree in electrical engineering from the University of Illinois.